THE UNIVERSE

ASTROPHYSICS AND SPACE SCIENCE LIBRARY

VOLUME 244

THE UNIVERSE

Visions and Perspectives

Edited by

NARESH DADHICH

Inter-University Centre for Astronomy and Astrophyics,
Pune, India

and

AJIT KEMBHAVI

Inter-University Centre for Astronomy and Astrophyics,
Pune, India

KLUWER ACADEMIC PUBLISHERS

DORDRECHT / BOSTON / LONDON

A C.I.P. Catalogue record for this book is available from the Library of Congress.

ISBN 0-7923-6210-1

Published by Kluwer Academic Publishers,
P.O. Box 17, 3300 AA Dordrecht, The Netherlands.

Sold and distributed in North, Central and South America
by Kluwer Academic Publishers,
101 Philip Drive, Norwell, MA 02061, U.S.A.

In all other countries, sold and distributed
by Kluwer Academic Publishers,
P.O. Box 322, 3300 AH Dordrecht, The Netherlands.

Courtesy of Prof. W.J. Couch, The University of New South Wales.

Printed on acid-free paper

Printed in the Netherlands.

Contents

Preface

It is with great joy that we present a collection of essays written in honour of Jayant Vishnu Narlikar, who completed 60 years of age on July 19, 1998, by his friends and colleagues, including several of his former students. Jayant has had a long research career in astrophysics and cosmology, which he began at Cambridge in 1960, as a student of Sir Fred Hoyle. He started his work with a big bang, expounding on the steady state theory of the Universe and creating a new theory of gravity inspired by Mach's principle. He also worked on action-at-a-distance electrodynamics, inspired by the explorations of Wheeler, Feynman and Hogarth in that direction. This body of work established Jayant's reputation as a bold and imaginative physicist who was ever willing to take a fresh look at fundamental issues, undeterred by conventional wisdom. This trait, undoubtedly inherited from his teacher and mentor, has always remained with Jayant. It is now most evident in his untiring efforts to understand anomalies in quasar astronomy, and to develop the quasi-steady state cosmology, along with a group of highly distinguished astronomers including Halton Arp, Geoffrey Burbidge and Fred Hoyle. In spite of all this iconoclastic activity, Jayant remains a part of the mainstream; he appreciates as well as encourages good work along conventional lines by his students and colleagues. This is clear from the range of essays included in this volume, and the variety and distribution of the essayists.

After a long stay in Cambridge, Jayant moved to the Tata Institute of Fundamental Research in Mumbai (then Bombay) in 1972. There he inspired several research students to work in gravitational theory and its many classical and quantum applications to cosmology and astrophysics, and established collaborations with his peers, which led to a fine body of work over the next 15 years. But perhaps his most enduring contribution of this period was to forge a link between distinguished senior relativists in India, and the younger generation of aspiring researchers. This has led to the formation of a warm and congenial community, spread throughout the country, working in relativity, cosmology and theoretical astrophysics. During this period Jayant also worked hard at the popularization of science, through the press, television and most importantly through talks to ever increasing audiences. This not only exposed people to good science, but it also helped to establish Jayant as one of the public figures of science in India. Jayant has used his formidable reputation and influence, developed during this period, for the advancement of science in India, always in a very quiet manner.

In 1988, inspired and aided by Professor Yashpal, then Chairman of the University Grants Commission, Jayant set up the Inter-University Centre for Astronomy and Astrophysics at Pune. Through this centre he has been able to open up for the university community avenues for excellent research in these areas. Jayant's broad vision, and his readiness to encourage every shade of opinion and to bring out the best in his colleagues, has enabled IUCAA to develop an international reputation. The centre is now seen as an example of how the energies of the research institutes and universities in India, usually considered disparate, could be harnessed together to excellent effect.

It is the general practice to list, in a volume of this kind, the scientific and other works of the person it seeks to honour. The list in the present case would have been rather unusually long, and we have therefore decided, in consultation with Jayant, that we will enumerate only his scientific books. These expose much of the work he has presented elsewhere in the form of research papers and review articles. They also present highly readable and often pedagogic accounts of modern astrophysics, and will surely continue to be read for a long time to come. Amongst the works that we will leave unlisted will be his contributions to the annals of science fiction, which have helped much to endear him to the general public. In this matter too Jayant has followed in the steps of Fred Hoyle.

Naresh Dadhich
Ajit Kembhavi

Acknowledgments

We wish to thank Professor K. S. V. S. Narasimhan for a careful reading of the manuscript.

Chapter 1

OBSERVATIONS AND THEORY[1]

Halton Arp
Max-Planck Institut für Astrophysik
Garching, Germany

1. INTRODUCTION

The most predictable observation concerning theories is that they will probably always turn out to be wrong. From Ptolemy to phlogisten these excercises have wasted untold model calculations and obsoleted endless sermons. Nevertheless, for the last 77 years, eschewing all humility, orthodox science has insisted on the theory that the entire universe was created instantaneously out of nothing. Observations for the last 33 years have shown this to be wrong - but these basic facts of science have been rejected on the grounds there was no theory to "explain" them.

Since 1977, however, there has not even been this feeble excuse for abandoning empiricism. That was the year in which Jayant Narlikar published a short paper in Annals of Physics (107, p325). The paper outlined how a more general solution of the equations of general relativity permitted matter to be "created" i. e. enter a black hole and remerge somewhere from a white hole without passing through a singularity where physics just broke down. This was not just another play with words because it turned out that the newly created matter would have to have a high intrinsic redshift. The latter is just what observations with optical and radio telescopes had been requiring since 1966!

As contradictory cases mounted over the years, the Big Bang theory had to be rescued by postulating an ever increasing number of adjustable parameters. As a consequence there is today a giant tsunami of evidence cresting above the Big Bang. It demonstrates continual creation of galaxies and evolution of intrinsic redshift in an indefinitely old and

[1]Editors' note: Dr. Halton Arp has requested that his contribution be presented as two separate articles, which we do in this chapter and the next.

N. Dadhich and A. Kembhavi (eds.), The Universe, 1–6.

large universe. By now we can start anywhere with this evidence so let us start with new results on a class of objects called active galaxies.

2. ACTIVE GALAXIES

In the preceding paper, preliminary investigation of two Ultra Luminous Infrared Galaxies (ULIRG"s) are reported. It is clear that these very disturbed objects are being torn apart during the process of ejecting high redshift quasars. Empirical evolutionary sequences show that the ULIRG's themselves are very active galaxies recently evolved from quasars. Therefore they also possess an appreciable component of intrinsic redshift. Conventionally this redshift gives too large a distance and this is why these objects are considered to be so "overluminous". As we shall comment later, however, they do not look at all like the most luminous galaxies of which we have certain knowledge. Instead they resemble small, active companion galaxies to larger, older parent galaxies. For example, Markarian 273 is an obvious companion to the large, nearby spiral, Messier 101.

The defining characteristic of active galaxies is that they show enormous concentrations of energy inside very small nuclei. They also show optical, radio and X-ray jets and plumes of material emerging from their centers. The latter is not surprising since the concentrated energy must expand and escape somehow. It has been accepted for about 40 years that active galaxies eject radio material so it is difficult to understand why the ejections associated with quasars are not recognized. But the expulsion of material is clearly responsible for the disrupted appearance of the active galaxies. Why then does conventional astronomy make an enormous industry out of a completely different, ad hoc explanation for morphologically disturbed galaxies - namely mergers!

3. MERGERS?

What is the conventional view of disturbed galaxies and ULIRG's? It is that two independent galaxies are merging. One galaxy sees another and heads directly for it. In its excitement it forgets about angular momentum and unerringly scores a direct hit. To judge how reasonable this is one could ask how many comets are perturbed into the solar system and proceed to plunge directly into the sun?

In all honesty, however, I must admit that my long term scorn for the merger scenario has been tempered by recent evidence on ejection from active galaxies. For many years it was clear that there was a tendency for galaxies to eject along their minor axes. But recently there have been some cases where ejection has been aligned with striking accuracy along the minor axis (6 quasars from NGC3516 , Chu et al. 1998, and five Quasars and four companion galaxies from NGC5985, Arp 1999). It is clear that proto galaxies ejected exactly out along the minor axis, and evolving into companion galaxies as they eventually fall back (Arp 1997;1998) will have little or no angular momentum and therefore move on plunging orbits. Their chances of colliding with the parent galaxy are therefore much greater than if they were field galaxies. So maybe there is some usefulness after all to those many detailed calculations which have been carried out on colliding galaxies.

But when the ejection of protogalactic material takes place in the plane or tries to exit through the substance of the parent galaxy then an entirely different scenario develops. Using the low mass creation theory, one can now begin to connect these events with previously uninterpretable observations.

4. SUPERFLUID

In 1957 the famous Armenian astronomer Ambartsumian concluded from looking at survey photographs that galaxies were formed by ejection from other galaxies. As an accomplished astrophysicist he realized that would require ejection in an initially non-solid form form but with properties different from a normal plasma. He called it "superfluid". In spite of general agreement that Ambartsumian was a great scientist his important conclusion about the formation of galaxies has been ignored.

But now the Hoyle-Narlikar variable mass theory is required to explain the high intrinsic redshifts of the quasars ejected from galaxies. The creation of mass in the centers of galaxies with this same variable mass theory then also solves the major problem which must have caused Ambartsumian to use the term "superfluid", namely that a normal, hot plasma expanding from the small dimensions of a galaxy nucleus would not have been able to condense into a new galaxy. In contrast, as the particles in the newly created plasma age they gain mass and, in order to conserve momentum, must slow their velocity. This means the plasma cools as it ages and also its self gravitation increases - both factors working in the direction of condensing the material into a proto galaxy.

The second major obstacle overcome by starting the particles off with near zero mass is the initial velocity of ejection. Observations have shown examples of ejected material in jets approaching closer and closer to the speed of light. Physicists believe that as a particle approaches the speed of light its mass must approach infinity. In other words one has to pump an enormous amount of energy into a huge number of particles to get the velocities (gamma factors) which are implied by the observations. If the particles are initially near zero mass, however, they are almost all energy and are emerging naturally with near the signal velocity, c.

In M87, the very strong radio galaxy in the Virgo Cluster, knots in the jet have been measured by their proper motion to have apparent outgoing velocities of 5 to 6 c. But further out along this jet we find quasars and companion galaxies which the knots must evolve into. Now, however, all the calculations based on the assumption that the knots consist of normal plasma will have to be redone with a low mass plasma, e.g. the calculations of supposed shock fronts and containment envelopes. (See Arp 1998,1999)

5. EXPLODING GALAXIES

There is a strong (and in some cases almost perfect) tendency for quasars to be initially ejected out along the minor axis and also ordered in descending redshift with angular separation. Nevertheless there are some cases where quasars are found close to their galaxy of origin but not ordered in redshift. The key to understanding this situation lies in the observation that the nearby galaxy of origin is usually spectacularly disrupted. What could cause this disruption? The obvious inference is that the process of ejection has, somehow, fragmented the galaxy when the ejection is not out along the minor axis.

At this particular point the usefulness of the variable mass theory becomes especially apparent. We are able to visualize a cloud of low particle mass material pushing out against the material of the galaxy, initially with velocity c. Low mass particle cross sections are large and eject and entrain the material of the galaxy into long, emerging jets. The initial pulse of energy concentrated at the center of the galaxy plus the sudden decentralization of mass explodes and tears asunder the parent galaxy. Moreover, the new material is retarded and fragmented so that it develops into many smaller new proto galaxies much closer to the, by now, thoroughly disrupted galaxy. This is the case where the new material does not exit along the minor axis.This is exactly what is observed as shown here in Figures 1 and 2.

Here the disrupted galaxy is 53W003 (a blue, radio, galaxy). As the picture shows it has been disrupted into at least three pieces. A pair of almost perfectly aligned quasars of $z = 2.389$ and $z = 2.392$ have apparently come out fairly unimpeded. (There are, as expected, brighter quasars of $z = 1.09$ and $z = 1.13$ about 7 arcmin further along in this direction). The rest of the quasars, about 18 similarly high redshift objects, have wound up in a cloud only about 1.5 arcmin from the disrupted

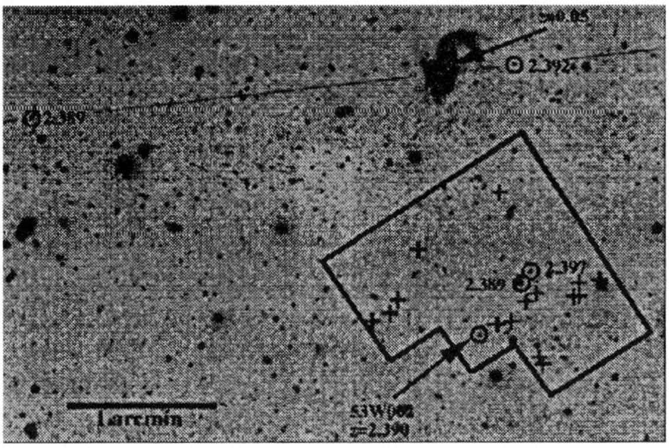

Figure 1.1 Part of a 4m PF-CCD field in the F410M filter (4150A, filter width 150A). The WFPC2 search fields are outlined - plus signs show non-AGN Ly α emitters. Quasars in the cluster are circled with z marked. From Keel et al. 1998.

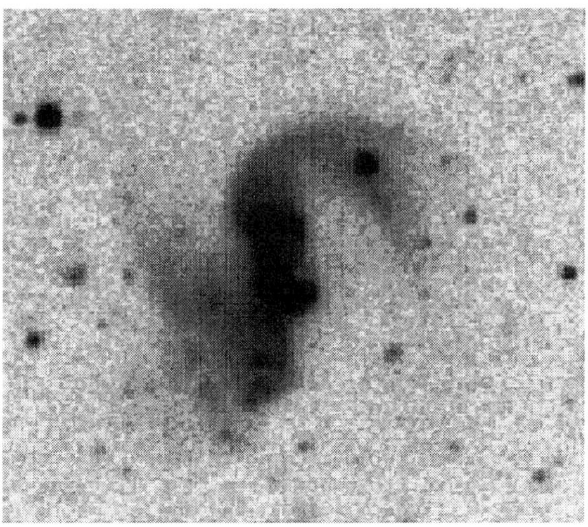

Figure 1.2 Enlargement of z = .05 galaxy in Fig.1. Note how this blue radio galaxy, 53W003, has multiple components. Image courtesy W. Keel.

galaxy. Evidently they represent some low mass plasma that was broken up into smaller clouds in its violent exit from the galaxy. In support of this scenario, high resolution, Hubble Telescope images of these high redshift objects show them to be blue and irregular. At their conventional redshift distance they would have absolute magnitudes of M = -24 mag. - well into the supposed quasar range of luminosity. Yet they have an extended morphology, whereas, in general, brighter quasars of the same redshift are point-like.

More broadly, this leads me to comment that the faint images in the famous Hubble Deep Field exposure which have such large redshifts are of predominantly blue, irregular morphology. At their conventional redshift distance they should be enormously luminous. But all our experience with genuinely luminous galaxies indicates such galaxies should be massive, relaxed, equilibrium forms - like E galaxies, or at least Sb's. These Hubble Deep Field objects have ragged, irregular looking dwarf morphology. Instead of a new kind of object suddenly discovered in the universe would it not be plausible that they are really relatively nearby dwarfs but simply have high redshifts because they are young?

6. A USEFUL THEORY

Speaking for myself, the Narlikar general solution of the relativistic field equations has been a salvation. It has opened up possibilities of understanding the observational facts - facts which must be accounted for if we are to have a science. In the dogma of current astronomy, evidence no matter how many times confirmed, cannot be accepted if it does not fit Big Bang assumptions. With the the variable mass theory, however, essentially all the salient observational facts can be related to each other in a physically understandable, reasonable way. Even if it is only a stepping stone to a future, deeper theory - I must say, thank you Jayant.

References

[1] Arp, H., 1997, *Journ. Astrophys. Astron.*, 393.

[2] Arp, H., 1998, *Seeing Red: Redshifts, Cosmology and Academic Science*, Apeiron, Montreal.

[3] Arp, H. 1999, *Astrophys. J.* , submitted.

Keel W., Windhorst R., Cohen S., Pascarelle S. & Holmes M. 1998, NOAO Newsletter **53**, 1.

Narlikar, J. V. 1977, *Ann. Phys.* **107**, 325.

Chapter 2

EJECTION FROM ULTRALUMINOUS INFRARED GALAXIES

Halton Arp
Max-Planck Institut für Astrophysik
Garching, Germany

Abstract

Active galaxies, particularly Seyferts, have been shown to eject material in various forms including quasars with high intrinsic redshifts. A class of active galaxy which has so far not been analyzed from this standpoint is the so called Ultra Luminous Infrared Galaxies (ULRIG's). Here we report the very beginning of an analysis of the three most luminous examples of such galaxies. Aided by the availability of the new VLA all sky radio surveys it is clear that these ULRIG's show especially strong evidence for ejection in optical, radio and X-ray wavelengths . These ejections are strikingly connected with adjacent quasars, both with those of known redshifts and those which are candidate quasars waiting to be confirmed.

1. MARKARIAN 273

This is a torn apart galaxy with a brilliant, long optical jet. At a conventional distance corresponding to its redshift (z = .038) it is one of the most luminous galaxies known in red wavelengths. Hence it is called an Ultra Luminous Infrared Galaxy (ULIRG). When observed in X-rays the galaxy has an active center. Only 1.3 arcmin NE, right at the end of a broad optical filament, lies another X-ray source (see Figure 2.1). When the spectrum of this companion (Mark273x) was taken it was reported as z = .038, the same as the central galaxy. Naturally this was interpreted as showing that Mark273x was a "dwarf" Seyfert interacting with Mark273. Fortunately the investigators checked the spectrum (Xia et al. [2], [3]). They found they had accidentally measured an HII region

N. Dadhich and A. Kembhavi (eds.), The Universe, 7–12.

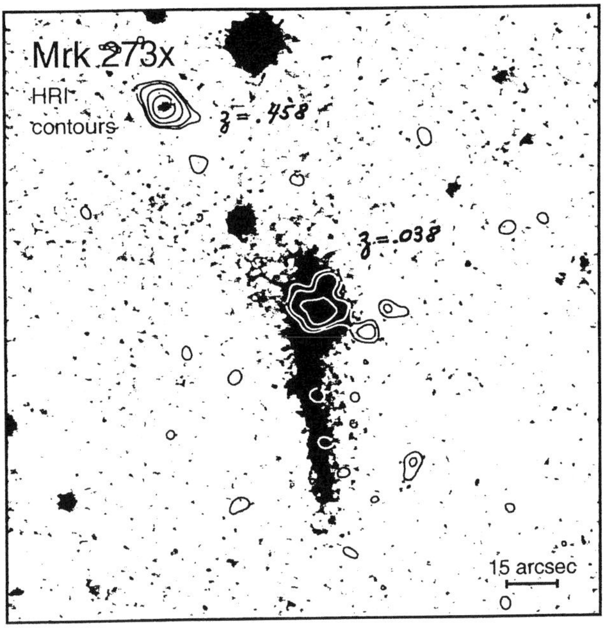

Figure 2.1 Copy of R film from POSSII. The X-ray contours around Mark273x (upper left) and Mark273 (center) are from Xia et al. [2]. Redshifts of each object as measured by Xia et al. [3]. Photographs to fainter surface brightnesses show luminous material extending in the direction of, and almost to, Mark273x.

in Mark273 and that Mark273x was actually a high redshift object of z = .458.

As in untold numbers of similar cases, as soon as the high redshift of the companion was discovered it was relegated to the background as an unassociated object. But, embarassingly, in this case it had already been claimed to be associated at the same distance. Tracking down the X-ray map of this system revealed at a glance that the z = .038 galaxy and the z = .458 companion were elongated toward each other! Moreover there was a significant excess of X-ray sources around the active central galaxy indicating further physically associated X-ray sources. Two of the brightest lay only 6.2 and 6.6 arcmin to the SE. The first was a catalogued quasar of z = .941 and the second an obvious quasar candidate whose redshift needs to be measured. As shown in Figure 2.2 there are both X-ray and radio jets emanating from Mark273 in the direction of these two additional quasars. Moreover the fainter radio emissions form two separate filaments leading directly to the two quasars. On a deep optical plate one can see the beginning of these two filaments starting SE from the strong optical jet which dominates

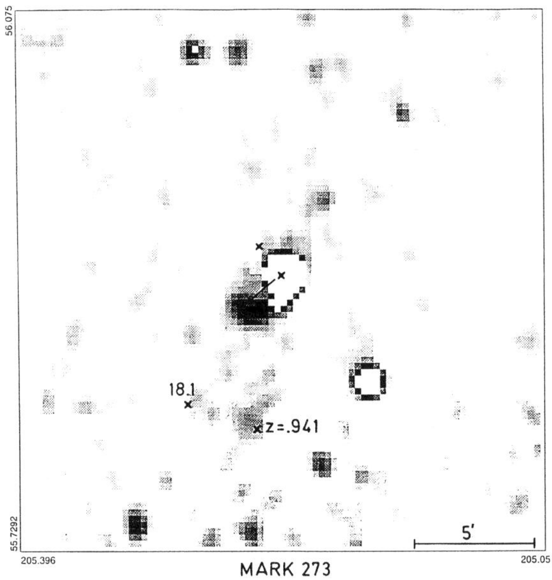

Figure 2.2 Radio map from the NRAO VLA Sky Survey (NVSS). The four brightest X-ray sources in the region are marked with X's. The direction of the X-ray jet from Mark273 is indicated by an arrow. Faint radio filaments lead southeastward to the quasar (z = 0.941) and the the quasar candidate (V = 18.1 mag.). This is generally along the line of the main radio and X-ray extensions from Mark273. Note also the exact alignment of Mark273x and the strong radio source to the SW across Mark273.

Mark273. (See deep R photograph of Mark 273 on web page of John Hibbard, www.cv.nrao.edu/ jhibbard)

This active galaxy appears to be ejecting optical, X-ray and radio material in two roughly orthogonal directions. (Note the exact alignment of 273x with the strong radio source to the SW of Mark273.) Associated with these ejections are high redshift quasars and quasar-like objects. Although all of these kinds of ejections have been observed many time before (see Arp [1] for a review), the ULIRG galaxies seem to be especially active. The authors of the original paper measuring Mark273x (Xia et al. [2]) report that in correlating ROSAT X-ray sources with ULIRG's: "...we find that some ULIRG's have soft X-ray companions within a few arcminutes of each source" and "This phenomenon was first mentioned by Turner, Urry and Mushotzky (1993)...". Later (Xia et al. [3]) state: "It is interesting to note in passing that the X-ray companions of the three nearest ULIGs (Arp 220, Mrk 273 and Mrk 231) are all background sources...".

Just a glance at two of the other most luminous ULIRG's (Mark231 and Arp220) shows similar evidence for ejection from these enormously

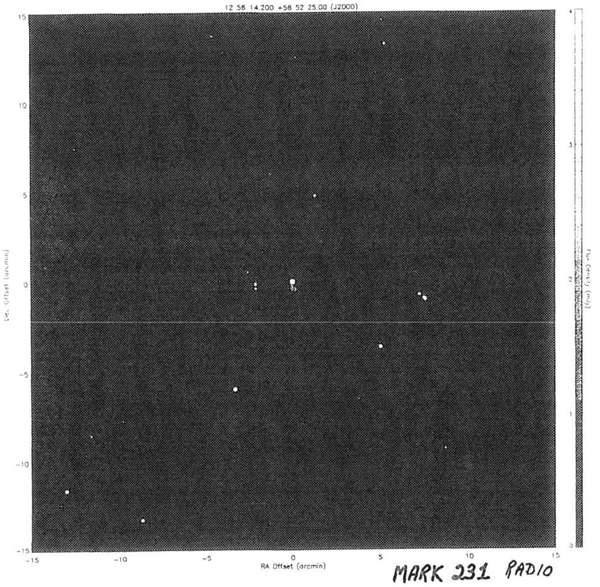

Figure 2.3 High resolution radio map centered on Mark231 (at 20 cm from VLA FIRST). Note puff of radio material just below the ULIRG and double nature of radio sources paired across Mark231.

disturbed galaxies. I will show now some preliminary evidence for Mark231 but it is already clear that there appear to be strong X-ray sources, radio ejections and physically associated high redshift objects connected to all three of these ULIRG's.

2. MARKARIAN 231

Figure 2.3shows a 30x30 arcmin radio map around Mark231. The images are high resolution 20cm from the VLA FIRST survey (www.nrao.edu). The brightest object in the center is Mark231. There is a puff of radio material immediately below the galaxy. Forming a striking pair across Mark231 are radio sources both of which are close doubles. The multiplicity of these flanking sources is unusual and suggests secondary ejection. At the least these radio sources are strongly indicated to be associated with the central, active galaxy.

Figure 2.4 shows an approximately 19x19 arcmin continuum radio map at lower resolution but fainter surface brightness. Here we see a continuous radio extension to the East of Mark231 including the multiple source seen previously on the higher resolution map. In addition we see a radio extension to the West, in the direction of the strong, close double source. There is also a string of small sources extending northward

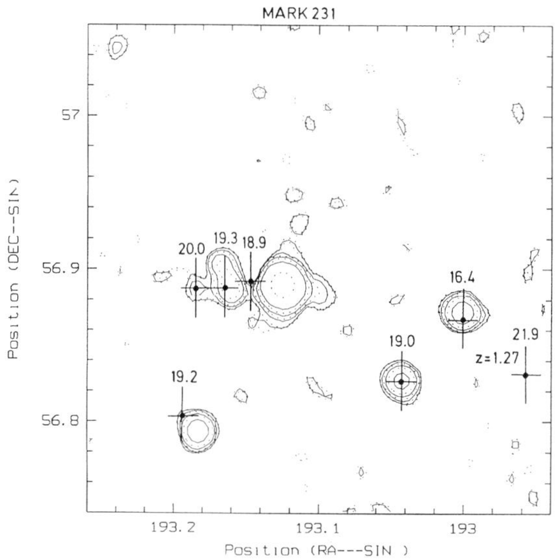

Figure 2.4 Contour maps of low surface brightness radio material around Mark231. Continuous connection of radio material to the east of the galaxy contains blue quasar candidates with the labeled, V apparent magnitudes. The remaining radio sources have quasar candidates at the marked positions. The strong double source to the west falls close to a quasar candidate of V = 16.4 mag. The only catalogued quasar in the field is faint and of z = 1.27.

from the central galaxy. We appear to be seeing another example of ejection in roughly orthogonal directions. (It is interesting to note that at FIRST resolution the strong radio source opposite Mark273x is also a close double.)

Two color APM finding charts have been centered at the positions of some of the radio sources indicated in Fig. 4. The charts reveal blue, candidate quasar images quite close to the radio positions. They are labeled in Fig. 4 with plus signs and the apparent visual magnitude of the candidate. They need to be analyzed spectroscopically but it can already be noted that the candidate at the position of the eastern radio lobe (V=19.3 mag.) is very blue and therefore highly probable. The strong western (double) source is close to a bright (V=16.4 mag.) candidate which has fainter candidates aligned across it - suggestive again of secondary ejection. The only catalogued quasar has z = 1.27 and is located in the direction of the western radio extension from Mark231. The X-ray maps are in the process of being analyzed and will undoubtedly add considerably to the understanding of the Mark231 region. Similarly, X-ray and radio maps of Arp220 are being analyzed and together with

Mark273 and Mark231 will form a representative sample of the most active infrared excess galaxies.

3. CONCLUSION

In the case of the tendency for long lines of ordered quasars to come out along the minor axes of disk galaxies [1] it was suggested that ejections encountered the least resistance along this spin axis. It is suggested here that if the ejections try to penetrate any appreciable material in the parent galaxy that they will expel and entrain gas and dust and dynamically rupture the galaxy. The production of new material in the centers of such galaxies would then then be responsible for the energetic X-ray and radio jets, the explosive morphology and the numbers of high energy, intrinsically redshifted quasars found nearby.

References

[1] Arp, H., 1998, *Seeing Red: Redshifts, Cosmology and Academic Science*, Apeiron, Montreal.

[2] Xia, X.-Y., Boller, T., Wu, H., Deng, Z.-G., Gao, Y., Zou, Z.-L., Mao, S. and Boerner, G., 1998, *Astrophys. J.* **496**, L99.

[3] Xia, X.-Y., Mao, S., Wu, H., Liu, X.-W., Deng Z.-G., Gao, Y., Zou, Z.-L., 1999, *Astrophys. J.* , in press.

Chapter 3

QUANTUM MECHANICS OF GEOMETRY

A. Ashtekar
Center for Gravitational Physics and Geometry
Department of Physics, The Pennsylvania State University
University Park, PA 16802, USA

It is a pleasure to dedicate this article to Professor Jayant Narlikar on the occasion of his 60th birthday.

—

Abstract

Over the past six years, a detailed framework has been constructed to unravel the quantum nature of the Riemannian geometry of physical space. A review of these developments is presented at a level which should be accessible to graduate students in physics. As an illustrative application, I indicate how some of the detailed features of the microstructure of geometry can be tested using black hole thermodynamics. Current and future directions of research in this area are discussed.

1. INTRODUCTION

During his Göttingen inaugural address in 1854, Riemann [1] suggested that geometry of space may be more than just a fiducial, mathematical entity serving as a passive stage for physical phenomena, and may in fact have direct physical meaning in its own right. General relativity provided a brilliant confirmation of this vision: curvature of space

N. Dadhich and A. Kembhavi (eds.), The Universe, 13–34.

now encodes the physical gravitational field. This shift is profound. To bring out the contrast, let me recall the situation in Newtonian physics. There, space forms an inert arena on which the dynamics of physical systems –such as the solar system– unfolds. It is like a stage, an unchanging backdrop for all of physics. In general relativity, by contrast, the situation is very different. Einstein's equations tell us that matter curves space. Geometry is no longer immune to change. It reacts to matter. It is dynamical. It has "physical degrees of freedom" in its own right. In general relativity, the stage disappears and joins the troupe of actors! Geometry is a physical entity, very much like matter.

Now, the physics of this century has shown us that matter has constituents and the 3-dimensional objects we perceive as solids are in fact made of atoms. The continuum description of matter is an approximation which succeeds brilliantly in the macroscopic regime but fails hopelessly at the atomic scale. It is therefore natural to ask: Is the same true of geometry? If so, what is the analog of the 'atomic scale?' We know that a quantum theory of geometry should contain three fundamental constants of Nature, $c, G, \hbar$, the speed of light, Newton's gravitational constant and Planck's constant. Now, as Planck pointed out in his celebrated paper that marks the beginning of quantum mechanics, there is a unique combination, $\ell_P = \sqrt{\hbar G/c^3}$, of these constants which has dimension of length. ($\ell_P \approx 10^{-33}$cm.) It is now called the Planck length. Experience has taught us that the presence of a distinguished scale in a physical theory often marks a potential transition; physics below the scale can be very different from that above the scale. Now, all of our well-tested physics occurs at length scales much bigger than than ℓ_P. In this regime, the continuum picture works well. A key question then is: Will it break down at the Planck length? Does geometry have constituents at this scale? If so, what are its atoms? Its elementary excitations? Is the space-time continuum only a 'coarse-grained' approximation? Is geometry quantized? If so, what is the nature of its quanta?

To probe such issues, it is natural to look for hints in the procedures that have been successful in describing matter. Let us begin by asking what we mean by quantization of physical quantities. Take a simple example –the hydrogen atom. In this case, the answer is clear: while the basic observables –energy and angular momentum– take on a continuous range of values classically, in quantum mechanics their eigenvalues are discrete; they are quantized. So, we can ask if the same is true of geometry. Classical geometrical quantities such as lengths, areas and volumes can take on continuous values on the phase space of general relativity. Are the eigenvalues of corresponding quantum operators discrete? If so, we would say that geometry is quantized and the precise eigenvalues and eigenvectors of geometric operators would reveal its detailed microscopic properties.

Thus, it is rather easy to pose the basic questions in a precise fashion. Indeed, they could have been formulated soon after the advent of quantum mechanics. Answering them, on the other hand, has proved to be surprisingly difficult. The main reason, I believe, is the inadequacy of standard techniques. More precisely, to examine the microscopic structure of geometry, we must treat Einstein gravity quantum mechanically, i.e., construct at least the basics of a quantum theory of the gravitational field. Now, in the traditional approaches to quantum field theory, one *begins* with a continuum, background geometry. To probe the nature of quantum geometry, on the other hand, we should *not* begin by assuming the validity of this picture. We must let quantum gravity decide whether this picture is adequate; the theory itself should lead us to the correct microscopic model of geometry.

With this general philosophy, in this article I will summarize the picture of quantum geometry that has emerged from a specific approach to quantum gravity. This approach is non-perturbative. In perturbative approaches, one generally begins by assuming that space-time geometry is flat and incorporates gravity –and hence curvature– step by step by adding up small corrections. Discreteness is then hard to unravel[1]. In the non-perturbative approach, by contrast, there is no background metric at all. All we have is a bare manifold to start with. All fields –matter as well as gravity/geometry– are treated as dynamical from the beginning. Consequently, the description can not refer to a background metric. Technically this means that the full diffeomorphism group of the manifold is respected; the theory is generally covariant.

As we will see, this fact leads one to Hilbert spaces of quantum states which are quite different from the familiar Fock spaces of particle physics. Now gravitons –the three dimensional wavy undulations on a flat metric– do not represent fundamental excitations. Rather, the fundamental excitations are *one* dimensional. Microscopically, geometry is rather like a polymer. Recall that, although polymers are intrinsically one dimensional, when densely packed in suitable configurations they can exhibit properties of a three dimensional system. Similarly, the familiar continuum picture of geometry arises as an approximation: one can regard the fundamental excitations as 'quantum threads' with which one can 'weave' continuum geometries. That is, the continuum picture arises upon coarse-graining of the semi-classical 'weave states'. Gravitons are no longer the fundamental mediators of the gravitational interaction. They now arise only as approximate notions. They represent perturbations of weave states and mediate the gravitational force only in the semi-classical approximation. Because the non-perturbative states are polymer-like, geometrical observables turn out to have discrete spectra. They provide a rather detailed picture of quantum geometry from which physical predictions can be made.

The article is divided into two parts. In the first, I will indicate how one can reformulate general relativity so that it resembles gauge theories. This formulation provides the starting point for the quantum theory. In particular, the one-dimensional excitations of geometry arise as the analogs of 'Wilson loops' which are themselves analogs of the line integrals $\exp i \oint A.d\ell$ of electro-magnetism. In the second part, I will indicate how this description leads us to a quantum theory of geometry. I will focus on area operators and show how the detailed information about the eigenvalues of these operators has interesting physical consequences, e.g., to the process of Hawking evaporation of black holes.

I should emphasize that this is *not* a technical review. Rather, it is written in the same spirit that drives Jayant's educational initiatives. I thought this would be a fitting way to honor Jayant since these efforts have occupied so much of his time and energy in recent years. Thus my aim is present to beginning researchers an overall, semi-quantitative picture of the main ideas. Therefore, the article is written at the level of colloquia in physics departments in the United States. I will also make some historic detours of general interest. At the end, however, I will list references where the details of the central results can be found.

2. FROM METRICS TO CONNECTIONS

2.1 GRAVITY VERSUS OTHER FUNDAMENTAL FORCES

General relativity is normally regarded as a dynamical theory of metrics —tensor fields that define distances and hence geometry. It is this fact that enabled Einstein to code the gravitational field in the Riemannian curvature of the metric. Let me amplify with an analogy. Just as position serves as the configuration variable in particle dynamics, the three dimensional metric of space can be taken to be the configuration variable of general relativity. Given the initial position and velocity of a particle, Newton's laws provide us with its trajectory in the position space. Similarly, given a three dimensional metric and its time derivative at an initial instant, Einstein's equations provide us with a four dimensional space-time which can be regarded as a trajectory in the space of 3-metrics [2].

However, this emphasis on the metric sets general relativity apart from all other fundamental forces of Nature. Indeed, in the theory of electro-weak and strong interactions, the basic dynamical variable is a (matrix-valued) vector potential, or a connection. Like general relativity, these theories are also geometrical. The connection enables one to parallel-transport objects along curves. In electrodynamics, the object is a charged particle such as an electron; in chromodynamics, it is a particle with internal color, such as a quark. Generally, if we move the object around a closed loop, we find that its state does not return to the initial value; it is rotated by an unitary matrix. In this case, the connection is said to have curvature and the unitary matrix is a measure of the curvature in a region enclosed by the loop. In the case of electrodynamics, the connection is determined by the vector potential and the curvature by the electro-magnetic field strength.

Since the metric also gives rise to curvature, it is natural to ask if there is a relation between metrics and connections. The answer is in the affirmative. Every metric defines a connection —called the Levi-Civita connection of the metric. The object that the connection enables one to parallel transport is a vector. (It is this connection that determines the geodesics, i.e. the trajectories of particles in absence of non-gravitational forces.) It is therefore natural to ask if one can not use this connection as the basic variable in general relativity. If so, general relativity would be cast in a language that is rather similar to gauge theories and the description of the (general relativistic) gravitational interaction would be very similar to that of the other fundamental interactions of Nature. It turns out that the answer is in the affirmative. Furthermore, both Einstein and Schrödinger gave such a reformulation of general relativity. Why is this fact then not generally known? Indeed, I know of no textbook on general relativity which even mentions it. One reason is that in their reformulation the basic equations are somewhat complicated —but not much more complicated, I think, than the standard ones in terms of the metric. A more important reason is that we tend to think of distances, light cones and causality as fundamental. These are directly determined by the metric and in a connection formulation, the metric is a 'derived' rather than a fundamental concept. But in the last few years, I have come to the conclusion that the real reason why the connection formulation of Einstein and Schrödinger has remained so obscure may lie in an interesting historical episode. I will return to this point at the end of this section.

2.2 METRICS VERSUS CONNECTIONS

Modern day researchers re-discovered connection theories of gravity after the invention and successes of gauge theories for other interactions. Generally, however, these formulations lead one to theories which are quite distinct from general relativity and the stringent experimental tests of general relativity often suffice to rule them out. There is, however, a reformulation of general relativity itself in which the basic equations are simpler than the standard ones: while Einstein's equations are non-polynomial in terms of the metric and its conjugate momentum, they turn out to be low order polynomials in terms of the new connection and its conjugate momentum. Furthermore, just as the simplest particle trajectories in space-time are given by geodesics, the 'trajectory' determined by the time evolution of this connection according to Einstein's equation turns out to be a geodesic in the configuration space of connections.

In this formulation, the phase space of general relativity is identical to that of the Yang-Mills theory which governs weak interactions. Recall first that in electrodynamics, the (magnetic) vector potential constitutes the configuration variable and the electric field serves as the conjugate momentum. In weak interactions and general relativity, the configuration variable is a matrix-valued vector potential; it can be written as $\mathbf{A}_i\tau_i$ where $\mathbf{A}_i$ is a triplet of vector fields and τ_i are the Pauli matrices. The conjugate momenta are represented by $\mathbf{E}_i\tau_i$ where $\mathbf{E}_i$ is a triplet of vector fields[3]. Given a pair $(\mathbf{A}_i, \mathbf{E}_i)$ (satisfying appropriate conditions as noted in footnote 2), the field equations of the two theories determine the complete time-evolution, i.e., a dynamical trajectory.

The field equations –and the Hamiltonians governing them– of the two theories are of course very different. In the case of weak interactions, we have a background space-time and we can use its metric to construct the Hamiltonian. In general relativity, we do not have a background metric. On the one hand this makes life very difficult since we do not have a fixed notion of distances or causal structures; these notions are to arise from the solution of the equations we are trying to write down! On the other hand, there is also tremendous simplification: Because there is no background metric, there are very few mathematically meaningful, gauge invariant expressions of the Hamiltonian that one can write down. (As we will see, this theme repeats itself in the quantum theory.) It is a pleasant surprise that the simplest non-trivial expression one can construct from the connection and its conjugate momentum is in fact the correct one, i.e., is the Hamiltonian of general relativity! The expression is at most quadratic in $\mathbf{A}_i$ and at most quadratic in $\mathbf{E}_i$. The similarity with gauge theories opens up new avenues for quantizing general relativity and the simplicity of the field equations makes the task considerably easier.

What is the physical meaning of these new basic variables of general relativity? As mentioned before, connections tell us how to parallel transport various physical entities around curves. The Levi-Civita connection tells us how to parallel transport vectors. The new connection, $\mathbf{A}_i$, on the other hand, determines the parallel transport of *left handed spin-$\frac{1}{2}$ particles* (such as the fermions in the standard model of particle physics) —the so called *chiral fermions*. These fermions are mathematically represented by spinors which, as we know from elementary quantum mechanics, can be roughly thought of as 'square roots of vectors'. Not surprisingly, therefore, the new connection is not completely determined by the metric alone. It requires additional information which roughly is a square-root of the metric, or a tetrad. The conjugate momenta $\mathbf{E}_i$ represent restrictions of these tetrads to space. They can be interpreted as spatial triads, i.e., as 'square-roots' of the metric of the 3-dimensional space. Thus, information about the Riemannian geometry of space is coded directly in these momenta. The (space and) time-derivatives of the triads are coded in the connection.

To summarize, there *is* a formulation of general relativity which brings it closer to theories of other fundamental interactions. Furthermore, in this formulation, the field equations simplify greatly. Thus, it provides a natural point of departure for constructing a quantum theory of gravity and for probing the nature of quantum geometry non-perturbatively.

2.3 HISTORICAL DETOUR

To conclude this section, let me return to the piece of history involving Einstein and Schrödinger that I mentioned earlier. In the forties, both men were working on unified field theories. They were intellectually very close. Indeed, Einstein wrote to Schrödinger saying that he was perhaps the only one who was not 'wearing blinkers' in regard to fundamental questions in science and Schrödinger credited Einstein for inspiration behind his own work that led to the Schrödinger equation. Einstein was in Princeton and Schrödinger in Dublin. But During the years 1946-47, they frequently exchanged ideas on unified field theory and, in particular, on the issue of whether connections should be regarded as fundamental or metrics. In fact the dates on their letters often show that the correspondence was going back and forth with astonishing speed. It reveals how quickly they understood the technical material the other hand sent, how they hesitated, how they teased each other. Here are a few quotes:

The whole thing is going through my head like a millwheel: To take Γ [the connection] *alone as the primitive variable or the* g*'s* [metrics] *and*

Γ's ? ...

—Schrödinger, May 1st, 1946.

How well I understand your hesitating attitude! I must confess to you that inwardly I am not so certain ... We have squandered a lot of time on this thing, and the results look like a gift from devil's grandmother.

—Einstein, May 20th, 1946

Einstein was expressing doubts about using the Levi-Civita connection alone as the starting point which he had advocated at one time. Schrödinger wrote back that he laughed very hard at the phrase 'devil's grandmother'. In another letter, Einstein called Schrödinger 'a clever rascal'. Schrödinger was delighted and took it to be a high honor. This continued all through 1946. Then, in the beginning of 1947, Schrödinger thought he had made a breakthrough. He wrote to Einstein:

Today, I can report on a real advance. May be you will grumble frightfully for you have explained recently why you don't approve of my method. But very soon, you will agree with me...

—Schrödinger, January 26th, 1947

Schrödinger sincerely believed that his breakthrough was revolutionary [4]. Privately, he spoke of a second Nobel prize. The very next day after he wrote to Einstein, he gave a seminar in the Dublin Institute of Advanced Studies. Both the Taoiseach (the Irish prime minister) and newspaper reporters were invited. The day after, the following headlines appeared:

Twenty persons heard and saw history being made in the world of physics. ... The Taoiseach was in the group of professors and students. ..[To a question from the reporter] *Professor Schrödinger replied "This is the generalization. Now the Einstein theory becomes simply a special case ..."*

—Irish Press, January 28th, 1947

Not surprisingly, the headlines were picked up by New York Times which obtained photocopies of Schrödinger's paper and sent them to prominent physicists –including of course Einstein– for comments. As Walter Moore, Schrödinger's biographer puts it, Einstein could hardly believe that such grandiose claims had been made based on a what was at best a small advance in an area of work that they both had been pursuing for some time along parallel lines. He prepared a carefully worded response to the request from New York Times:

It seems undesirable to me to present such preliminary attempts to the public. ... Such communiqués given in sensational terms give the lay public misleading ideas about the character of research. The reader gets the impression that every five minutes there is a revolution in Science, somewhat like a coup d'état in some of the smaller unstable republics. ...

Einstein's comments were also carried by the international press. On seeing them, Schrödinger wrote a letter of apology to Einstein citing his desire to improve the financial conditions of physicists in the Dublin Institute as a reason for the exaggerated account. It seems likely that this 'explanation' only worsened the situation. Einstein never replied. He also stopped scientific communication with Schrödinger for three years.

The episode must have been shocking to those few who were exploring general relativity and unified field theories at the time. Could it be that this episode effectively buried the desire to follow up on connection formulations of general relativity until an entirely new generation of physicists who were blissfully unaware of this episode came on the scene?

3. QUANTUM GEOMETRY

3.1 GENERAL SETTING

Now that we have a connection formulation of general relativity, let us consider the problem of quantization. Recall first that in the quantum description of a particle, states are represented by suitable wave functions $\Psi(\mathbf{x})$ on the classical configuration space of the particle. Similarly, quantum states of the gravitational field are represented by appropriate wave functions $\Psi(\mathbf{A}_i)$ of connections. Just as the momentum operator in particle mechanics is represented by $\hat{P} \cdot \Psi_I = -i\hbar\,(\partial\Psi/\partial x_I)$ (with $I = 1, 2, 3$), the triad operators are represented by $\hat{\mathbf{E}}_i \cdot \Psi = -i\hbar G(\delta\Psi/\delta\mathbf{A}_i)$. The task is to express geometric quantities, such as lengths of curves, areas of surfaces and volumes of regions, in terms of triads using ordinary differential geometry and then promote these expressions to well-defined operators on the Hilbert space of quantum states. In principle, the task is rather similar to that in quantum mechanics where we first express observables such as angular momentum or Hamiltonian in terms of configuration and momentum variables $\mathbf{x}$ and $\mathbf{p}$ and then promote them to quantum theory as well-defined operators on the quantum Hilbert space.

In quantum mechanics, the task is relatively straightforward; the only potential problem is the choice of factor ordering. In the present case, by contrast, we are dealing with a *field theory*, i.e., a system with an infinite number of degrees of freedom. Consequently, in addition to factor ordering, we face the much more difficult problem of regularization. Let me explain qualitatively how this arises. A field operator, such as the triad mentioned above, excites infinitely many degrees of freedom. Technically, its expectation values are distributions rather than smooth fields. They don't take precise values at a given point in space. To obtain numbers, we have to integrate the distribution against a test function, which extracts from it a 'bit' of information. As we change our test or smearing field, we get more and more information. (Take the familiar Dirac δ-distribution $\delta(x)$; it does not have a well-defined value at $x = 0$. Yet, we can extract the full information contained in $\delta(x)$ through the formula: $\int \delta(x) f(x) dx = f(0)$ for all test functions $f(x)$.) Thus, in a precise sense, field operators are distribution-valued. Now, as is well known, product of distributions is not well-defined. If we attempt naively to give meaning to it, we obtain infinities, i.e., a senseless result. Unfortunately, all geometric operators involve rather complicated (in fact non-polynomial) functions of the triads. So, the naive expressions of the corresponding quantum operators are typically meaningless. The key problem is to regularize these expressions, i.e., to extract well-defined operators from the formal expressions in a coherent fashion.

This problem is not new; it arises in all physically interesting quantum field theories. However, as I mentioned in the Introduction, in other theories one has a background space-time metric and it is invariably used in a critical way in the process of regularization. For example, consider the electro-magnetic field. We know that the energy of the Hamiltonian of the theory is given by $H = \int (\mathbf{E} \cdot \mathbf{E} + \mathbf{B} \cdot \mathbf{B})\, d^3x$. Now, in the quantum theory, $\hat{\mathbf{E}}$ and $\hat{\mathbf{B}}$ are both operator-valued distributions and so their square is ill-defined. But then, using the background flat metric, one Fourier decomposes these distributions, identifies creation and annihilation operators and extracts a well-defined Hamiltonian operator by normal ordering, i.e., by physically moving all annihilators to the right of creators. This procedure removes the unwanted and unphysical infinite zero point energy form the formal expression and the subtraction makes the operator well-defined. In the present case, on the other hand, we are trying to construct a quantum theory of geometry/gravity and do not have a flat metric –or indeed, any metric– in the background. Therefore, many of the standard regularization techniques are no longer available.

3.2 GEOMETRIC OPERATORS

Fortunately, between 1992 and 1995, a new functional calculus was developed on the space of connections $\mathbf{A}_i$ —i.e., on the configuration space of the theory. This calculus is mathematically rigorous and makes no reference at all to a background space-time geometry; it is generally covariant. It provides a variety of new techniques which make the task of regularization feasible. First of all, there is a well-defined integration theory on this space. To actually evaluate integrals and define the Hilbert space of quantum states, one needs a measure: given a measure on the space of connections, we can consider the space of square-integrable functions which can serve as the Hilbert space of quantum states. It turns out that there is a preferred measure, singled out by the physical requirement that the (gauge covariant versions of the) configuration and momentum operators be self-adjoint. This measure is diffeomorphism invariant and thus respects the underlying symmetries coming from general covariance. Thus, there is a natural Hilbert space of states to work with[5]. Let us denote it by $\mathcal{H}$. Differential calculus enables one to introduce physically interesting operators on this Hilbert space and regulate them in a generally covariant fashion. As in the classical theory, the absence of a background metric is both a curse and a blessing. On the one hand, because we have very little structure to work with, many of the standard techniques simply fail to carry over. On the other hand, at least for geometric operators, the choice of viable expressions is now severely limited which greatly simplifies the task of regularization.

The general strategy is the following. The Hilbert space $\mathcal{H}$ is the space of square-integrable functions $\Psi(\mathbf{A}_i)$ of connections $\mathbf{A}_i$. A key simplification arises because it can be obtained as the (projective) limit of Hilbert spaces associated with systems with only a finite number of degrees of freedom. More precisely, given any graph γ (which one can intuitively think of as a 'floating lattice') in the physical space, using techniques which are very similar to those employed in lattice gauge theory, one can construct a Hilbert space $\mathcal{H}_\gamma$ for a quantum mechanical system with $3N$ degrees of freedom, where N is the number of edges of the graph[6]. Roughly, these Hilbert spaces know only about how the connection parallel transports chiral fermions along the edges of the graph and not elsewhere. That is, the graph is a mathematical device to extract $3N$ 'bits of information' from the full, infinite dimensional information contained in the connection, and $\mathcal{H}_\gamma$ is the sub-space of $\mathcal{H}$ consisting of those functions of connections which depend only on these $3N$ bits. (Roughly, it is like focusing on only $3N$ components of a vector with an infinite number of components and considering functions which depend only on these $3N$ components, i.e., are constants along the orthogonal directions.) To get the full information, we need all possible graphs. Thus, a function of connections in $\mathcal{H}$ can be specified by fixing a function in $\mathcal{H}_\gamma$ for *every* graph γ in the physical space. Of course, since two distinct graphs can share edges, the collection of functions on $\mathcal{H}_\gamma$

must satisfy certain consistency conditions. These lie at the technical heart of various constructions and proofs.

The fact that $\mathcal{H}$ is the (projective) limit of $\mathcal{H}_\gamma$ breaks up any given problem in quantum geometry into a set of problems in quantum mechanics. Thus, for example, to define operators on $\mathcal{H}$, it suffices to define a *consistent family of* operators on $\mathcal{H}_\gamma$ for each γ. This makes the task of defining geometric operators feasible. I want to emphasize, however, that the introduction of graphs is only for technical convenience. Unlike in lattice gauge theory, we are not *defining* the theory via a continuum limit (in which the lattice spacing goes to zero.) Rather, the full Hilbert space $\mathcal{H}$ of the continuum theory is already well-defined. Graphs are introduced only for practical calculations. Nonetheless, they bring out the one-dimensional character of quantum states/excitations of geometry: It is because 'most' states in $\mathcal{H}$ can be realized as elements of $\mathcal{H}_\gamma$ for some γ that quantum geometry has a 'polymer-like' character.

Let me now outline the result of applying this procedure for geometric operators. Suppose we are given a surface S, defined in local coordinates by x_3 = const. The classical formula for the area of the surface is: $A_S = \int d^2x \sqrt{E_i^3 E_i^3}$, where E_i^3 are the third components of the vectors $\mathbf{E}_i$. As is obvious, this expression is non-polynomial in the basic variables $\mathbf{E}_i$. Hence, off-hand, it would seem very difficult to write down the corresponding quantum operator. However, thanks to the background independent functional calculus, the operator can in fact be constructed rigorously.

To specify its action, let us consider a state which belongs to $\mathcal{H}_\gamma$ for *some* γ. Then, the action of the final, regularized operator $\hat{A}_S$ is as follows. If the graph has no intersection with the surface, the operator simply annihilates the state. If there are intersections, it acts at each intersection via the familiar angular momentum operators associated with $SU(2)$. *This simple form is a direct consequence of the fact that we do not have a background geometry*: given a graph and a surface, the diffeomorphism invariant information one can extract lies in their intersections. To specify the action of the operator in detail, let me suppose that the graph γ has N edges. Then the state Ψ has the form: $\Psi(\mathbf{A}_i) = \psi(g_1, ...g_N)$ for some function ψ of the N variables $g_1, ..., g_N$, where g_k ($\in SU(2)$) denotes the spin-rotation that a chiral fermion undergoes if parallel transported along the k-th edge using the connection $\mathbf{A}_i$. Since g_k represent the possible rotations of spins, angular momentum operators have a natural action on them. In terms of these, we can introduce 'vertex operators' associated with each intersection point

v between S and γ:

$$\hat{O}_v \cdot \Psi(A) = \sum_{I,L} k(I,L) \mathbf{J}_I \cdot \mathbf{J}_L \cdot \psi(g_1, ..., g_N) \tag{3.1}$$

where I, L run over the edges of γ at the vertex v, $k(I,J) = 0, \pm 1$ depending on the orientation of edges I, L at v, and $\mathbf{J}_I$ are the three angular momentum operators associated with the I-th edge. (Thus, $\mathbf{J}_I$ act only on the argument g_I of ψ and the action is via the three left invariant vector fields on $SU(2)$.) Note that the *the vertex operators resemble the Hamiltonian of a spin system, $k(I,L)$ playing the role of the coupling constant.* The area operator is just a sum of the square-roots of the vertex operators:

$$\hat{A}_S = \frac{G\hbar}{2c^3} \sum_v |O_v|^{\frac{1}{2}} \tag{3.2}$$

Thus, the area operator is constructed from angular momentum-like operators. Note that the coefficient in front of the sum is just $\frac{1}{2}\ell_P^2$, the square of the Planck length. This fact will be important later.

Because of the simplicity of these operators, their complete spectrum –i.e., full set of eigenvalues– is known explicitly: Possible eigenvalues a_S are given by

$$a_S = \frac{\ell_P^2}{2} \sum_v \left[2j_v^{(d)}(j_v^{(d)}+1) + 2j_v^{(u)}(j_v^{(u)}+1) - j_v^{(d+u)}(j_v^{(d+u)}+1) \right]^{\frac{1}{2}} \tag{3.3}$$

where v labels a finite set of points in S and $j^{(d)}, j^{(u)}$ and $j^{(d+u)}$ are non-negative half-integers assigned to each v, subject to the usual inequality

$$j^{(d)} + j^{(u)} \geq j^{(d+u)} \geq |j^{(d)} - j^{(u)}| . \tag{3.4}$$

from the theory of addition of angular momentum in elementary quantum mechanics. Thus the entire spectrum is discrete; *areas are indeed quantized!* This discreteness holds also for the length and the volume operators. Thus the expectation that the continuum picture may break down at the Planck scale is borne out fully. Quantum geometry is *very* different from the continuum picture. This may be the fundamental reason for the failure of perturbative approaches to quantum gravity.

Let us now examine a few properties of the spectrum. The lowest eigenvalue is of course zero. The next lowest eigenvalue may be called the *area gap.* Interestingly, area-gap is sensitive to the topology of the surface S. If S is open, it is $\frac{\sqrt{3}}{4}\ell_P^2$. If S is a closed surface –such as a 2-torus in a 3-torus– which fails to divide the spatial 3-manifold into an 'inside' and an 'outside' region, the gap turns out to be larger, $\frac{2}{4}\ell_P^2$. If S is a closed surface –such as a 2-sphere in R^3– which divides space into an 'inside' and an 'outside' region, the area gap turns out to be even larger; it is $\frac{2\sqrt{2}}{4}\ell_P^2$. Another interesting feature is that in the large area limit, the eigenvalues crowd together. This follows directly from the form of eigenvalues given above. Indeed, one can show that for large eigenvalues a_S, the difference Δa_S between consecutive eigenvalues goes as $\Delta a_S \leq (exp - \sqrt{a_S/\ell_P^2})\ell_P^2$. Thus, Δa_S goes to zero very rapidly. (The crowding is noticeable already for low values of a_S. For example, if S is open, there is only one non-zero eigenvalue with $a_S < 0.5{\ell_P}^2$, seven with $a_S < \ell_P^2$ and 98 with $a_S < 2\ell_P^2$.) Intuitively, this explains why the continuum limit works so well.

3.3 PHYSICAL CONSEQUENCES: DETAILS MATTER!

However, one might wonder if such detailed properties of geometric operators can have any 'real' effect. After all, since the Planck length is so small, one would think that the classical and semi-classical limits should work irrespective of, e.g., whether or not the eigenvalues crowd. For example, let us consider not the most general eigenstates of the area operator $\hat{A}_S$ but –as was first done in the development of the subject– the simplest ones. These correspond to graphs which have simplest intersections with S. For example, n edges of the graph may just pierce S, each one separately, so that at each one of the n vertices there is just a straight line passing through. For these states, the eigenvalues are $a_S = (\sqrt{3}/2)n\ell_P^2$. Thus, here, the level spacing Δa_S is uniform, like that of the Hamiltonian of a simple harmonic oscillator. If we restrict ourselves to these simplest eigenstates, even for large eigenvalues, the level spacing does not go to zero. Suppose for a moment that this is the *full* spectrum of the area operator. wouldn't the semi-classical approximation still work since, although uniform, the level-spacing is so small?

Surprisingly, the answer is in the negative! What is perhaps even more surprising is that the evidence comes from unexpected quarters: the Hawking evaporation of *large* black holes. More precisely, we will see that if Δa_S had failed to vanish sufficiently fast, the semi-classical approximation to quantum gravity, used in the derivation of the Hawking process, must fail in an important way. The effects coming from area quantization would have implied that even for large macroscopic black holes of, say, a thousand solar masses, we can not trust semi-classical arguments.

Let me explain this point in some detail. The original derivation of Hawking's was carried out in the framework of quantum field theory in curved space-times which assumes that there is a specific underlying continuum space-time and explores the effects of curvature of this space-time on quantum matter fields. In this approximation, Hawking found that the classical black hole geometrics arc such that there is a spontaneous emission which has a Planckian spectrum at infinity. Thus, black-holes, seen from far away, resemble black bodies and the associated temperature turns out to be inversely related to the mass of the hole. Now, physically one expects that, as it evaporates, the black hole must lose mass. Since the radius of the horizon is proportional to the the mass, the area of the horizon must decrease. Thus, to describe the evaporation process adequately, we must go beyond the external field approximation and take in to account the fact that the underlying space-time geometry is in fact dynamical. Now, if one treated this geometry classically, one would conclude that the process is continuous. However, since we found that the area is in fact quantized, we would expect that the black hole evaporates in discrete steps by making a transition from one area eigenvalue to another, smaller one. The process would be very similar to the way an excited atom descends to its ground state through a series of discrete transitions.

Let us look at this process in some detail. For simplicity let us use units with $c=1$. *Suppose, to begin with, that the level spacing of eigenvalues of the area operator is the naive one, i.e. with* $\Delta a_S = (\sqrt{3}/2)\ell_P^2$. Then, the fundamental theory would have predicted that the smallest frequency, ω_o, of emitted particles would be given by $\hbar\omega_o$ and the smallest possible change ΔM in the mass of the black hole would be given by $\Delta M = \hbar\omega_o$. Now, since the area of the horizon goes as $A_H \sim G^2M^2$, we have $\Delta M \sim \Delta a_H/2G^2M \sim \ell_P^2/G^2M$. Hence, $\hbar\omega_o \sim \hbar/GM$. Thus, the 'true' spectrum would have emission lines only at frequencies $\omega = N\omega_o$, for $N = 1, 2, ...$ corresponding to transitions of the black hole through N area levels. How does this compare with the Hawking prediction? As I mentioned above, according to Hawking's semi-classical analysis, the spectrum would be the same as that of a black-body at temperature T given by $kT \sim \hbar/GM$, where k is the Boltzmann constant. Hence, the peak of this spectrum would appear at ω_p given by $\hbar\omega_p \sim kT \sim \hbar/GM$. But this is precisely the order of magnitude of the minimum frequency ω_o that would be allowed if the area spectrum were the naive one. Thus, in this case, a more fundamental theory would have predicted that the spectrum would not resemble a black body spectrum. The most probable transition would be for $N = 1$ and so the spectrum would be peaked at ω_p as in the case of a black body. However, there would be no emission lines at frequencies low compared with ω_p; this part of the black body spectrum would be simply absent. The part of the spectrum for $\omega > \omega_p$ would also not be faithfully reproduced since the discrete lines with frequencies $N\omega_o$, with $N = 1, 2, ...$ would *not* be sufficiently near each other –i.e. crowded– to yield an approximation to the continuous black-body spectrum.

The situation is completely different for the correct, full spectrum of the area operator if the black hole is macroscopic, i.e., large. Then, as I noted earlier, the area eigenvalues crowd and the level spacing goes as $\Delta a_H \leq (\exp -\sqrt{a_H/\ell_P^2})\ell_P^2$. As a consequence, as the black hole makes transition from one area eigenvalue to another, it would emit particles at frequencies equal to or larger than $\sim \omega_p \exp -\sqrt{a_H/\ell_P^2}$. Since for a macroscopic black-hole the exponent is very large (for a solar mass black-hole it is $\sim 10^{38}$!) the spectrum would be well-approximated by a continuous spectrum and would extend well below the peak frequency. Thus, the precise form of the area spectrum ensures that, for large black-holes, the potential problem with Hawking's semi-classical picture disappears. Note however that as the black hole evaporates, its area decreases, it gets hotter and evaporates faster. Therefore, a stage comes when the area is of the order of ℓ_P^2. Then, there *would* be deviations from the black body spectrum. But this is to be expected since in this extreme regime one does not expect the semi-classical picture to continue to be meaningful.

This argument brings out an interesting fact. There are several iconoclastic approaches to quantum geometry in which one simply begins by postulating that geometric quantities should be quantized. Then, having no recourse to first principles from where to derive the eigenvalues of these operators, one simply postulates them to be multiples of appropriate powers of the Planck length. For area then, one would say that the eigenvalues are integral multiples of ℓ_P^2. The above argument shows how this innocent looking assumption can contradict semi-classical results *even for large black holes.* In the present approach, we did not begin by postulating the nature of quantum geometry. Rather, we *derived* the spectrum of the area operator from first principles. As we see, the form of these eigenvalues is rather complicated and could not have been guessed a priori. More importantly, the detailed form does carry rich information and in particular removes the conflict with semi-classical results in macroscopic situations.

3.4 CURRENT AND FUTURE DIRECTIONS

Exploration of quantum Riemannian geometry continues. Last year, it was found that geometric operators exhibit certain unexpected non-commutativity. This reminds one of the features explored by Alain Connes in his non-commutative geometry. Indeed, there are several points of contact between these two approaches. For instance, the Dirac operator that features prominently in Connes' theory is closely related to the connection $\mathbf{A}_i$ used here. However, at a fundamental level, the two approaches are rather different. In Connes' approach, one constructs a non-commutative analog of entire differential geometry. Here, by contrast, one focuses only on Riemannian geometry; the underlying manifold structure remains classical. In three space-time dimensions, it is possible to get rid of this feature in the final picture and express the theory in purely combinatorial fashion. Whether the same will be possible in four dimensions remains unclear. However, combinatorial methods continue to dominate the theory and it is quite possible that one would again be able to present the final picture without any reference to an underlying smooth manifold.

Perhaps the most striking application of quantum geometry has been to black hole thermodynamics. We saw in section 3.3 that the Hawking process provides a non-trivial check on the level spacing of the eigenvalues of area operators. Conversely, the discrete nature of these eigenvalues provides a statistical mechanical explanation of black hole entropy. To see this, first recall that for familiar physical systems —such as a gas, a magnet, or a black body— one can arrive at the expression of entropy by counting the number of micro-states. The counting in turn requires one to identify the building blocks that make up the system. For a gas, these are atoms; for a magnet, electron spins and for the radiation field in a black body, photons. What are the analogous building blocks for a large black hole? They *can not* be gravitons because the gravitational fields under consideration are static rather than radiative. Therefore, the elementary constituents must be non-perturbative in nature. In our approach they turn out to be precisely the quantum excitations of the geometry of the black hole horizon. The polymer-like one dimensional excitations of geometry in the bulk pierce the horizon and endow it with its area. It turns out that, for a given area, there are a specific number of permissible bulk states and for each such bulk state, there is a precise number of permissible surface states of the intrinsic quantum geometry of the horizon. Heuristically, the horizon resembles a pinned balloon —pinned by the polymer geometry in the bulk— and the surface states describe the permissible oscillations of the horizon subject to the given pinning. A count of all these quantum states provides, in the usual way, the expression of the black hole entropy.

Another promising direction for further work is construction of better candidates for 'weave states', the non-linear analogs of coherent states approximating smooth, macroscopic geometries. Once one has an 'optimum' candidate to represent Minkowski space, one would develop quantum field theory on these weave quantum geometries. Because the underlying basic excitations are one-dimensional, the 'effective dimension of space' for these field theories would be less than three. Now, in the standard continuum approach, we know that quantum field theories in low dimensions tend to be better behaved because their ultra-violet problems are softer. Hence, there is hope that these theories will be free of infinities. If they are renormalizable in the continuum, their predictions at large scales can not depend on the details of the behavior at very small scales. Therefore, one might hope that quantum field theories on weaves would not only be finite but also agree with the renormalizable theories in their predictions at the laboratory scale.

A major effort is being devoted to the task of formulating and solving quantum Einstein's equations using the new functional calculus. Over the past two years, there have been some exciting developments in this area. The methods developed there seem to be applicable also to supergravity theories. In the coming years, therefore, there should be much further work in this area. More generally, since quantum geometry does not depend on a background metric, it may well have other applications. For example, it may provide a natural arena for other problem such as that of obtaining a background independent formulation of string theory.

So far, I have focussed on theoretical ideas and checks on them have come from considerations of consistency with other theoretical ideas, e.g., those in black hole thermodynamics. What about experimental tests of predictions of quantum geometry? An astonishing recent development suggests that direct experimental tests may become feasible in the near future. I will conclude with a summary of the underlying ideas. The approach one takes is rather analogous to the one used in proton decay experiments. Processes potentially responsible for the decay come from grand unified theories and the corresponding energy scales are very large, 10^{15} GeV —only four orders of magnitude below Planck energy. There is no hope of achieving these energies in particle accelerators to actually create in large numbers the particles responsible for the decay. Therefore the decays are very rare. The strategy adopted was to carefully watch a *very* large number of protons to see if one of them decays. These experiments were carried out and the (negative) results actually ruled out some of the leading candidate grand unified theories.

Let us return to quantum geometry. The naive strategy of accelerating particles to Planck energy to directly 'see' the Planck scale geometry is hopeless. However, as in proton decay experiments, one can let these minutest of effects accumulate till they become measurable. The laboratory is provided by the universe itself and the signals are generated by the so-called γ-ray bursts. These are believed to be of cosmological origin. Therefore, by the time they arrive on earth, they have traveled extremely large distances. Now, if the geometry is truly quantum mechanical, as I suggested, the propagation of these rays would be slightly different from that on a continuum geometry. The difference would be minute but could accumulate on cosmological distances. Following this strategy, astronomers have already put some interesting limits on the possible 'graininess' of geometry. Now the challenge for theorists is to construct realistic weave states corresponding to the geometry we observe on cosmological scales, study in detail propagation of photons on them and come up with specific predictions for astronomers. The next decade should indeed be very exciting!

Acknowledgments

The work summarized here is based on contributions from many researchers especially John Baez, Alejandro Corichi, Roberto DePitri, Rodolfo Gambini, Chris Isham, Junichi Iwasaki, Jerzy Lewandowski, Renate Loll, Don Marolf, Jose Mourao, Jorge Pullin, Thomas Thiemann, Carlo Rovelli, Steven Sawin, Lee Smolin and José-Antonio Zapata. Special thanks are due to Jerzy Lewandowski for long range collaboration. This work was supported in part by the NSF Grant PHY95-14240 and by the Eberly fund of the Pennsylvania State University.

Notes

1. The situation can be illustrated by a harmonic oscillator: While the exact energy levels of the oscillator are discrete, it would be very difficult to "see" this discreteness if one began with a free particle whose energy levels are continuous and then tried to incorporate the effects of the oscillator potential step by step via perturbation theory.

2. Actually, only six of the ten Einstein's equations provide the evolution equations. The other four do not involve time-derivatives at all and are thus constraints on the initial values of the metric and its time derivative. However, if the constraint equations are satisfied initially, they continue to be satisfied at all times.

3. As usual, summation over the repeated index i is assumed. Also, technically each $\mathbf{A}_i$ is a 1-form rather than a vector field. Similarly, each $\mathbf{E}_i$ is a vector density of weight one, i.e., natural dual of a 2-form.

4. The 'breakthrough' was to drop the requirement that the (Levi-Civita) connection be symmetric, i.e., to allow for torsion.

5. This is called the kinematical Hilbert space; it enables one to formulate the quantum Einstein's (or supergravity) equations. The final, physical Hilbert space will consist of states which are solutions to these equations.

6. The factor 3 comes from the dimension of the gauge group SU(2) which acts on Chiral spinors. The mathematical structure of the gauge-rotations induced by this SU(2) is exactly the same as that in the angular-momentum theory of spin-$\frac{1}{2}$ particles in elementary quantum mechanics.

References

[1] Riemann, B., 1854, Über die Hypothesen, welche der Geometrie zugrunde liegen.

Monographs and Reviews on Non-perturbative Quantum Gravity:

[2] Ashtekar, A., 1991, *Lectures on Non-perturbative Canonical Gravity*, Notes prepared in collaboration with R.S. Tate. (World Scientific, Singapore).

[3] Gambini, R., Pullin, J., 1996 *Loops, Knots, Gauge Theories and Quantum Gravity*, Cambridge University Press, Cambridge.

[4] Ashtekar, A., 1995, *Gravitation and Quantizations*, ed B. Julia and J. Zinn-Justin Elsevier, Amsterdam.

[5] Rovelli, C., 1998, *Gravitation and Relativity: At the Turn of the Millennium*, ed N. Dadhich and J. Narlikar, IUCAA, Pune.

Background-independent Functional Calculus:

[6] Ashtekar, A., Isham, C.J., 1992, Class. & Quan. Grav. **9**, 1433.

[7] Ashtekar, A., Lewandowski, J., 1994, *Representation theory of analytic holonomy $C^\star$ algebras*, in *Knots and Quantum Gravity*, ed Baez, J., Oxford University Press, Oxford.

[8] Baez, J. 1994, *Diffeomorphism invariant generalized measures on the space of connections modulo gauge transformations*, Lett. Math. Phys., **31**, 213, hep-th/9305045, in *The Proceedings of the Conference on Quantum Topology*, ed D. Yetter, World Scientific, Singapore.

[9] Ashtekar, A., Lewandowski, J., 1995, J. Math. Phys. **36**, 2170.

[10] Marolf, D., Mourão, J.M., 1995, Commun. Math. Phys. **170**, 583.

[11] Ashtekar, A. Lewandowski, J., 1995 J. Geo. & Phys. **17**, 191.

[12] Baez, J., 1996, Adv. Math. **117**, 253 (1996); "Spin networks in non-perturbative quantum gravity," gr-qc/9504036 in *The Interface of Knots and Physics*, ed L. Kauffman (American Mathematical Society, Providence.

[13] Rovelli, C., Smolin, L., 1995, Phys.Rev. **D52**, 5743.

[14] Ashtekar, A., Lewandowski, J., Marolf, D., Mourão, J., Thiemann, T., 1995, J. Math. Phys. **36**, 6456.

[15] Baez, J., Sawin, S., 1997, J. Funct. Analysis **150**, 1 .

[16] Zapata, J.A., 1998, Gen.Rel.Grav. **30**, 1229.

Geometric Operators

[17] Ashtekar, A., Rovelli, C., Smolin, L., 1992, Phys. Rev. Lett. **69**, 237.

[18] Iwasaki, J., Rovelli, C., 1993 Int. J. Modern. Phys. **D1**, 533; Class. Quant. Grav. **11**, 2899 (1994).

[19] Rovelli, C. Smolin, L., 1995, Nucl. Phys. B**442**, 593 .

[20] Ashtekar, A., Lewandowski, J., Marolf, D., Mourão, J., Thiemann, T., 1996, J. Funct. Analysis, **135**, 519.

[21] Loll, R., 1995, Phys. Rev. Lett. **75**, 3084.

[22] Ashtekar, A., Lewandowski, J., 1997, Class. & Quant. Grav. **14**, A55-A81.

[23] Loll, R., 1997, Nucl.Phys. **B500** 405.

[24] Thiemann, T., 1997, J.Math.Phys. **39**, 3372.

[25] Ashtekar, A., Lewandowski, J., 1997, Adv. Theo. Math. Phys. **1**, 388.

[26] Ashtekar, A., Corichi, A., Zapata, J.A., 1998, Class. & Quant. Grav. **15**, 2955.

Black Hole Thermodynamics:

[27] Bekenstein, J.D., 1973, Phys. Rev. **D7**, 2333; Phys. Rev. **D9**, 3292 (1974).

[28] Bardeen, J.W., B. Carter, Hawking, S.W., 1973, Commun. Math. Phys. **31**, 161.

[29] Hawking, S.W., 1975, Comun. Math. Phys. **43**, 199.

[30] Bekenstein, J., Mukhanov, V.F., 1995, Phys. Lett. **B360**, 7.

[31] Fairhurst, S., 1996, Properties of the Spectrum of the Area Operator (unpublished Penn State Report).

[32] Rovelli, C., 1996, Helv.Phys.Acta **69**, 582.

[33] Ashtekar, A., 1997, *Geometric issues in quantum gravity* , in: *The Geometric Universe*, ed S. Hugget, L. Mason, K.P. Tod, S.T. Tsou and N.M.J. Woodhouse, Oxford University Press, Oxford, .

[34] Ashtekar, A., Baez, J., Corichi, A., Krasnov, K., 1998, Phys. Rev. Lett. **80**, 904.

[35] Ashtekar, A., Krasnov, K., 1998, *Quantum geometry and black holes* in *Black Holes, Gravitational Radiation and the Universe*, ed B. Bhawal and B. K. Iyer, Kluwer, Dodrecht.

[36] Ashtekar, A., Beetle, C., Fairhurst, S., *A generalization of black hole mechanics*, gr-qc/9812065, Class. & Quant. Grav.,in press.

[37] Ashtekar, A., Corichi, A., Krasnov, K., 1998, *Isolated horizons: Classical Phase space* CGPG pre-print.

[38] Ashtekar, A., Baez, J., Krasnov, K., 1998, *Quantum geometry of isolated horizons and black hole entropy* CGPG pre-print.

Experimental tests

[39] Amelino-Camelina, G., Ellis, J., Marvomatos,N., Nanopoulos, D., Sarkar, S., 1998, Nature **393**, 763.

[40] Gambini, R., Pullin, J., *Nonstandard optics from quantum space-time*, gr-qc/9809038.

[41] Biller, S.D., Breslin, A.C., Buckley, J., Catanese, M., Carson, M., Carter-Lewis, D.A., Cawley, M.F. , Fegan,D.J., Finley,J., Gaidos, J.A. , Hillas, A.M. , Krennrich, F., Lamb, R.C. , Lessard,R. , Masterson, C. , McEnery, J.E., McKernan, B. , Moriarty, P., Quinn, J., Rose, H.J., Samuelson, F., Sembroski, G., Skelton, P., Weekes, T.C., "Limits to Quantum Gravity Effects from Observations of TeV Flares in Active Galaxies", gr-qc/9810044

Chapter 4

QUANTUM MECHANICS AND RETROCAUSALITY

D. Atkinson
Institute for Theoretical Physics
University of Groningen
NL-9747 AG Groningen
The Netherlands

Abstract The classical electrodynamics of point charges can be made finite by the introduction of effects that temporally precede their causes. The idea of retrocausality is also inherent in the Feynman propagators of quantum electrodynamics. The notion allows a new understanding of the violation of the Bell inequalities, and of the world view revealed by quantum mechanics.

1. INTRODUCTION

Dirac was never happy with quantum electrodynamics, although it was in large part his own creation. In old age, during an after-dinner seminar in 1970 that I attended in Austin, Texas, he lambasted such upstarts as Feynman, Schwinger, Tomonaga, and their ilk, under the dismissive collective term 'people'. These "People neglect infinities in an arbitrary way. This is not sensible mathematics. Sensible mathematics involves neglecting a quantity when it is small — not neglecting it just because it is infinitely great and you do not want it." A timorous spirit among the chastened listeners asked: "But, Professor Dirac, what about $g-2$?", referring of course to the g-factor in the expression for the magnetic moment of the electron. Dirac's own equation had predicted that this factor should be precisely 2, and the highly accurate quantum electrodynamical calculation of its deviation from 2 was, and is, one of the tours de force of modern physics. The agreement with painstaking experimental measurement of this quantity is phenomenal (the Particle Data Group gives on the World Wide Web ten digits of agreement after

N. Dadhich and A. Kembhavi (eds.), The Universe, 35–50.

the decimal point[1]). But the old maestro had his own views about this remarkable result: "It might just be a coincidence," he remarked evenly.

Quantum mechanics, married to electromagnetism, has produced a very successful theory, as measured by its empirical adequacy. The matter is not so adequate, however, at a conceptual level. There are still many competing *interpretations* of what quantum mechanics is telling us about the nature of the world. Despite the early preoccupation with the breakdown of determinism, the serious difficulties have to do rather with causality, which is by no means the same thing. Classical electromagnetic theory is in fact not immune to such problems either: the only known way to remove disastrous infinities in the theory of point charges interacting through the electromagnetic field is by the introduction of retrocausal effects. Quantum electrodynamics inherits the diseases of causality and of divergence from both of its parents. Their nature is pervasive, the cure unknown.

2. ADVANCED POTENTIALS

An electrically neutral particle, of mass m, subject to a force $\mathbf{F}$, satisfies Newton's second law of motion, which may be expressed in the form

$$m\mathbf{a} = \mathbf{F}\,, \tag{4.1}$$

where $\mathbf{a} = \ddot{\mathbf{r}}$ is the acceleration, on condition that $|\dot{\mathbf{r}}| << c$, so that relativistic corrections may be neglected. A similar charged particle cannot satisfy the same equation, because an accelerated charge emits electromagnetic waves, losing energy in the process. Newton's law may be repaired by adding an effective radiative damping force that accounts for this extra source of energy loss to space:

$$m\mathbf{a} = \mathbf{F} + \mathbf{F}_{\text{rad}}\,, \tag{4.2}$$

where one finds, for a point charge e,

$$\mathbf{F}_{\text{rad}} = \frac{2e^2}{3c^3}\dot{\mathbf{a}}\,. \tag{4.3}$$

We may rewrite Eq.(4.2)-(4.3) in the form

$$m(\mathbf{a} - \tau\dot{\mathbf{a}}) = \mathbf{F}\,, \tag{4.4}$$

where

$$\tau = \frac{2e^2}{3mc^3}\,,$$

is called the Abraham-Lorentz relaxation time. For an electron it is about 6×10^{-24} sec., in which time light travels only about 10^{-13} cm., the size of a proton.

The general solution of Eq.(4.4) is

$$\mathbf{a}(t) = \frac{1}{m\tau} \int_t^c dt' \, e^{(t-t')/\tau} \mathbf{F}(t') \,,$$

where c is an integration constant. Clearly $\mathbf{a}(t)$ blows up exponentially as $t \to \infty$, the so-called runaway solution, unless $c = \infty$. Accordingly, we choose this latter value, and find we can rewrite the solution in the form

$$m\mathbf{a}(t) = \int_0^\infty ds \, e^{-s} \mathbf{F}(t + \tau s) \,, \tag{4.5}$$

from which we derive the following Taylor series in τ:

$$m\mathbf{a}(t) = \sum_{n=0}^{\infty} \tau^n \mathbf{F}^{(n)}(t) \,. \tag{4.6}$$

The Newton law Eq.(4.1), as it applies to a neutral particle, corresponds to the zeroth term only. From Eq.(4.5), the acceleration at time t is determined not only by the value of the applied force at time t, but also by the force at all times *later* than t.

For a simple force, one can evaluate Eq.(4.5) explicitly. For example, if a force is turned on at time $t = 0$, after which it remains constant, i.e. $\mathbf{F}(t) = 0$ for $t < 0$ and $\mathbf{F}(t) = \mathbf{K}$ for $t \geq 0$, then we find $m\mathbf{a}(t) = \mathbf{K}$ for $t \geq 0$, as we would for a neutral particle, but surprisingly $m\mathbf{a}(t) = \mathbf{K}\, e^{t/\tau}$ for $t < 0$. This preacceleration violates a naïve notion of causality, according to which a cause *precedes* its effect, whereas here the force, which is not applied before time $t = 0$, produces (has already produced!) an acceleration *before* $t = 0$.

Consider next a universe consisting of many particles, at positions $x_a, x_b, \ldots$ with masses $m_a, m_b, \ldots$ and charges $e_a, e_b, \ldots$ For particle a, the relativistic generalization of Eq.(4.2) for the four-momentum p_a^μ is

$$\frac{dp_a^\mu}{d\tau_a} = e_a \left[F^\mu{}_\nu + R^\mu{}_\nu \right] \frac{dx_a^\nu}{d\tau_a} \,. \tag{4.7}$$

Here τ_a is the proper time of particle a, and $F^\mu{}_\nu$ is the retarded field tensor that gives rise to the usual Lorentz force. It may be written

$$F^\mu{}_\nu = \sum_{b \neq a} F_b^{\mathrm{ret}\,\mu}{}_\nu \,,$$

where the sum is over all the contributions to the field from the particles *other* than a itself: there is no self-interaction. The term $R^{\mu}{}_{\nu}$ is the radiation damping tensor: it corresponds to $\mathbf{F}_{\text{rad}}$ in the nonrelativistic approximation (4.3). Dirac deduced the explicit form of this tensor and showed that it can be written

$$R^{\mu}{}_{\nu} = \tfrac{1}{2}\left[F_a^{\text{ret}\,\mu}{}_{\nu} - F_a^{\text{adv}\,\mu}{}_{\nu}\right] . \tag{4.8}$$

It is very interesting that this expression involves the advanced, as well as the retarded fields arising from particle a. For the point particles that we are considering, these fields are separately singular on the world-line of a itself, but their difference (4.8) is finite.

To simplify the notation, we will henceforth suppress the Lorentz indices. It is important to distinguish the sum $\sum_{b\neq a}$, in which one sums over all particles *except* a, in order to calculate the influence of the rest of the universe on particle a, and the sum $\sum_b$, in which a is also included, giving a quantity that refers to the universe in its entirety.

$$\begin{aligned} F + R &= \sum_{b\neq a} F_b^{\text{ret}} + \tfrac{1}{2}\left[F_a^{\text{ret}} - F_a^{\text{adv}}\right] \\ &= \sum_{b} F_b^{\text{ret}} - F_a^{\text{ret}} + \tfrac{1}{2}\left[F_a^{\text{ret}} - F_a^{\text{adv}}\right] \\ &= \sum_{b} F_b^{\text{ret}} - \tfrac{1}{2}\left[F_a^{\text{ret}} + F_a^{\text{adv}}\right] \end{aligned} \tag{4.9}$$

The essential assumption of Wheeler and Feynman[2] is that the universe is a perfect absorber: all radiation is absorbed somewhere and none escapes to infinity. Since a radiation field is of order $1/r$ for large distances r, to eliminate energy loss by radiation it is enough to require

$$\sum_b F_b^{\text{ret}} = o\left(r^{-1}\right) \qquad \sum_b F_b^{\text{adv}} = o\left(r^{-1}\right) ,$$

for all times, i.e. the sum of all retarded (advanced) fields is assumed always to vanish faster than $1/r$ at spatial infinity. However, $\sum_b F_b^{\text{ret}}$ and $\sum_b F_b^{\text{adv}}$ each satisfies Maxwell equations with the same sources and sinks (the charges). They are indeed two independent solutions of the same second-order equations. Hence their difference,

$$\sum_b \left[F_b^{\text{ret}} - F_b^{\text{adv}}\right] , \tag{4.10}$$

satisfies a homogeneous system of equations, i.e. a system without sources or sinks. Such a system possesses nontrivial solutions, but they are radiation fields that decrease like r^{-1} at spatial infinity: there are no $o\left(r^{-1}\right)$ nontrivial solutions. Thus the difference (4.10) is not merely zero at spatial infinity, it must be identically zero everywhere. Hence

$$\sum_b F_b^{\mathrm{ret}} = \sum_b F_b^{\mathrm{adv}} = \tfrac{1}{2} \sum_b \left[F_b^{\mathrm{ret}} + F_b^{\mathrm{adv}}\right], \tag{4.11}$$

for all times.

On combining this result with Eq.(4.9), we obtain

$$\begin{aligned} F + R &= \tfrac{1}{2} \sum_b \left[F_b^{\mathrm{ret}} + F_b^{\mathrm{adv}}\right] - \tfrac{1}{2}\left[F_a^{\mathrm{ret}} + F_a^{\mathrm{adv}}\right] \\ &= \tfrac{1}{2} \sum_{b \neq a} \left[F_b^{\mathrm{ret}} + F_b^{\mathrm{adv}}\right] \end{aligned} \tag{4.12}$$

This is a stunning result: it says that to calculate the response of a charged particle to all the other charged particles in the universe, one has to sum over the fields emanating from all those other particles, *on condition that one uses the time-symmetric solution of the Maxwell equation.* In this approach there is no need, nor room, to add a further radiation damping term: it is all contained in the average of the retarded and advanced solutions of Maxwell's equations. Turning the argument around, one can say that the time-symmetric form is equivalent to, and so validates, the conventional calculation in which a retarded solution is supplemented, in a somewhat *ad hoc* manner, by a radiation damping field.

It must not be thought that we have hereby forged an arrow of time from a time-symmetric theory. This can be seen by complementing Eq.(4.9) by

$$\begin{aligned} F + R &= \sum_b F_b^{\mathrm{adv}} - \tfrac{1}{2}\left[F_a^{\mathrm{ret}} + F_a^{\mathrm{adv}}\right] \\ &= \sum_{b \neq a} F_b^{\mathrm{adv}} + \tfrac{1}{2}\left[F_a^{\mathrm{adv}} - F_a^{\mathrm{ret}}\right]. \end{aligned} \tag{4.13}$$

This is an equally valid *modus operandi*, involving the full advanced potential, supplemented by a radiation damping term, but since it is precisely minus the corresponding term in the first line of Eq.(4.9), it might better be called a radiation boosting term.

3. BELL INEQUALITY

Let us turn now to the Einstein-Podolsky-Rosen scenario[3] in its modern experimental avatar[4]. We will see that the violation of the

Bell inequality loses much of its impact once we entertain the notion of advanced fields.

Briefly, two photons are prepared with opposed spins by the sequential decay of a calcium atom from an excited S state, through an intermediate P state, to the ground state, which is also S. The state of linear polarization of one photon is measured by means of a birefringent calcite crystal and a photo-detector at location A, and that of the other photon by a similar arrangement at location B. The separation of A and B is several metres, and the measurement events are contained within small space-time hypervolumes that have a mutual spacelike separation. Thus the measurement events at A and B are independent of one another in the sense that no information about the result of the measurement at A can be transmitted to B in time to influence the result of the measurement there (and vice versa). This is true *only if we limit ourselves to the usual retarded fields.* The two photons are not independent, however, in the sense that their spins are correlated because of their common genesis in an atomic decay. The polarizations have, in the locution of Reichenbach, a common cause[5].

If the optical axes of the calcite crystals at A and B are parallel, then whenever a photon at A is found to go in the direction of the ordinary ray, the same is found at B. Similarly, there is perfect correlation in the case that the photons are deflected along the extraordinary ray directions. The more general situation, in which the optical axis at A is at an angle α to the vertical, and the optical axis at B is at an angle β to the vertical, leads to the following joint probabilities:

$$\begin{aligned} P_{oo}(\alpha,\beta) &= \tfrac{1}{2}\cos^2(\alpha-\beta) &= P_{ee}(\alpha,\beta) \\ P_{oe}(\alpha,\beta) &= \tfrac{1}{2}\sin^2(\alpha-\beta) &= P_{eo}(\alpha,\beta)\,. \end{aligned} \tag{4.14}$$

Here P_{oo} is the probability that the photons at A and B both go into the ordinary rays, P_{ee} that both photons go into the extraordinary rays, P_{oe} is the probability that the photon at A goes into the ordinary ray but the photon at B goes into the extraordinary ray, and finally P_{eo} is the probability that the photon at A goes into the extraordinary ray but the photon at B goes into the ordinary ray. The results Eq.(4.14) are predicted by quantum mechanics and confirmed by experiment.

The correlation coefficient is defined as follows:

$$C(\alpha,\beta) = P_{oo}(\alpha,\beta) + P_{ee}(\alpha,\beta) - P_{eo}(\alpha,\beta) - P_{oe}(\alpha,\beta) = \cos 2(\alpha-\beta)\,. \tag{4.15}$$

If we suppose, with Bell[6], that the joint probabilities, and hence the correlation coefficient, are separable, in the sense of classical probability theory, then we can write, for this correlation coefficient,

$$C(\alpha,\beta) = \sum_{\lambda} \rho(\lambda) C(\alpha|\lambda) C(\beta|\lambda)\,, \tag{4.16}$$

where λ are *hidden variables* that account for the correlations between the two photon polarizations: they arise from the birth of the twin photons in the de-exciting calcium atom. The weight $\rho(\lambda)$ is supposed to be positive and normalized; and $C(\alpha|\lambda)$ is the correlation coefficient

$$C(\alpha|\lambda) = P_o(\alpha|\lambda) - P_e(\alpha|\lambda)$$

at location A, *conditioned by the hidden variable* λ. Similarly, $C(\beta|\lambda)$ is the conditional correlation coefficient at location B. Clearly each conditional correlation coefficient, being the difference between two probabilities, lies in the interval $[-1,1]$.

The Bell coefficient is defined as the following combination of four correlation coefficients:

$$B = C(\alpha,\beta) + C(\alpha',\beta) + C(\alpha',\beta') - C(\alpha,\beta')\,. \tag{4.17}$$

It can be measured by combining the results of four separate runs of the experiment, with a choice of two possible orientations (α or α') of the calcite optical axis at A, and two possible orientations (β or β') at B. One can show, under the assumption of separability, and

$$\sum_{\lambda} \rho(\lambda) = 1\,, \tag{4.18}$$

with $\rho(\lambda) \geq 0$, that

$$|B| \leq 2\,. \tag{4.19}$$

However, by choosing the angles α, β, α' and β' suitably, one can arrange that quantum mechanics yields $B = 2\sqrt{2} > 2$. However,

$$C(\alpha,\beta) = \cos 2\alpha \cos 2\beta + \sin 2\alpha \sin 2\beta\,,$$

so the normalization Eq.(4.18) is ruined[1] — on the right-hand side of Eq.(4.18) we obtain 2 instead of 1! We must conclude that something is amiss; and we seem to have (at least) the following options:

1. No hidden variables can be found that screen off the common cause.

2. Classical probability theory is simply inapplicable in the quantum domain, in particular Kolmogorov's definition of stochastic independence is inappropriate[7].

3. Advanced as well as retarded fields are present.

In this paper we will concentrate on the third possibility. If the absorption of the photon at A, after its passage through the calcite crystal at A, is accompanied by an advanced, as well as a retarded field, then *information about the interaction of the photon at A, in particular details about the polarizer orientation at the moment of measurement, will ride the advanced wave back to the genesis of the photon pair, arriving at the calcium atom just at the moment that it de-excites.* We can understand how, even if the orientation of the A polarizer is changed at the last moment before the polarization measurement, still the interaction can carry information back about the measurement configuration. This way of speaking about information being carried back and forth, as if there were a sort of internal biological time of the sort that science fictional time travellers seem to carry about with themselves, is imprecise and may be confusing. It is better to say that, in the advanced field approach, one has a self-consistent picture in which the state of the photon's polarization is correlated to its future, as well as to its past interactions. The notions of 'cause' and 'information' are replaced by that of 'correlation'.

In one variant of Aspect's experiment, the selection between the angles α and α' at A, and β and β' at B, was changed randomly by two independent oscillators every few nanoseconds. Still the predictions of quantum mechanics were borne out and the Bell inequality violated. Most people interpret this as a demonstration of nonlocality (more soberly of nonseparability). With option 3 we can retain Lorentz covariance while achieving action at a distance. Is this action local or nonlocal? In a sense it is a semantic matter. It is not usual to call conventional retarded field theory nonlocal, the idea being that a particle is only influenced by a distant causal agent in the particle's past light cone. This influence is fleshed out by imputing a real existence to the field (in quantum theory to the field quanta). In this way the field serves as a messenger from afar, bringing influence and information at no more than light speed and delivering it in the vicinity of the particle. One might describe advanced action also as being local in an analogous manner: an influence is transmitted by the advanced field, also within the light cone, arriving in the vicinity of the particle to deliver its information, much on a par with the retarded case. However, this account, even after deanthropomorphization in terms of correlations rather than of causes and of influences, is incomplete. Since correlations can be established forwards and backwards in time, really the only logical requirement is one of consistency. The theory need only be such that it is impossible for an event in a space-time hypervolume both to occur and not to occur[2].

4. RETROCAUSALITY

According to David Hume, causality is based on nothing more than the observed constant conjunction of two or more kinds of events, say A and B. It is a mere habit we have to call the earlier of the occurrences, say A, the cause, and the later, B, the effect; no relation of necessity, nor even of likelihood, of a B's succeeding an A in the future can be deduced. If we replace the word 'habit' by 'theory', then we may reconstrue Hume's admonition as the trite Scottish verity that we have no proof that a theory, based on the results of observations in the past, will yield reliable predictions in the future, no matter how numerous the observations in question are. Indeed, we neither have, nor expect to be able to provide, such a proof concerning empirical matters. Moreover, if it is a mere habit, a mere linguistic convention, to call the temporal antecedent a cause, and the successor an effect, why should we not expand our horizons, generalize our theories, and envisage causes that can occur *later* than their hypothesized effects?

In his intriguing article "Bringing About the Past", Michael Dummett has indeed claimed that the temporal asymmetry of the causal relation is contingent rather than necessary[8]. He describes two situations in which one might speak of a voluntary action performed with the intention of bringing about a past event. Nevertheless, stringent conditions must be satisfied to ensure the coherence of such a standpoint. In particular, Dummett claims that it is *incoherent* to hold all of the following claims:

1. There is a positive correlation between an agent's performing an action of type A at time t_A and the occurrence of an event of type B at time t_B, where $t_A > t_B$.

2. It is entirely within the power of the agent to perform A at time t_A, if he so chooses.

3. It is possible for the agent to find out, at time t_A, whether B has or has not already occurred, independently of his performing A.

One of the two examples that Dummett describes concerns a tribe that has the following custom: "Every second year the young men of the tribe are sent, as part of their initiation ritual, on a lion hunt: they have to prove their manhood. They travel for two days, hunt lions for two days, and spend two days on the return journey; ... While the young men are away from the village the chief performs ceremonies—dances, let us say—intended to cause the young men to act bravely. We notice that he continues to perform these dances for the whole six days that the party is away, that is to say, for two days during which the events that the dancing is supposed to influence have already taken place. Now there is generally thought to be a *special* absurdity in the idea of affecting the past, much greater than the absurdity of believing that the performance of a dance can influence the behavior of a man two days' journey away; ... " Ref.[8], pages 348-9. In physicists' terms, retrocausality seems even more absurd than action at a distance.

The chief is a wise and rational man: he believes the first of the above-mentioned three claims, at any rate as a statement of the significant statistical efficacity of his magic dancing. Let us further suppose that he does not believe that he is somehow hindered from dancing, or perhaps caused to dance inadequately, during the last two days, in the case that his young men have been cowardly. Then he must deny the third claim: he must assume that there is no way that he can find out, during the crucial days 5 and 6, what in fact has happened during days 3 and 4. For if it were possible to find it out, he could *bilk* the correlation. That is to say, he could choose to dance properly if, and only if, he knew that his men had not been brave. Then there would not be a positive correlation of the sort envisaged in claim 1.

It seems that we, as anthropologists, would at any rate accept claim 3, and thus conclude incoherence. With the aid of radio communication and a field worker, we could always arrange a bilking scenario, so that A could not count, even stochastically, as a cause of the earlier event B. But is there a situation in which claim 3 could defensibly be denied? There seem indeed to be such cases in subatomic physics. For example, the state of polarization of a photon, which has passed through one polarizer, and will pass through a second polarizer, is a property that we can only test by passing it through the next polarizer that it will encounter. If we choose to insert a calcite crystal in the path of the photon in such a way that it effects a polarization measurement, then *this crystal is the next polarizer.* If it be claimed that the state of polarization of a photon is correlated, not only with the orientation of the polarizer in its past, but also with that of the polarizer in its future trajectory, no bilking of the claim is possible. Here is indeed a clear candidate for retrocausal effects.

5. THE VIEW FROM NOWHEN

Is there a way to fit the notion of retrocausality into a general theoretical framework, rather than merely to permit its fugitive occurrence when all bilking scenarios are impossible? The Australian philosopher Huw Price elaborates a Weltanschauung that he calls the view from nowhen[9]. His point of departure is the time reversal (T) invariance of microscopic processes[3]. When two inert gases of different colours, initially segregated and at different temperatures, are allowed to mix, the approach to an equilibrium mixture, of an intermediate colour and at an intermediate temperature, is irreversible, although the dynamics of the molecular collisions is T-invariant. A reversed video recording of the process would not look queer at the level of individual collisions, seen one by one, but it would appear odd at the macro-level, where it would show an apparently spontaneous segregation of the two gaseous components. It is generally agreed that the Stoßzahlansatz of Boltzmann, an example of what Price calls PI^3, or the principle of the independence of incoming influences, is not acceptable as an *explanation* of the irreversibility in question. For if PI^3 holds, why should not $PIOI$ hold, the principle of the independence of outgoing influences? If one suggests that $PIOI$ breaks down because correlations are generated by a collision, then one must ask whether after all PI^3 is justified. That is, if correlations are generated in a collision process, may they not be present *before* as well as after the scattering? There seems in fact to be no good reason for adopting a double standard in this matter. Indeed, to do so in the search for a thermodynamic arrow of time is a flagrant example of *petitio principii.*

A convincing case can be made that the the master arrow of time is cosmological, and the major task lies in explaining why the cosmos had such a low entropy in what for us is the distant past. The thermodynamic arrow follows readily: there is no need for an *ad hoc* PI^3 without a $PIOI$. The Wheeler-Feynman time symmetric treatment of electromagnetic radiation implicitly appeals ultimately to cosmology, for the effective retardation arises from the assumption of perfect future absorption. This absorption is treated as a matter of irreversible thermodynamics, in terms in fact of a phenomenological absorptive (complex) refraction index. The thermodynamic arrow is tied to the cosmological one, and Wheeler and Feynman reason that radiation appears to us to be retarded because of thermodynamic processes in the future universe. The reason for the direction of the thermodynamic arrow itself seems to lie in the statistical properties of the *early* universe, i.e. in the fact that it was in such a low entropy condition.

If the arrow of radiation ultimately derives from cosmological considerations, it would be desirable to show this directly, in terms of the properties of a cosmological model, rather than indirectly, via thermodynamics. This is precisely what Hoyle and Narlikar have done[10]. Suppose that the future is *not* a perfect absorber, but only works at efficiency f, in the sense that the reaction of the universe, on particle a, is not the full Dirac radiation damping of $\frac{1}{2}\left[F_a^{\mathrm{ret}} - F_a^{\mathrm{adv}}\right]$, but only

f times this quantity. Analogously, suppose that the past is also not perfect as an absorber, but has efficiency p. That is, the boosting is not minus the Dirac term, but rather $-p$ times that quantity. Let us write the symmetric sum over all the fields acting on particle a as a general linear superposition of retarded and advanced contributions, each with its damping or boosting terms:

$$A\left\{\sum_{b\neq a} F_b^{\text{ret}} + \tfrac{1}{2}f\left[F_a^{\text{ret}} - F_a^{\text{adv}}\right]\right\} + B\left\{\sum_{b\neq a} F_b^{\text{adv}} - \tfrac{1}{2}p\left[F_a^{\text{ret}} - F_a^{\text{adv}}\right]\right\}, \tag{4.20}$$

with $A + B = 1$. This leads to

$$\begin{aligned}(1-2A)\sum_b F_b^{\text{ret}} + (1-2B)\sum_b F_b^{\text{adv}} = \\ (1-2A+Af-Bp)F_a^{\text{ret}} + (1-2B-Af+Bp)F_a^{\text{adv}}\end{aligned} \tag{4.21}$$

The system is consistent if the coefficients of F_a^{ret} and F_a^{adv} vanish:

$$\begin{aligned} A &= \frac{1-p}{2-f-p} \\ B &= \frac{1-f}{2-f-p}, \end{aligned} \tag{4.22}$$

and this is indeed consistent with $A + B = 1$.

The Hoyle-Narlikar relation Eq.(4.22) is interesting. Unless the past and the future are *both* fully absorbing, the values of A and B are uniquely defined. For $p < 1$ and $f < 1$, since neither A nor B is zero, the radiation from an accelerated charge is effectively neither retarded nor advanced, but a superposition of the two, and the radiation damping is a definite fraction of the Dirac value. The special case in which the future is a perfect, but the past an imperfect absorber, $f = 1$ but $p < 1$, leads to $A = 1$ and $B = 0$, which is the empirically satisfactory situation of effectively retarded radiation, together with the full strength Dirac radiation damping. With $p = 1$ but $f < 1$, on the other hand, we obtain $B = 1$ and $A = 0$. That is, in the situation in which the big bang acts as a perfect absorber but the future is not fully absorbing—in an open Friedmann model, for instance—one finds the unacceptable effectively advanced solution, with a radiation boosting term, i.e. minus the Dirac radiation damping. The main point to be made here is that, while the basic emission is time symmetric, the effective radiation is not symmetric if and only if $p \neq f$. That is, the radiative temporal symmetry is broken by an asymmetry in the absorptive properties of the past and future universe, in short by a cosmological asymmetry.

It seems that Feynman himself, after he had elaborated quantum electrodynamics (QED) in the form that we still use today, rejected only part of the credo of symmetric action at a distance[11]: "It was based on two assumptions:

1. Electrons act only on other electrons
2. They do so with the mean of retarded and advanced potentials

The second proposition may be correct but I wish to deny the correctness of the first." The reason given for accepting that a charged particle can interact with its own field was precisely the success of the calculation of the anomalous magnetic moment of the electron—the famous $g-2$ to which we alluded at the beginning.

The close similarity between the Wheeler-Feynman account of radiation and that given in QED—and also the crucial difference—can be appreciated by looking at the Green's functions of the theories. The electromagnetic field tensor may be expressed in terms of the four-potential, $A^\mu(x)$, by

$$F_{\mu\nu} = \partial_\mu A_\nu - \partial_\nu A_\mu\,,$$

and the Maxwell equations can be written

$$\partial^2 A_\mu = j_\mu\,, \tag{4.23}$$

in the Lorentz gauges, for which $\partial_\mu A^\mu = 0$. Here j_μ is the four-current density. A solution of Eq.(4.23) is expressible as an integral,

$$A_\mu(x) = \int d^4y D_{\mu\nu}(x-y) j^\nu(y)\,,$$

where $D_{\mu\nu}$ is a Green's function that satisfies

$$\partial^2 D_{\mu\nu}(x) = g_{\mu\nu}\delta^4(x)$$

The relations between the different theories can be appreciated by comparing the various choices of Green's function. The standard classical choice is the retarded one:

$$D^{\text{ret}}_{\mu\nu}(x) = -\frac{g_{\mu\nu}}{(2\pi)^4}\int d^4p \frac{\mathrm{e}^{-ipx}}{(p_0+i\epsilon)^2 - \mathbf{p}.\mathbf{p}} = \frac{g_{\mu\nu}}{2\pi}\theta(x_0)\delta(x^2)\,.$$

The $i\epsilon$ prescription means that the Green's function is to be interpreted as a distribution on a space of analytic functions: the implicit limit $\epsilon \to 0$ through positive values is equivalent to a small deformation of the k_0-integration contour in the appropriate direction. The advanced Green's function is obtained from the above by changing the sign of ϵ, which implies that $\theta(x_0)$ is replaced by $\theta(-x_0)$. The Green's function of the Wheeler-Feynman theory is

$$\begin{aligned} D^{\mathrm{WF}}_{\mu\nu}(x) &= \tfrac{1}{2}\left[D^{\mathrm{ret}}_{\mu\nu}(x) + D^{\mathrm{adv}}_{\mu\nu}(x)\right] \\ &= -\frac{g_{\mu\nu}}{(2\pi)^4}\int d^4p\,\mathrm{e}^{-ipx}\frac{P}{p^2} = \frac{g_{\mu\nu}}{4\pi}\delta(x^2)\,, \end{aligned} \tag{4.24}$$

where P means the principal value in the sense of Cauchy.

The QED Feynman propagator, defined through the vacuum expectation value of the time ordered product of two fields, is in QED

$$D^{\mathrm{F}}_{\mu\nu}(x) = -\frac{g_{\mu\nu}}{(2\pi)^4}\int d^4p\frac{\mathrm{e}^{-ipx}}{p^2+i\epsilon} = \frac{g_{\mu\nu}}{4i\pi^2}\frac{1}{x^2-i\epsilon}\,.$$

Now we can write

$$D^{\mathrm{F}}_{\mu\nu}(x) = -\frac{g_{\mu\nu}}{(2\pi)^4}\int d^4p\,\mathrm{e}^{-ipx}\left[\frac{P}{p^2} - i\pi\delta(p^2)\right]\,.$$

On comparing this with Eq.(4.24), we see that the Wheeler-Feynman Green's function is the real part of the Feynman Green's function. The extra piece, the imaginary part of the Feynman propagator, corresponds to the mass-shell contribution in momentum space, and has to do with the self-interaction of a charged particle that is coupled to the electromagnetic field. It guarantees the meromorphy of scattering amplitudes on the principal sheet of a suitably cut p^2-plane.

Microcausality, as it is now understood in quantum field theory, is expressed by the vanishing of (anti-)commutators of fields outside the light-cone; and this leads to analyticity of scattering amplitudes with respect to momenta. However, this new style causality is perfectly consistent with, indeed requires, retrocausality on the same footing as ordinary (Humean) causality. However, the heavy price that we must pay is the introduction of self interaction. This gives rise to divergences that are only provisionally hidden in the renormalization programme. Feynman was not satisfied with what he had achieved[11]: "I invented a better way to figure, but I hadn't fixed what I wanted to fix ... The problem was how to make the theory finite ... I wasn't satisfied at all."

Hoyle and Narlikar also add a self-action term to their quantized action at a distance theory, almost as an afterthought, and clearly against their better inclination[10]. As Dirac had done before them, they simply introduce an ultraviolet cut-off that breaks Lorentz covariance. Dirac writes, at the end of the fourth edition of his classic book, Quantum Mechanics[12]: "It would seem that we have followed as far as possible the path of logical development of the ideas of quantum mechanics as they are at present understood. The difficulties, being of a profound character, can be removed only by some drastic change in the foundations of the theory, probably a change as drastic as the passage from Bohr's orbit theory to the present quantum mechanics."

Could it be that the change to the view from nowhen, following in the footsteps of Wheeler, Feynman, Hoyle, Narlikar and Price, is sufficiently drastic to cure the malaise of electromagnetism and of quantum mechanics? As we have shown, retrocausality was built into the very foundations of QED. Yet the T-symmetry of quantum mechanics is routinely squandered in the projection postulate, with its attendant mystique of the measurement process. Might a rigorously atemporal viewpoint lead to a physical picture closer to Einstein's than to Bohr's, and might it be that the infinite self interaction is somehow a mistake induced by our time-bound viewpoint?

Notes

1. $\rho(\lambda) = 1$, $\lambda = \{1, 2\}$. $C(\gamma|1) = \cos 2\gamma$, $C(\gamma|2) = \sin 2\gamma$, $\gamma = \{\alpha, \beta\}$.

2. We leave out of consideration the science fiction scenario of many worlds. This option is logically flabby and it carries moreover an unwieldly metaphysical baggage.

3. This must be generalized to PCT invariance for some electroweak interactions, for example those responsible for K^0-decay.

References

[1] http://pdg.lbl.gov/1998/contents_tables.html
Kinoshita, T. 1981, Phys. Rev. Lett. **47** 1573.

[2] Wheeler, J.A., Feynman, R.P. 1945, 1949, Rev. Mod. Phys. **17** 157; **21** 425.

[3] Einstein, A., B. Podolsky, B., Rosen, N. 1935,Phys. Rev. **47** 777.

[4] Aspect, A., Dalibard, J., Roger, G. 1982, Phys. Rev. Lett. **49** 1804.

[5] Reichenbach, H. 1956, *The Direction of Time,* Univ. of Calif. .

[6] Bell, J.S. 1987, *Speakable and Unspeakable in Quantum Mechanics,* Cambridge Univ. Press.

[7] Atkinson, D. 1998, Dialectica **52** 103.

[8] Dummett, M.A.E. 1964, Philos. Rev. **73** 338.

[9] Huw Price, 1996, *Time's Arrow and Archimedes' Point,* Oxford.

[10] Hoyle, F., Narlikar, J.V. 1974, 1995, *Action at a Distance in Physics and Cosmology,* Freeman and Co. ; Rev. Mod. Phys. **67** 113.

[11] Schweber, S.S., 1986, Rev. Mod. Phys. **58** 449; see page 501.

[12] Dirac, P.A.M. 1958, *Quantum Mechanics,* Oxford (4th edition); see page 310.

[13] Dirac, P.A.M. 1978, *Directions in Physics,* Wiley, New York.

Chapter 5

INSTANTONS FOR BLACK HOLE PAIR PRODUCTION

Paul M. Branoff and Dieter R. Brill
Department of Physics, University of Maryland
College Park, MD 20742, USA

Abstract Various ways are explored to describe black hole pair creation in a universe with a cosmological constant that do not rely on an intermediate state of "nothing".

1. INTRODUCTION

Of Jayant Narlikar's many important contributions to astrophysics and cosmology, none is more creative and imaginative than the idea, developed with Fred Hoyle, that particles may be created as the universe expands. Stated long before quantum effects of gravity could be treated, this proposal has new meaning today. Methods are now available to analyze quantum particle production in dynamic spacetimes, and even black hole creation can be understood semiclassically as a tunneling process. The latter process is the main subject of this paper.

Although a complete theory of quantum gravity does not yet exist, examples of gravitational tunneling have been studied for a number of years, including such processes as pair creation of black holes and vacuum decay of domain walls. In each case the treatment is based on an instanton (solution of the Euclidean field equations) that connects the states between which tunneling is taking place. However, there are some nucleation processes of interest where the standard instanton method is not effective, for example because no solutions exist to the Euclidean Einstein equations that smoothly connect the spacelike sections representing the initial and final states of the tunneling process. It is therefore an interesting challenge to adapt the "bounce" method, most suitable for vacuum decay calculations, to deal with non-static initial states and

N. Dadhich and A. Kembhavi (eds.), The Universe, 51–68.

background fields such as a positive cosmological constant or domain walls typically present when particle-like states are created.

A positive cosmological constant (and other strong gravitational sources, such as a positive energy density domain wall) acts to increase the separation of timelike geodesics. It is therefore expected to "pull particles out of the vacuum" by favoring creation of pairs over their annihilation. The analogous creation of black hole pairs in de Sitter space can be treated in WKB approximation by the "no boundary" realization of quantum cosmology [1]. The first (and usually only) step in such a treatment consists of finding a solution of the Euclidean field equations containing the initial state (pure de Sitter universe) and the final state (Schwarzschild-de Sitter space) as totally geodesic boundaries. Such a solution exists only if we accept it in two disconnected pieces. If the cosmological constant is large enough one then obtains an appreciable probability of creating in each Hubble volume a pair of black holes comparable to the volume's size; if these break up into smaller ones (see, for example, Gregory and Laflamme [2]) one has, within pure gravity, a model of continuous creation not too far removed in spirit from that of Hoyle and Narlikar.

This model is, however, not fully satisfactory in several respects. For example, it is not clear how to calculate the "prefactor" of the exponential in the transition probability, which would define the dimensionful rate of the process. When it can be calculated from the fluctuations about the instanton [3], a "negative mode" is necessary for a non-vanishing rate. But this negative mode would have to connect the two parts of the instanton, and therefore cannot be treated as a small perturbation. A discontinuous instanton is of course also conceptually unsatisfactory, because the usual composition rules assume that histories are continuous.

Each of the two parts of the disconnected instanton has the universe's volume reaching zero. By forbidding arbitrarily small volumes one can connect the two parts. The exploration of modifications of Einstein gravity in which this is possible is still in its infancy. For example, Bousso and Chamblin [4] have used virtual domain walls to construct interpolating instantons. A similar technique using 'pseudomanifolds' has also been used to construct such solutions [5].

Modifications of Einstein's theory that have been proposed in other contexts may also give continuous instantons, if the change from Einstein's theory becomes important at small volumes. For this reason it is natural to consider higher curvature gravity theories.

Another promising modification of Einstein gravity is Narlikar's C-field [6]. This field can describe reasonable energetics of particle production in a context where quantum mechanics plays no essential role, and it is therefore interesting to explore, as we will below, whether it can also solve the disconnectedness problem in the instanton treatment.

So we ask whether these modifications of the Einstein-Hilbert action allow continuous paths from an initial to final cosmological state when calculating amplitudes for cosmological black hole production in the context of closed universes. We will outline a modified version of the calculation of Bousso and Chamblin concerning the use of virtual domain walls in constructing interpolating instantons. We next discuss the existence of continuous instantons in higher curvature gravity theories whose Lagrangians are nonlinear in the Ricci scalar. Finally, we consider the case of general relativity with a cosmological constant and a Narlikar C-field.

2. GRAVITATIONAL TUNNELING

Processes such as black hole pair creation can be analyzed semiclassically through the use of instanton methods. One can think of such a process as a tunneling phenomenon. The initial state consists of a universe with some background metric and no black holes, and the final state consists of a universe with two black holes supplementing the background metric. Classical dynamics is prevented from connecting the two states by a generalized potential barrier. The quantum process can "penetrate" the barrier with some probability, and the same barrier makes it improbable for the final state, once created, to "annihilate" back to the initial state. In problems that can be treated by instantons, the non-classical transition from initial to final state can be described approximately as an excursion in imaginary time. A solution that goes from the initial state to the final state and back again is called a bounce solution; an instanton is a solution which goes from the initial state to the final state, i.e., half a bounce. In the WKB interpretation the excursion into imaginary time simply signifies an exponentially decreasing wavefunction that is large only near configurations contained in the instanton. In the sum over histories interpretation the instanton is a saddle point by means of which the propagator is to be be evaluated.

The exponential of the instanton's classical Euclidean action is the dominant factor in the transition probability, provided it is normalized so that the action vanishes when there is no transition. That is, we are really comparing two instantons, one corresponding to the background alone in which initial and final states are the same, and the instanton of the bounce, in which they are different. If the initial state is static, it is typically approached asymptotically by the bounce, and therefore the normalization of the action can be achieved by a suitable surface term. If the initial state is only momentarily static, as in the case of the de Sitter universe, we must find the two instantons explicitly and evaluate their actions. In the context of the disconnected instanton the background instanton corresponds to two disconnected halves of a 4-sphere: a de Sitter space fluctuating into nothing and back again. A first test whether a modification of Einstein's theory can have connected instantons is therefore to see whether the background instanton can be connected (Fig. 1).

The rate of processes like black hole pair creation is calculated by subtracting from the action of the bounce, $I^{\rm bc}$, the action corresponding to the background state, $I^{\rm bg}$. The pair creation rate is then given as

$$\Gamma = A \exp\left[-(I^{\rm bc} - I^{\rm bg})\right], \tag{5.1}$$

where A is a prefactor, which is typically neglected in most calculations because it involves fluctuations about the classical instantons that are difficult to calculate. Without this dimensionful prefactor one can find the relative transitions to different final states, but the actual the number of transitions per spacetime volume to a given final state can only be estimated, for example as 1/(instanton four-volume) for finite volume instantons.

The connected background instanton as described above is closely related to a Euclidean wormhole, or birth of a baby universe [7]: if the two parts are connected across a totally geodesic 3-surface, we can, according to the usual rules, join a Lorentzian space-time at that surface, passing back to real time. An instanton with this surface as the final state would then describe the fluctuation of a large universe into a small one, with probability comparable to that of the creation of a black hole pair. Thus whatever process provides a connected instanton is likely to lead not only to the pair creation but also to formation of a baby universe. (In section 5 we will see how the latter can be avoided)

An instanton calculation has been used by a number of authors to find the pair creation of black holes on various backgrounds (see, for example, Garfinkle et al [8]). The instantons involved a continuous interpolation between an initial state without black holes and a final state with a pair of black holes. By contrast, in cosmological scenarios where the universe closes but Lorentzian geodesics diverge, as in the presence of a positive cosmological constant or a domain wall, there are Lorentzian solutions to Einstein's equations with and without black holes (such as de Sitter and Schwarzschild-de Sitter spacetimes, respectively), but there are no Euclidean solutions that connect the spacelike sections of these geometries [4]. (For the related case of baby universe creation the absence of such solutions is understood, for it is necessary that the Ricci tensor have at least one negative eigenvalue [9].)

The No-Boundary Proposal of Hartle and Hawking [1] can be modified to provide answers in these cases. The original proposal was designed to eliminate the initial and final singularities of cosmological models by obtaining the universe as a sum of regular histories, which may include intervals of imaginary time. One can think of the Euclidean sector of the dominant history as an instanton that mediates the creation of a (typically totally geodesic) Lorentzian section from nothing. By calculating the action corresponding to these instantons, one can calculate the wave function for this type of universe, i.e.,

$$\Psi(\mathcal{G}) = e^{-I_{\text{inst}}(\mathcal{G})} \tag{5.2}$$

where $I_{\text{inst}}(\mathcal{G}) = \frac{1}{2} I^{\text{bc}}$ is the action corresponding to a saddlepoint solution of the Euclidean Einstein equations whose only boundary is the 3-dimensional geometry $\mathcal{G}$. The probability measure associated with this universe is then given by

$$P = \Psi^* \Psi = e^{-2I_{\text{inst}}} \tag{5.3}$$

To relate the probability measure to the pair creation rate of black holes given in Eq. (5.1) one writes

$$\Gamma = \frac{P_{\text{bh}}}{P_{\text{bg}}} = \exp[-(2I^{\text{bh}}_{\text{inst}} - 2I^{\text{bg}}_{\text{inst}})] \tag{5.4}$$

so the ratio of the probability of a universe with black holes to the probability of a background universe without black holes is taken to be also the rate at which an initial cosmological state can decay into a final cosmological state, that is, the pair creation rate. In the latter sense the two disconnected instantons together describe the tunneling process.

Although this formalism allows one to calculate, in principle, the rates of nucleation processes, there is no well-justified reason why Eq. (5.4) should be identified with this quantity. The straightforward interpretation of the instanton concerns the probability for one universe to annihilate to nothing and for a second universe to be nucleated from nothing. This second universe can either contain a pair of black holes, or it can be identical to the initial universe, but it retains no "memory" of the initial state. It would clearly be preferable to have a continuous interpolation between the initial and final states. (This would allow degrees of freedom that interact only weakly with the dynamics of gravity to act as a memory that survives the pair creation.) In the following sections we will consider several ways in which this continuity of spacetime can be achieved, the first of which involves matter fields that can form virtual domain walls.

3. CONTINUOUS INSTANTONS VIA VIRTUAL DOMAIN WALLS

In this section, we will consider the method by which the authors of [4] use virtual domain walls to construct continuous paths between two otherwise disconnected instantons. They illustrated the method for the nucleation of magnetically charged Reissner-Nordström black holes in the presence of a domain wall. We will confine attention to nucleation of uncharged black holes in a universe with a cosmological constant. The initial state is the de Sitter universe and the final state is the extremal form of a Schwarzschild-de Sitter universe known as the Nariai universe [10], which is dictated by the requirement that the Euclidean solution be non-singular. To understand virtual domain walls we will need some elementary properties of real domain walls. These have been discussed extensively in [4, 11, 12, 13, 14, 15].

3.1 BRIEF OVERVIEW OF DOMAIN WALLS

A vacuum domain wall is a $(D-2)$-dimensional topological defect in a D-dimensional spacetime that forms as a result of a field ϕ undergoing the spontaneous breaking of a discrete symmetry. If we let $\mathcal{M}$ denote the manifold of vacuum expectation values of the field ϕ, then a necessary condition for a domain wall to form is that the vacuum manifold is not connected ($\pi_0(\mathcal{M}) \neq 0$). An example of a potential energy function $U(\phi)$ of the field ϕ giving rise to domain walls is the double-well potential.

Throughout this section, we will be dealing with a domain wall in the "thin-wall" approximation, which means that the thickness of the domain wall is negligible compared to its other dimensions, and it is homogeneous and isotropic in its two spacelike dimensions, so that the spatial section of the wall can be treated as planar, and the spacetime geometry as reflection symmetric with respect to the wall.

The action of a real scalar field ϕ, interacting with gravity, that may form a domain wall is given by

$$I_{\rm dw} = \int d^4x\sqrt{-g}\left[L_{\rm mat} + \frac{R-2\Lambda}{16\pi}\right] \tag{5.5}$$

with matter Lagrangian

$$L_{\rm mat} = -\frac{1}{2}\,g^{\mu\nu}\,\partial_\mu\phi\partial_\nu\phi - U(\phi) \tag{5.6}$$

and stress-energy tensor

$$T_{\mu\nu} = \partial_\mu\phi\,\partial_\nu\phi - g_{\mu\nu}\left[\frac{1}{2}\,g^{\alpha\beta}\partial_\alpha\phi\,\partial_\beta\phi + U(\phi)\right]. \tag{5.7}$$

Here $U(\phi)$ is a potential function with two degenerate minima ϕ_- and ϕ_+, at which $U = 0$; g is the determinant of the 4-metric $g_{\mu\nu}$; and R is the Ricci scalar. (We have neglected boundary terms in the action since the instantons we will be considering are compact and have no boundary.)

The trace of the Einstein equations (resulting from the variation of $I_{\rm dw}$ with respect to $g_{\mu\nu}$) gives

$$\frac{R - 4\Lambda}{8\pi} = g^{\mu\nu}\, \partial_\mu\phi\, \partial_\nu\phi + 4U(\phi)\,, \tag{5.8}$$

which can be used to simplify the action (5.5) when evaluated on a solution:

$$I_{\rm dw} = \int \left[U(\phi) + \frac{\Lambda}{8\pi} \right] \sqrt{-g}\, d^4x\,. \tag{5.9}$$

The ϕ-field is essentially constant away from a domain wall, with $\phi = \phi_-$ on one side and $\phi = \phi_+$ on the other. In Gaussian normal coordinates (ζ^i, z) with the wall at $z = 0$, ϕ depends only on z, and the field equation for ϕ implies that T_{zz} of Eq. (5.7) is negligible. The rest of the components of the stress-energy tensor differ from zero only near the wall, where ϕ changes rapidly from ϕ_- to ϕ_+:

$$T^\mu_\nu = \sigma\delta(z)\mathrm{diag}(1, 1, 1, 0) \tag{5.10}$$

where σ can be related via the ϕ-field equation to the ϕ-potential alone,

$$\sigma = \int 2U(\phi(z))dz\,. \tag{5.11}$$

Thus σ is the surface energy density of the wall. For such surface distributions the Israel matching condition imply that the intrinsic geometry h_{ij} of the domain wall is continuous, and that the extrinsic curvature jumps according to

$$K^+_{ij} - K^-_{ij} = 4\pi\sigma h_{ij}\,. \tag{5.12}$$

Here the normal with respect to which K_{ij} is defined points from the + side of the surface to the − side. Outside the wall we have the sourceless Einstein equations.

3.2 JOINING INSTANTONS BY DOMAIN WALLS

The jump (5.12) in extrinsic curvature across a domain wall can be used to join the two parts of a disconnected instanton (Fig. 1) by "surgery": We remove a small 4-ball of radius η from each instanton.

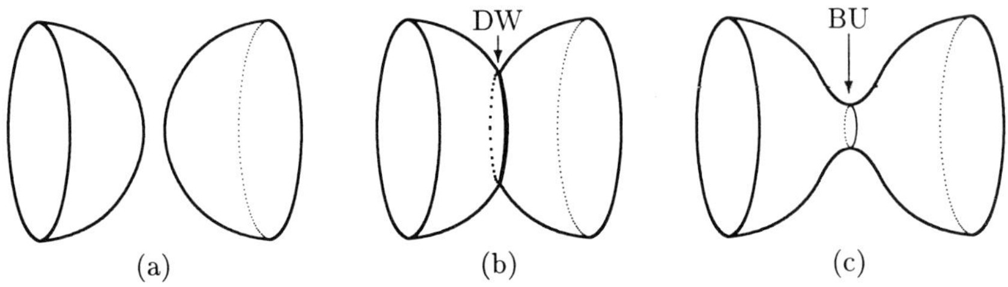

Figure 5.1 Two-dimensional analog of de Sitter instanton. Imaginary time runs horizontally. Because no significant change can be shown in two dimensions, this is a "background" instanton with identical initial and final states. (a) The disconnected instanton. (b) "Yoyo" instanton connected by domain wall (heavy curve labeled DW). (c) Instanton connected by a "virtual baby universe" (BU).

Their two 3-surface boundaries have the same intrinsic geometry, and their extrinsic curvatures are proportional to the surface metric. They can therefore be joined together in such a way as to satisfy the Israel matching conditions, Eq. (5.12), thereby inserting a domain wall.

However, the surface energy density $\tilde{\sigma}$ of the domain wall used to join the instanton must be negative: As we approach the domain wall from the initial state, heading towards annihilation, successive 3-spheres are shrinking, $K_{ij}^+ < 0$. After we pass through the domain wall, successive 3-spheres are expanding, $K_{ij}^- > 0$. Because of the negative energy density the authors of [4] call this a virtual domain wall, but it is not virtual in the sense that it corresponds to a Euclidean solution of the equations of section 3.1, for the σ of Eq. (5.11) remains positive when passing to imaginary time. Within this scheme the only way to achieve a "yoyo" instanton as a saddle point of the Euclidean action is to have a scalar field with a negative energy also in the real domain, that is, a Lagrangian with the opposite sign as that of Eq. (5.6). As we will see in section 5, in that case a plain scalar field, without the domain-wall-forming potential $U(\phi)$, will do as well and is preferable.

By how much does the Euclidean action change when we introduce a domain wall whose radius η is small compared to the radius $\sqrt{3/\Lambda}$ of the instanton itself? The extrinsic curvature of the connecting 3-sphere is then nearly the same as what it would be in flat space, $K_{ij} = h_{ij}/\eta$, and the jump in curvature is twice that; hence the size of the domain wall is determined from Eq. (5.12),

$$\eta = -\frac{1}{2\pi\tilde{\sigma}} \,. \tag{5.13}$$

The Euclidean version of Eq. (5.9) is

$$I_{\text{dw}} = -\int \left[U(\phi) + \frac{\Lambda}{8\pi} \right] \sqrt{g}\, d^4x\,. \tag{5.14}$$

We have taken a 4-ball with scalar curvature $R = 4\Lambda$ away from each part of the original instanton, for a total change in action (including a boundary term) by $3\pi\eta^2/2 - \Lambda\pi\eta^4/8$; this is comparable to that due to the added domain wall with action given by Eqs. (5.14) and (5.11), $I_{\text{dw}} = -\pi^2\tilde{\sigma}\eta^3 = \frac{1}{2}\pi\eta^2$, which is small compared to the total action $-3\pi/\Lambda$. Thus the Euclidean action increases when we add the domain wall, and the connected instanton therefore has a relatively smaller probability measure (although the difference is small compared to the total action), and the disconnected instanton will dominate. If the path integral is extended over continuous histories only, the domain wall provides the only saddle point, with action very close to what the discontinuous history would have given, thus justifying the calculation using the discontinuous history alone. But in that case a path integral without a domain-wall-forming scalar field gives a very similar value of the action, as shown in [4].

Introducing this scalar field may therefore be considered a high price to pay for gaining a saddle point, particularly because it entails other, less desirable processes. For example, the "center" $z = 0$ of the domain wall is totally geodesic with $\partial\phi/\partial z = 0$, that is, a possible place to revert from imaginary time back to real time. This corresponds to the formation of a baby universe of size comparable to η and smaller Euclidean action than that for the black hole formation.

If a field exists that can form small domain walls, any two instanton parts can be connected by such surgery across one or several small 3-spheres, with a change in action as estimated above for each; the dominant history will have the fewest connections.

Finally, recall that the periodicity in imaginary time of each part of the disconnected instanton is well defined by the requirement that conical singularities should be absent from each part. If the parts are connected where there would otherwise be a conical singularity, one such requirement is eliminated. Thus there are connected instantons for which the final state is not Nariai but Schwarzschild-de Sitter geometry with black hole and cosmological horizons of unequal size.

4. CONTINUOUS INSTANTONS IN HIGHER CURVATURE THEORIES

Higher curvature theories have a long history and have been proposed in several different contexts. For example, they arise naturally in theories describing gravity by an effective action [16, 17].

In this section we will explore whether higher order theories can pass the "first test" of Section 2, namely whether there is a continuous instanton describing the annihilation and rebirth of de Sitter space (generalized to these theories). Adding higher order terms to the action does not, however, immediately eliminate disconnected instantons; for example, de Sitter space (that is, a spacetime of constant curvature) is a solution of many higher-order theories. In fact, if the universe without and with black holes can originate by tunneling from nothing, a disconnected instanton will also exist. Therefore connected instantons may again co-exist with the de Sitter-like, disconnected instantons.

The Euclidean action we will be considering has the form

$$I = -\frac{1}{16\pi}\int f(R)\sqrt{g}\,d^4x \tag{5.15}$$

where

$$f(R) = R - 2\Lambda + \alpha R^2 + \gamma R^3 + \cdots , \tag{5.16}$$

R is the Ricci scalar, Λ is the cosmological constant, and α, γ, etc., are coupling constants whose value we leave unspecified for the moment. The metric has the Euclidean Robertson-Walker form appropriate to three-dimensional space slices of constant positive curvature:[1]

$$ds^2 = N^2(\tau)d\tau^2 + a^2(\tau)d\Omega_3^2 . \tag{5.17}$$

Here τ is imaginary time determined from the analytic continuation $t \to i\tau$, N is the lapse function, a is the universe radius and $d\Omega_3^2$ is the metric on the unit three-sphere. Having the metric depend on N and a allows us to obtain all the independent Einstein equations by varying only these functions in the action (5.15): variation with respect to a gives us the one independent spacelike time development equation, and variation with respect to N yields the timelike constraint equation, as in ordinary Einstein theory. A further variation that is easily performed is a *conformal* change of the metric, giving the trace of the field equations, which is not independent of the other equations but involves only the function $f(R)$:

$$\frac{\partial}{\partial a}(fNa^3) - \frac{d}{d\tau}\left[\frac{\partial}{\partial \dot{a}}(fNa^3)\right] + \frac{d^2}{d\tau^2}\left[\frac{\partial}{\partial \ddot{a}}(fNa^3)\right] = 0 \tag{5.18}$$

$$
\begin{aligned}
a^3 f + \frac{\partial f}{\partial N} N a^3 - \frac{d}{d\tau}\left[\frac{\partial}{\partial \dot{N}}(f N a^3)\right] &= 0 & (5.19)\\
2Rf' + 6\nabla^2 f' - 4f &= 0 & (5.20)
\end{aligned}
$$

where

$$
\nabla^2 = \frac{d^2}{d\tau^2} + \frac{3\dot{a}}{a}\frac{d}{d\tau}, \tag{5.21}
$$

a dot denotes $d/d\tau$, and a prime denotes d/dR.

Equation (5.18) is a fourth order ordinary differential equation, and Eq. (5.19) is a third order first integral of this equation. The trace equation (5.20) shows that we can regard R as an independent variable, satisfying a second order equation. In this view Eq. (5.20) replaces Eq. (5.18) (to which it is equivalent), and a also satisfies a second-order differential equation, namely its definition in terms of R,

$$
R = -6\left(\frac{\ddot{a}}{aN^2} - \frac{\dot{a}\dot{N}}{aN^3} + \frac{\dot{a}^2}{a^2N^2} - \frac{1}{a^2}\right) \tag{5.22}
$$

In addition we still have the constraint, Eq. (5.19), a first order relation between a and R.

A general Hamiltonian analysis (c.f. [18] and references therein), not confined to the symmetry of Eq. (5.17), bears out the idea that, as a second-order field theory, this is Einstein theory coupled to a non-standard scalar field [19]. For example, for a Lagrangian quadratic in the Ricci scalar with no cosmological constant, the relationship between R and the non-standard scalar field ϕ is given by [20]

$$
\phi = \sqrt{\frac{3}{4\pi}}\alpha R \tag{5.23}
$$

where the ϕ-field has the standard stress energy tensor multiplied by $(1 + 4(\pi/3)^{1/2}\phi)^{-2}$.

Can this effective scalar field form a domain wall in four dimensions? If the coefficients up to γ in Eq. (5.16) are non-zero, then the "force term" $Rf' - 2f$ occurring in the trace equation (5.20) vanishes at three equilibrium states for R, where $\nabla^2 f'(R) = 0$, approximately at $R = \pm 1/\sqrt{\gamma}$ and at $R = 4\Lambda$, for small γ. But in order to have a macroscopic universe on either side of the wall we need $R = 4\Lambda$ on either side, so the usual wall formation where the scalar field changes from one equilibrium to another is unsuitable in this case. A solution of the bounce type may appear possible, since the equilibrium at $R = 4\Lambda$ is unstable. At the turning point the time-dependent R would then have to "overshoot" the stable equilibrium near $R = -1/\sqrt{\gamma}$. A negative R is required there so that the universe radius can turn around at the same moment. This synchronization, if possible at all (numerical calculations have failed to reveal it to us; see however ref. [21]), would require fine tuning that does not appear natural in this context. Furthermore, if we had a bounce for both a and R, half of it would be an instanton describing the formation of a baby universe of size $\sim \gamma^{-1/4}$, which would then continue to collapse classically, and this process would be exponentially more probably than the black hole formation. For these reasons the effective scalar field that derives from higher curvature Lagrangians of the form (5.15) does not appear promising for connected instantons.

We therefore consider solutions to Eqs (5.18) - (5.20) when R is constant, $R = R_0$. To allow a continuous transition to imaginary time at $\tau = 0$ we make the usual ansatz that all odd time derivatives of $a(\tau)$ vanish at $\tau = 0$. With the choice $N = 1$, the above equations at $\tau = 0$ take the form

$$\begin{aligned} R_0 f' - 2f &= 0 \qquad (5.24)\\ a_0 f + 6\ddot{a}_0 f' &= 0 \qquad (5.25)\\ a_0^2 f - 2f'(a_0\ddot{a}_0 - 2) &= 0\,. \qquad (5.26) \end{aligned}$$

Eliminating f from these equations we get the condition

$$(a_0^2 R_0 - 12) f' = 0\,. \qquad (5.27)$$

Thus, we have two classes of solutions. The first class is described by the condition $R_0 = 12/a_0^2$. The second class is described by the condition $f = f' = 0$.

If $R = R_0 = 12/a_0^2$ and $\dot{a}_0 = 0$ then the unique regular solution of Eq. (5.22) is a de Sitter-like solution, $a(\tau) = a_0 \cos(\tau/a_0)$, leading only to the disconnected instanton.

The second condition indeed allows periodic, non-collapsing solutions with any amplitude A of the form

$$a^2 = \frac{6}{R_0} + A\cos\sqrt{\frac{R_0}{3}}\tau \qquad \text{where} \qquad f(R_0) = 0 = f'(R_0)\,. \qquad (5.28)$$

If we want this R_0 to be close to that of the de Sitter universe, $R_0 = 4\Lambda$, then at least one of the higher-order coefficients (α, γ ...) in $f(R)$ of Eq. (5.16) has to be large and rather fine tuned. Furthermore, because the action for all of these solutions vanishes, we should integrate over all values of A, which includes some disconnected instantons, so this problem is not really avoided by these solutions. (They appear pathological also in other ways, for example they would allow production of baby universes of any radius. They would also tend to be unstable in the

Lorentzian sector, although this can be confined to the largest scale by judiciously choosing $f(R) = R - 2\Lambda$ except near $R_0 \sim 2\Lambda$.)

5. CONTINUOUS INSTANTONS IN C-FIELD THEORY

Except for boundary terms, which describe classical matter creation and which we neglect in the present context, the C-field Lagrangian is similar to the usual scalar field Lagrangian without self-interaction (Eq. (5.6) with $U = 0$), but with the important difference that the coupling constant $-f$ of the C-field has the opposite sign from the usual one [6]. Thus the total action of gravity with cosmological constant and C-field has the form, for Lorentzian geometries

$$I_C = \int d^4x\sqrt{-g}\left[\frac{1}{2}f\, g^{\mu\nu}\,\partial_\mu C \partial_\nu C + \frac{R-2\Lambda}{16\pi}\right]. \tag{5.29}$$

(We have not included ordinary matter fields here because we are confining attention to pair production of black holes as purely geometrical objects.)

5.1 SOURCELESS C-FIELD IN LORENTZIAN COSMOLOGY

The field equations that follow from this action by varying C and $g_{\mu\nu}$ are, for the C-field:

$$\triangle C = C^{\mu}_{\ ;\mu} = 0 \quad (\text{where } C_\mu = C_{,\mu}) \tag{5.30}$$

and for the geometry,

$$G_{\mu\nu} + \Lambda g_{\mu\nu} = T^C_{\mu\nu} \quad \text{where} \quad T^C_{\mu\nu} = -f\left(C_\mu C_\nu - \frac{1}{2}g_{\mu\nu}C^\alpha C_\alpha\right). \tag{5.31}$$

The stress-energy tensor $T^C_{\mu\nu}$ gives a negative energy density (for $f > 0$). Narlikar [6] has given reasons why this violation of the energy condition is not an objection when the C-field is coupled to Einstein gravity of an expanding universe.

For Lorentzian cosmology we make a Robertson-Walker ansatz analogous to (5.17),

$$ds^2 = -N^2(t)dt^2 + a^2(t)d\Omega_3^2\,. \tag{5.32}$$

In agreement with the homogeneous nature of this geometry we assume that C is homogeneous in space and hence depends only on t. The field

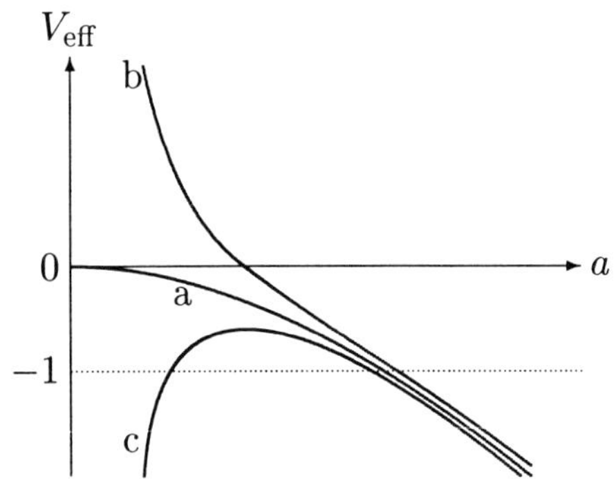

Figure 5.2 The effective potential for de Sitter-like universes: a universe with only cosmological constant (curve a), one with a real C-field (curve b), and one with a virtual C-field (curve c).

equations, derived by varying a, N, and C, and then setting $N = 1$, are

$$2\frac{\ddot{a}}{a} + \frac{\dot{a}^2 + 1}{a^2} - \Lambda = 4\pi f \dot{C}^2 \tag{5.33}$$

$$\dot{a}^2 + 1 - \frac{\Lambda}{3}a^2 = -\frac{4\pi f}{3}\dot{C}^2 a^2 \tag{5.34}$$

$$\frac{d(a^3\dot{C})}{a^3\,dt} = 0\,. \tag{5.35}$$

The second equation, as usual, is a first integral of the first (time development) equation, and it implies the latter except for extraneous solutions $a = \text{const}$. The third equation has the integral

$$\dot{C} = \frac{K}{a^3} \tag{5.36}$$

where K is a constant. By eliminating $\dot{C}$ we obtain an equation of the "conservation of energy" type for a:

$$\dot{a}^2 + V_{\text{eff}} = \dot{a}^2 - \frac{\Lambda}{3}a^2 + \frac{4\pi f K^2}{3a^4} = -1\,. \tag{5.37}$$

This is the usual de Sitter equation supplemented by a term in $1/a^4$, which is unimportant at late times when a is large and does not change the qualitative Lorentzian time development at any time (Fig. 2).

5.2 SOURCELESS C-FIELD IN EUCLIDEAN COSMOLOGY

The effective potential in Eq. (5.37) increases monotonically as a decreases below the minimum classically allowed value. The corresponding Euclidean motion in such a potential therefore does not bounce; instead, a would continue to decrease and reach $a = 0$ in a finite Euclidean time. This is a geometrical singularity if $K \neq 0$ because, for example, it follows from Eqs. (5.33) and (5.36) that $R = 4\Lambda + 8\pi f(K^2/a^6)$.

However, a different potential is obtained if the motion of both a and C is continued to imaginary time,[2] thereby describing a virtual process that involves both of these variables, so that we take into account fluctuations in C as well as in a. Then the K of Eq. (5.36) becomes imaginary, $K = ik$, and the Euclidean "conservation of energy" equation becomes

$$-\dot{a}^2 - \frac{\Lambda}{3}a^2 - \frac{4\pi f k^2}{3a^4} = -1\,. \tag{5.38}$$

It is easily seen that, for sufficiently small k, this equation does have bounce solutions, with a turning point at $a \sim k^{1/2} f^{1/4}$ (Fig. 2). Thus the C-field theory passes the "first test": it has a continuous instanton describing a fluctuation with identical initial and final state. It is reasonable to suppose that the theory will also have continuous instantons describing the creation of a black hole pair, because for small k the turning point occurs at small a, so that two disconnected instantons can be joined by surgery similar to that of section 3.

It is essential that the fluctuation of the C-field be virtual, that is, that the coupling constant f have the opposite sign from the usual, positive energy density scalar field. If the C-field were real, time could revert to real values at the minimum radius of the bounce and continue in a small, Lorentzian universe [23] that we have above described as a baby universe. This transition would be the most probable if allowed. By contrast, in the case of the virtual C-field this transition is not allowed, The reason is that at the moment of the bounce, the C-field's effective potential dominates. A return to real time (K changing from imaginary to real) would make a large change in V_{eff} of Eq. (5.37), violating this Lorentzian Hamiltonian constraint. A much smaller violation is involved at the first change to imaginary time, at large a. This can occur if the background is not exactly de Sitter-like, but contains some gravitational wave excitation that can supply the necessary small energy difference in the local region where the black hole will form. Thus the C-field makes a continuous instanton possible, but avoids forming a baby universe.[3]

5.3 BLACK HOLES IN C-FIELD COSMOLOGY

As a final step we exhibit as an endstate of the particle creation instanton an expanding universe in C-field theory of spatial topology $S^1 \times S^2$. This describes a universe with an extremal black hole pair in the same sense that the Nariai solution [10, 25] describes such a universe in Einstein's theory. The metric has the homogeneous form

$$ds^2 = -dt^2 + a^2(t)d\chi^2 + b^2(t)(d\theta^2 + \sin^2\theta d\phi^2) \tag{5.39}$$

where χ has periodicity appropriate to S^1, θ and ϕ are coordinates on S^2, and a and b are functions only of t. The C-field likewise is a function only of t and therefore obeys the conservation law analogous to (5.36),

$$\dot{C} = \frac{K}{ab^2}\,. \tag{5.40}$$

The field equations then take the form

$$G^t_t + \Lambda = -\frac{2\dot{a}\dot{b}}{ab} - \frac{\dot{b}^2+1}{b^2} + \Lambda \quad = \quad \frac{4\pi f K^2}{a^2 b^4} \tag{5.41}$$

$$G^\chi_\chi + \Lambda = -\frac{2\ddot{b}}{b} - \frac{\dot{b}^2+1}{b^2} + \Lambda \quad = \quad -\frac{4\pi f K^2}{a^2 b^4} \tag{5.42}$$

$$G^\theta_\theta + \Lambda = -\frac{\ddot{a}}{a} - \frac{\dot{a}\dot{b}}{ab} - \frac{\ddot{b}}{b} + \Lambda \quad = \quad -\frac{4\pi f K^2}{a^2 b^4}\,. \tag{5.43}$$

If the universe volume expands similar to the Nariai solution, the effects the C-field will become negligible at late times. It is therefore reasonable to solve the field equations with the condition that the solution be asymptotic to the Nariai universe, $a(t) = (1/\sqrt{\Lambda})\cosh\sqrt{\Lambda}t$, $b(t) = 1/\sqrt{\Lambda}$. We also require a moment of time-symmetry (to enable the transition from imaginary time). The solution to first order in $\varepsilon = 4\pi f K^2 \Lambda^{3/2}$ is

$$a(t) \quad = \quad \frac{1}{\sqrt{\Lambda}}\cosh\sqrt{\Lambda}t - \frac{\varepsilon}{3}\ln(2\cosh\sqrt{\Lambda}t) - \frac{\varepsilon}{8}e^{-2\sqrt{\Lambda}t} + \cdots \tag{5.44}$$

$$b(t) \quad = \quad \frac{1}{\sqrt{\Lambda}} + \frac{\varepsilon}{6}e^{-2\sqrt{\Lambda}t} + \cdots\,. \tag{5.45}$$

These functions do not differ much from those for the Nariai solution for any time t. However, the differences would become large in the continuation to imaginary time, as the volume decreases. In order to reach a minimum volume we again need an imaginary K (virtual C-field). This minimum volume, like all $t =$ const. surfaces, has topology

$S^1 \times S^2$ and would therefore not fit directly on the minimum-a surface of a de Sitter-like metric, Eq. (5.17); a solution with less symmetry in both spaces would be needed to make the match.

6. CONCLUSIONS

In Einstein's theory of gravity with a cosmological constant, typical Euclidean solutions describe a universe originating from "nothing," or decaying into nothing, but there are no equally simple solutions corresponding to quantum processes, such as creation of a pair of black holes, which change a universe that is already present. According to the simple interpretation of Euclidean solutions in Einstein's theory, the most probable path to black hole creation is discontinuous via nothing as an intermediate state. In the present paper we have considered several modifications of Einstein's theory that allow continuous histories as saddle points of the Euclidean action between two finite universes. Considered as a matter source, these modifications involve extreme forms of the stress-energy tensor because the Ricci tensor will typically have at least one negative eigenvalue. Therefore the formation of baby universes is a possible competing process.

A matter field that can form sufficiently small domain walls is a universal connector, replacing the intermediate state of nothing with at least a small three-sphere. Higher-order Lagrangians in the scalar curvature have to be fine tuned to allow the desired continuous histories. In many ways the most successful solution involves a scalar C-field of negative (but small) coupling constant.

Notes

1. The cases of zero or negative curvature present additional normalization problems because the naive Euclidean action would be infinite. Therefore we confine attention to the positive curvature case.

2. We assume that this transition is the most probable; this would not be so if a transition were possible in the potential of Eq. (5.37). For example in penetrating radially a spherically symmetric potential barrier the most likely transition maintains the real angular momentum [22].

3. We also note that, as remarked in [24], a real change in C (if K were real) during the instanton could be interpreted as a change in the gravitational constant after the pair creation, which would be undesirable.

References

[1] J. B. Hartle and S. W. Hawking, *Phys. Rev.* D **28**, 2960 (1983).

[2] R. Gregory and R. Laflamme, *Phys. Rev. Lett.* **70**, 2837 (1993); *Nucl. Phys.* B**428**, 399 (1994).

[3] C. G. Callan, Jr., and S. Coleman, *Phys. Rev.* D **16**, 1762 (1977).

[4] R. Bousso and A. Chamblin, "Patching up the No-Boundary Proposal with Virtual Euclidean Wormholes" gr-qc/9803047.

[5] E. Farhi, A. H. Guth, and J. Guven, *Nucl. Phys.* B **339**, 417 (1990).

[6] J. V. Narlikar, *J. Astrophys. Astr.* **5**, 67 (1984); J. V. Narlikar and T. Padmanabhan, *Phys. Rev.* D **32**, 1928 (1985).

[7] S. B. Giddings and A. Strominger, *Nucl. Phys.* B **306**, 890 (1988).

[8] D. Garfinkle, S. B. Giddings, and A. Strominger, *Phys. Rev.* D **49**, 958 (1994).

[9] J. Cheeger and D. Grommol, *Ann. Math.* **96**, 413 (1972).

[10] H. Nariai, *Sci. Rep. Tohoku Univ.* Series 1 **34**, 160 (1950); *ibid.* **35**, 62 (1951).

[11] A. Vilenkin and E. P. S. Shellard, *Cosmic Strings and Other Topological Defects*, Cambridge University Press, Cambridge (1994).

[12] R. R. Caldwell, A. Chamblin, and G. W. Gibbons, *Phys. Rev.* D **53**, 7103 (1996).

[13] S. J. Kolitch and D. M. Eardley, *Phys. Rev.* D **56**, 4651, (1997).

[14] A. Vilenkin, *Phys. Lett.* B **133**, 177 (1983).

[15] J. Ipser and P. Sikivie, *Phys. Rev.* D **30**, 712 (1984).

[16] R. C. Myers, *Nucl. Phys.* B **289**, 701 (1987).

[17] S. Deser and N. Redlich, *Phys. Lett.* **176B**, 350 (1986).

[18] Y. Ezawa, M. Kajihara, M. Kiminami, J. Soda, and T. Yano, "On the Canonical Formalism for a Higher-Curvature Gravity" gr-qc/9801084.

[19] J. D. Barrow and S. Cotsakis, *Phys. Lett.* B **214**, 515 (1988); K. Maeda, *Phys. Rev.* D **39**, 3159 (1989).

[20] S. W. Hawking and J. C. Luttrell, *Nucl. Phys.* B **247**, 250 (1984).

[21] H. Fukutata, K. Ghoroku, and K. Tanaka, *Phys. Lett.* B **222**, 191 (1990); O. Bertolami, *ibid.* **234**, 258 (1990).

[22] K. Lee, *Phys. Rev. Lett.* **61**, 263 (1988) and *Phys. Rev.* D **48**, 2493 (1993).

[23] J. D. Brown, *Phys. Rev.* D **41**, 1125 (1990).

[24] S. Cotsakis, P. Leach, and G. Flessas, *Phys. Rev.* D **49**, 6489 (1994).

[25] R. Bousso and S. W. Hawking, *Phys. Rev.* D **54**, 6312 (1996).

Chapter 6

THE ORIGIN OF HELIUM AND THE OTHER LIGHT ELEMENTS

G. Burbidge
Department of Physics and Center for Astrophysics and Science
University of California, La Jolla, CA 92093, USA.

F. Hoyle
102 Admiral's Walk
Bournemouth BH2 5HF, Dorset, UK

Abstract The energy released in the synthesis of cosmic ^{4}He from hydrogen is almost exactly equal to the energy contained in the cosmic microwave background radiation. This result strongly suggests that the ^{4}He was produced by hydrogen burning in stars and not in the early stages of a big bang. In addition, we show that there are good arguments for believing that the other light isotopes, D, ^{3}He, ^{6}Li, ^{7}Li, ^{9}Be, ^{10}B and ^{11}B were also synthesized in processes involving stars. By combining these results with the earlier, much more detailed work of Burbidge et al. and of Cameron, we can finally conclude that all of the chemical elements were synthesized from hydrogen in stars over a time of about 10^{11} yr.

1. INTRODUCTION

There are more than 320 stable isotopes in the periodic table. In our original work ([1], hereafter B^{2}FH; see also [11]), we showed that nearly all of them, with the possible exception of the helium isotopes and D, Li, Be, and B, were synthesized by nuclear processes in stellar interiors. In the 1950s, there appeared to be several problems associated with explaining the observed abundances of these remaining nuclides, which we discuss in turn. We shall show here that another approach leads

N. Dadhich and A. Kembhavi (eds.), The Universe, 69–76.

to the conclusion that very likely all of them have been synthesized in processes involving stars.

2. ^{4}HE

In the 1950s, it appeared to us that there were two problems associated with explaining the origin of helium in its measured abundance through hydrogen burning. Assuming that the time-scale of the universe is $\sim H_0^{-1}$, there was not enough time for a ^{4}He/H ratio of about 0.24 to be built up, if the luminosities of the galaxies remained at normal levels for 10^{10} yr. Second, there appeared to be no evidence that the energy released by this amount of hydrogen burning was present. The energy density of starlight of about $10^{-14}\,\mathrm{erg\,cm^{-3}}$ is well below the energy released in hydrogen burning, which, for a ^{4}He/H ratio of 0.243 [33, 25] that we assume to be universal, is $4.37\times10^{-13}\,\mathrm{erg\,cm^{-3}}$. In deriving this quantity, we have taken the mean density of baryonic matter associated with galaxies to be $4.31\times10^{-31}\,\mathrm{gm\,cm^{-3}}$. This number has been obtained from the counts of galaxies, and we assume that baryonic dark matter in the form of massive halos, etc. (with 10 times the visible mass), is present. Here we have put $H_0 = 60\,\mathrm{km\,sec^{-1}\,Mpc^{-1}}$.

In the 1950s, Bondi, Gold & Hoyle [5] argued that the large amount of undetected energy, which must be present if the helium has been synthesized in stars, must reside in the far-infrared spectrum, while Burbidge [8] speculated that perhaps there was an earlier short-lived phase in the evolution of galaxies in which they were much more luminous, or else possibly the true helium abundance was lower than 0.24, because most of the mass is tied up in low-mass stars in which He/H < 0.24.

Of course, the solution to the He problem that became popular was that which Gamow, Alpher, & Herman proposed earlier [1]), that the helium was made in a hot big bang some 10^{10} yr ago. Several calculations following this work and starting with Hoyle & Tayler [24], Peebles [34], and Wagoner, Fowler, & Hoyle [46] demonstrated this. We have now reached the stage where it is argued that the existence of He and the other light isotopes is taken, together with the microwave background radiation, as primary evidence in favor of the standard, hot, big bang cosmological model. However, this argument is only powerful if there is no other way to explain the helium abundance and the microwave background radiation.

In 1941, McKellar [30] showed that there must be a radiation field present in the Galaxy with a temperature between 1.8 and 3.4 K. Penzias & Wilson's [35] measurements, followed by others and culminating in the COBE observations by J. Mather and his colleagues [16] have shown that the cosmic microwave background (CMB) has a blackbody form at least out to radio wavelengths with $T = 2.728\,\mathrm{K}$. The hot big bang cosmological model is not able to predict the temperature [43]. But what is remarkable about the result that we have described here is that the energy density of the *observed* blackbody radiation is extremely close to the energy density expected from the production of helium from hydrogen burning. We showed earlier that this energy is $4.37\times10^{-13}\,\mathrm{erg\,s^{-1}\,cm^{-3}}$ and when this energy is thermalized, the temperature turns out to be $T = 2.76\,\mathrm{K}$.

While the value of the baryonic density in galaxies and their environs is not known with anything like the precision with which the blackbody temperature is measured, it is clearly not very different from $\rho = 3\times10^{-31}\,\mathrm{gm\,cm^{-3}}$ ($H_0 = 60\,\mathrm{km\,sec^{-1}\,Mpc^{-1}}$, and dark-to-luminous baryon ratio ~ 10) and of course, the calculated temperature is only proportional to $\rho^{1/4}$. Indeed, it might be argued that the CMB temperature gives a more precise measure of the true mass density of baryonic matter in the universe than can be obtained by estimating the mass in galaxies.

We conclude that this result, based on two simple observational arguments, strongly suggests that the helium and the CMB were produced by hydrogen burning in stars. This requires a time much greater than 10^{10} yr, and there must be a physical mechanism operating that is able to thermalize the radiation that is initially released through hydrogen burning as ultraviolet photons from hot stars in starburst situations in galaxies. We have shown elsewhere that both of these conditions are fulfilled within the framework of the quasi-steady state cosmology (QSSC) (Hoyle, Burbidge, & Narlikar [20, 21, 22, 23]). In the QSSC, the universe is in a sequence of oscillations of period Q superposed on a general universal expansion of period P. In our model $Q \approx 10^{11}$ yr and $P \approx 10^{12}$ yr. These timescales correspond to the lifetimes of main-sequence dwarf stars with masses less than $0.7M_\odot$ and $0.4M_\odot$, respectively, thereby greatly enhancing the importance of dwarf stars in cosmogony. We conclude that ^{4}He in the cosmos is most likely a result of stellar nucleosynthesis. Given that this most abundant nucleus among the light elements is a result of stellar activity, it is then natural to ask whether the other light isotopes can also be due to processes involving stars.

3. ^{6}LI, ^{7}LI, ^{9}BE, ^{10}B AND ^{11}B

Much work has been done on these nuclides in recent years. It is generally accepted that ^{6}Li, ^{9}Be, ^{10}B and ^{11}B were produced in spallation reactions of high-energy protons on ^{12}C and ^{16}O with energy ultimately coming from galactic processes as we originally proposed B^{2}FH. Reeves, Fowler & Hoyle [37] showed that galactic cosmic rays are an important ingredient. The most modern work shows that it is the C and O that bombard the protons and α-particles. The Be and B abundances are proportional to the Fe/H ratio is subdwarfs, and Vangioni-Flam et al. [45] have shown that spallation by high-energy C and O can account for this. The high-energy C and O nuclei are ejected in the winds from massive stars and supernovae.

What about ^{7}Li? The early suggestion [37] that spallation is responsible gives a $^6\mathrm{Li}/sevenli$ ratio ~ 1, but in the solar system, $^6\mathrm{Li}/^7\mathrm{Li} \simeq 10$. This is one of the reasons why it has been argued that ^{7}Li at least is due to big bang nucleosynthesis. This argument has been supported by the claim that there is a "plateau" at $^7\mathrm{Li/H} = 1.7\times10^{-10}$ in a sample of Population II stars that are $> 10^{10}$ yr old [41]. However, it is now known that this plateau is breached and that several stars have $^7\mathrm{Li/H} < 10^{-10}$ [6]. Ryan et al. [38] conclude that there is an intrinsic spread in the ^{7}Li abundance due to influences other than uniform nucleosynthesis in a big bang. We must also not forget that while it is generally believed that susceptibility to destruction prevents ^{7}Li from being synthesized in stars, the observation that there is a class of lithium-rich supergiants (cf. WZ Cas; [30]) shows that stellar processes may be responsible, as was suggested in a complicated scenario by Cameron & Fowler [12].

Boesgaard & Tripicco [4] looked at the Li abundance as a function of [Fe/H] for both Population I and old disk stars. They found that the Li abundance could be very different in stars where the [Fe/H] abundance has the solar value but that there is an absence of stars that are Li rich but have low values of [Fe/H] (see also [36, 3]). The abundances and isotope ratios of Li in the interstellar gas have been determined most recently by Lemoine, Ferlet, & Vidal-Madjar [27]. They have concluded that there must be an extra source of ^{7}Li in the Galaxy. It is now clear from the observations that there may be at least three possible effects that have contributed to the observe Li abundance. They are (a) stellar processing, which tends to deplete Li, (b) galactic production which tends to build Li and (c) big bang nucleosynthesis. From the observations, the relative importance of (a), (b),and (c) is not yet clear. However, in view of our earlier arguments concerning the origin of ^{4}He, we consider it likely that (c) is not operating. Thus, we believe that (a) and (b) alone can explain the Li abundance and that further observational investigations will show this.

4. D AND ^{3}HE

The light isotope ^{3}He is produced in large quantities in dwarf stars where the masses are not large enough for it to be destroyed by ^{3}He(^{3}He, $2p$)^{4}He. It is also the case that there is a class of stars in which it has been shown from measurements of the isotope shift that most of the helium in their atmospheres is He. These stars include 21 Aquilae, three Centaurus A and several others [1, 39, 18, 42]. The stars are peculiar A, F, and B stars having He/H abundance that are 1/10 of the normal helium abundance. The ^{3}He/^{4}He ratio can range from 2.7 to 0.5. These stars occupy a narrow strip in the $(\log g, T_{\rm eff})$ plane between the B stars with strong helium lines that shows no evidence for the presence of ^{3}He. However, the detection of ^{3}He from the isotope shift will fail if the ^{3}He/^{4}He ratio is ≤ 0.1. Thus, many of the weak helium-line stars may well have ^{3}He/^{4}He abundance ratios far higher than abundance ratio that is normally assumed to be present, namely, $^3\text{He}/^4\text{He} \approx 2\times10^{-4}$. The high abundance of He in these stars has been attributed by G. Michaud and his colleagues to diffusion ([32] 1979 and earlier references). Whether or not this is the correct explanation, what these results do tell us is that stellar winds from such stars will enrich the interstellar gas with ^{3}He in large amounts. This ^{3}He is in addition to the ^{3}He that will be injected from dwarf stars. The final abundance required is $^3\text{He/H} \approx 2\times10^{-5}$. It has been argued by those who believe that ^{3}He is a product of big bang nulceosynthesis that there has not been time to build up the required abundance by astrophysical processes. However, not only do we not know what the rate of injection from stars is, but in the QSSC, the time scale for all of this stellar processing is $\sim 10^{11}$ rather than $H_0^{-1} \approx 10^{10}$ yr. Thus, we believe that ^{3}He may very well have been produced by stellar processes.

We turn finally to the production of deuterium. It has been arguead that D cannot be synthesized by spallation or photo-disintegration in supernova outbursts [15, 40]. Recently, however, Fuller & Shi [17] have argued that antineutrinos $\overline{\nu}_e$ can give rise to deuterons through $\overline{\nu}_e + p \rightarrow n + e^+$ followed by $n(p,\gamma)$D -reactions in the collapse of supermassive stars ($M \geq 5\times10^4 M_\odot$) in the early history of galaxies. This mechanishm may be important but in view of the fact that the ^{3}He/H and D/H ratios are very similar, and because we believe that the ^{3}He is likely to be produced by low-mass stars, we believe that the most likely source of the cosmic deuterium is the dwarf stars.

It is known that the dwarf M stars are a major constituent of normal galaxies. They have extensive convective envelopes, and thus they are likely to have outer layers in which extensive flare activity takes place. A very good example is the large UV flare in the red dwarf AU Microscopii, which has just been reported [26]. In our view, it is the cumulative effect of stellar winds and flares from these low-mass stars that has led to the build up of the deuterium.

It is easily shown that the smount of energy required to generate a D/H ratio $\sim 10^{-5}$ through flaring and ejection from dwarf stars is not very large. The energy required to produce D in stellar flares through the generation of neutrons and the subsequent capture by protons turns out to be close to 6×10^{18} erg gm^{-1} D, which is much the same as the energy release involved in hydrogen burning to ^{4}He. For a universal mass density of 3×10^{-31} gm cm^{-3}, the energy requirement is then 1.8×10^{-17} erg cm^{-3}. This is very small compared with the energy of starlight, which at present, is $\sim 10^{-14}$ erg cm^{-3} and which, in the QSSC will build up to $\sim 10^{-13}$ erg cm^{-3} in the full cycle. Thus, the energy requirement in the production of D is for a small fraction of the available energy that is to into the generation of neutrons.

Deuterium is known to be produced in solar flares [13, 2] and early work by Coleman & Worden [14] has shown how much mass can be ejected from the dwarf stellar component. They estimated that for a typical galaxy containing $10^{10} - 10^{11}$ dwarf M stars, the mass-loss rate will amount to about $0.1M_{\odot}$ yr-1 from the dwarfs. If we add to this the fact that the programs now underway to detect faint stars through microlensing are now showing that the number of dwarf stars is very large, and the fact that in the QSSC cosmology, the timescale for the buildup of D in the interstellar gas is much greater than 10^{10} yr, a large amount of interstellar gas that is enriched in deuterium will be produced in a timescale corresponding to a cycle of oscillation Q in the QSSC i.e. in 10^{11} yr

Of course, in the same period, the deuterium contained in gas that is recondensed into stars will be destroyed, so that the final abundance will depend on how much uncondensed gas remains. More measurements are required of D/H both in the gas in our Galaxy [28, 29] and elsewhere. Much has been made recently of the D/H ratio determined in the absorption-line spectra of QSOs with large redshifts. The value obtained by D. Tytler and his colleagues [44, 10], D/H $\lesssim 2\times10^{-5}$, is the best estimate that has been made so far for extragalactic material, and this has been discussed only in the context of big bang cosmology. In the QSSC, the absorbing clouds that give rise to the absorbtion spectrum may also lie at an earlier epoch in the cycle. However, as we have discussed elesewhere [19], there is independent evidence that many QSOs may not lie at the distances indicated by their redshifts, so the epoch to which these values of D/H correspond is not clear. Our prediction is that with the deuterium made largely in stellar flares, there will be a range of values of the D/H ratio. With values of D/H $\sim 10^{-5}$ at the high end. We do not expect that the D/H ratio will have a constant value throughout an individu al galaxy or throughout a cycle of the QSSC. Thus, a possible test is to look for difference in the D/H ratio both inside and outside our Galaxy.

5. CONCLUSION

We have shown that there are good reasons to argue that ^{4}He has been produced by astrophysical processes following stellar activity. Thus, provided that a timescale much greater than H_0^{-1} is available, as is the case in the QSSC, all of the chemical elements may well have been synthesized in stellar processes. The fact that the great majority of the 320 stable isotopes have been generated astrophysically has always made the idea that all of the isotopes were made this way very attractive.

References

[1] Alpher, R. A. & Herman R. 1950, *Rev. Mod. Phys.* **22**. 153.

[2] Anglin J. D. Dietrich, W. & Simpson J. 1973, *Astrophys. J.* **186**, L41.

[3] Balachandran S. 1990, *Astrophys. J.* **354**, 310.

[4] Boesgaard, A. & Tripicco, M. J. 1986, *Astrophys. J.* **303**, L42.

[5] Bondi, H., Gold, T. & Hoyle F. 1955, *The Observatory***75**, 80.

[6] Bonifacio, P. & Molero P. 1997, *Mon. Not. Roy. astr. Soc.* **285**, 547

[7] Burbidge E. M., Burbidge G., Fowler, W. A. & Hoyle F. 1957, *Rev. Mod. Phys.* **29**, 547 (B^2FH).

[8] Burbidge G.1958, *Proc. Astr. Soc. Pacific* **70**, 83.

[9] Burbidge M. & Burbidge G. 1956, *Astrophys. J.* **124**, 655.

[10] Burles, S. & Tytler, D. 1996, *Astrophys. J.* **460**, 584.

[11] Cameron, A. G. W. 1957, Chalk River Rep. CRL - 41.

[12] Cameron, A. G. W. & Fowler, W.A. 1971. *Astrophys. J.* **164**, 111.

[13] Chupp, E. et al. 1973, *Nature* **241**, 333.

[14] Coleman, G. & Worden P. 1976, *Astrophys. J.* **205**, 475.

[15] Epstein R. Lattimer J. & Schramm,D. 1976, *Nature* **263**, 198

[16] Fixsen D. et al 1996, *Astrophys. J.* **473**, 576.

[17] Fuller F. & Shi, X 1997, *Astrophys. J.* **487**, L25.

[18] Hartoog M. & Cowley A. 1979, *Astrophys. J.* **228**, 229.

[19] Hoyle F., Burbidge G. & 1996, *Astro. Astrophys.* **309**, 335.

[20] Hoyle F., Burbidge, G. & Narlikar, J. V. 1993, *Astrophys. J.* **410**, 437.

[21] Hoyle F., Burbidge, G. & Narlikar, J. V. 1994a, *Mon. Not. Roy. astr. Soc.* **267**, 1007.

[22] Hoyle F., Burbidge, G. & Narlikar, J. V. 1994b, *Astro. Astrophys.* **289**, 729.

[23] Hoyle F., Burbidge, G. & Narlikar, J. V. 1995, *Proc. Roy. Soc. Lond. A* **448**, 191.

[24] Hoyle F. & Tayler R. 1964, *Nature* **203**, 1108.

[25] Isotov, Y., Thuan, T. & Lipovetsky, V. 1997, *Astrophys. J. Suppl.* **108**, 1.

[26] Katsova M., Drake, J. & Livshits, M. 1998 CFA preprint 4704.

[27] Lemoine M., Ferlet. R. & Vidal-Madjar, A. 1995, *Astro. Astrophys.* **298**, 879.

[28] Linsky J., et al 1993, *Astrophys. J.* **402**, 695.

[29] Linsky J., et al 995 *Astrophys. J.* **451**, 335.

[30] McKellar A. 1940 *Proc. Astr. Soc. Pacific* **52**, 407.

[31] McKellar A. 1941, Publ Dom, Astrophysics Obs. Victoria **7**,15.

[32] Michaud G. et al. 1979, *Astrophys. J.* **234**, 206

[33] Pagel, B.E.J.1997, in *The Universe at Large*, ed. G. Munch, A. Mampaso, & F. Sanchez, Cambridge Univ. Press, 343.

[34] Peebles, P.J.E. 1966, ApJ, 146, 542.

[35] Penzias, A., & Wilson, R. 1965, *Astrophys. J.* **142**, 419.

[36] Rebolo, R., et al. 1988, *Astro. Astrophys.* **193**, 193.

[37] Reeves, H., Fowler, W.A., & Hoyle, F. 1970, *Nature* **226**, 727.

[38] Ryan, S. et al. 1996, *Astrophys. J.* **458**, 543.

[39] Sargent, W.L.W. & Jugaku, J. 1961, *Astrophys. J.* **134**, 777.

[40] Sigl, G., Jedamsik, K., Schramm, D., & Berezinsky, V. 1995, *Phys. Rev. D* **52**, 6682.

[41] Spite, M., & Spite, F. 1985, *Ann. Rev. Astron. Astrophys.* **23**, 225.

[42] Stateva, I., Ryabchikov, T, T., & Iliev, I. 1998, Preprint.

[43] Turner, M., 1993, *Science* **262**, 861.

[44] Tytler, D., Fan, X.M., & Burles, S. 1996, *Nature* **381**, 207.

[45] Vangioní-Flam, et al. , 1996, *Astrophys. J.* **468**, 199.

[46] Wagoner, R., Fowler, W.A., & Hoyle, F. 1967. *Astrophys. J.* **148**, 3.

Chapter 7

SUPERLUMINAL MOTION AND GRAVITATIONAL LENSING

S. M. Chitre
Tata Institute of Fundamental Research
Homi Bhabha Road, Bombay 400 005

Abstract The role of gravitational bending of light in generating observed apparent superluminal motions of VLBI components in the compact cores of some of the AGNs and quasars is highlighted.

1. INTRODUCTION

In the early part of 1970s, the very long base line interferometry (VLBI) enabled radio astronomers to probe the internal structure of radio sources at milliarcsecond scales. There was an understandable feeling of disbelief, therefore, when several radio sources monitored with VLBI over a number of years, revealed components in their nuclei separating at speeds exceeding that of light. The first hint of a superluminal motion in quasars was contained in observations of the sizes of variable components in quasars 3C273 and 3C279 ([1], [2]). More observational evidence for such motions accumulated through the 1970s when two distinct components apparently separating with a linear speed, $\beta_{app} = v_{app}/c \geq 5 - 10$, over a period of a few months, were detected ([3], [4], [5]). It has now been convincingly demonstrated for several dozens of sources, from the high-resolution VLBI observations, that the compact radio sources in the active galactic nuclei exhibit striking superluminal motion associated with several components ([6], [7], [8]. Since their discovery, the superluminal sources have remained one of the most intriguing themes in radio astronomy.

Even prior to the detection of apparent superluminal motion, the observations of some quasars had indicated the presence of fast bulk motions through their rapid intensity variations ([9]). The feasibility of

N. Dadhich and A. Kembhavi (eds.), The Universe, 77–86.

superluminal motion was in fact, suggested by Rees ([10]) in a prescient paper where he had argued that the relativistic expansion of a source at speed v can result in its size increasing at an apparent speed of $\gamma v(\gamma \equiv (1-v^2/c^2)^{-1/2})$. The arrival time differences of the signals from different parts of the source can then lead to the apparent size expansion at a transverse speed $\sim \gamma c(\gamma \gg 1)$. The VLBI measurements of the compact core regions of quasars suggest a typical Lorentz factor, $\gamma \leq 10$ for the relativistically moving components, corresponding to a typical proper motion of $\leq$ 1 milliarcsec yr^{-1}. Over the past quarter of a century, the superluminal sources have been observed with VLBI and VLBA to establish a number of striking features ([8]):

(i) The superluminal motion appears to be common amongst the brighter radio sources and generally exhibits properties such as rapid variability of intensity and polarization, although, not all well-surveyed sources display superlight motion (e.g. 3C84).

(ii) The expansion speeds are on the average larger for the core-dominated sources compared to the lobe-dominated sources.

(iii) The compact sources exhibit superluminally expanding relative motion of the components, with the emergence of new components from the core.

(iv) The superluminal motion is largely uniform, but there are cases of acceleration and deceleration, and in some cases there are instances of bent trajectories as well.

(v) The VLBI jets associated with the superluminal sources are invariably curved and misaligned with the large-scale symmetry axis of the extended lobes.

2. THEORETICAL SCENARIOS

There are two ways in which it is possible to account for the observed superluminal motion of VLBI components in the nuclei radio sources. The obvious way out to explain the velocities of components apparently exceeding the speed of light, c is to argue that these radio sources are located at distances considerably smaller than what is implied by Hubble's law. Since all the observations depend on measurements of angular separation between components, their conversion into linear transverse motion would necessarily require a knowledge of the distance to the objects. Should the sources be situated closer than what is indicated by the Hubble interpretation of their redshifts, the observed motion would turn out to be subluminal after all ([11], [12]). Indeed, the AGNs and quasars are now widely accepted to be located at cosmological distances. It, therefore, becomes necessary to imagine that the observed superluminal speeds are not physical, but rather, the result of cosmic illusions.

A number of ingenious proposals, mostly based on the kinematics of the source have been advanced for explaining such cosmic illusions. These include:

(a) Christmas-tree model proposed by Dent ([13]) invokes independent flares erupting all over at random locations in the source. Such random flaring could mimic a regular superlight motion, though it was realized that the observed motions were highly systematic and indicated only expanding motions of the components ([14]).

(b) Light-echo model of Lynden-Bell ([15]) attributed the superluminal motion to an outward propagating relativistic blast curve that can cause a progressive brightening of the region of the source with increasingly large size. If such an oppositly directed signal along an axis making a small angle with the sight-line can lead to a superluminal expansion. The model does not seem to be compatible with the observed core-jet structures in these sources.

(c) Gravitational screen model was proposed by Chitre and Narlikar ([16], [17]) as a plausible explanation of superluminal motion in AGNs, prior to the discovery of the first gravitational lens system, the twin quasar 0957 + 561 A,B, by Walsh, Carswell & Weymann ([18]). This model envisages the presence of a gravitational screen in the form of an intervening galaxy or a cluster of galaxies, between the source and the observer. Owing to the gravitational bending produced by the deflecting mass en route, the observer would see the components in the nucleus of the background source, not in their real positions, but at virtual transversely separated locations, thus creating an illusion of superlight motion.

The effect is due to the differential gravitational deflection caused by the intervening mass with the increasing impact parameter distance, from the centre, of the light rays emanating from the background source. For a spherically symmetric matter distribution in a galaxy, G, with mass, M and radius, R, the external gravitational bending of a typical ray is given by

$$\Delta = \frac{4GM}{c^2 r}, \quad \text{for } r \geq R.$$

It turns out the interesting effects are produced from light paths that go through the inner regions of G. It can be demonstrated ([17]) that value of the relativistic bending angle is exactly twice the Newtonian value in the case of weak gravity. The important feature of the bending angle is that $\Delta'(a) > 0$ in the inner regions of most physical mass-distributions associated with galaxies or galaxy-clusters. For external bending, on the other hand, $\Delta'(a) < 0$. The most striking aspect is the effect of gravitational bending on the apparent velocity of separation. For this purpose let us denote by D_d, D_{ds} and D_s the distances between the observer and the deflector, the deflector and the source, and the observer and the source respectively. Let $v_\perp$ be the transverse speed of separation between two components in the nuclear region of a stationary background source. Then, the apparent separation velocity as seen by the observer is

$$v_{app} = \frac{v_\perp}{1 - \frac{D_d Dds}{Ds}\Delta'(a)}.$$

It is clear that we can get a large magnification of the real transverse velocity provided, $\Delta'(a) > 0$ and $\frac{D_d Dds}{D_s}\Delta'(a) \simeq 1$, a condition that is satisfied when the source and the observer are situated at conjugate points with respect to the deflector.

It is, thus, essential for the manifestation of apparent superluminal motions to have a suitably placed gravitating intervenor between the source and the observer. The presence of an intervening deflector for producing the superluminally separating images is a requirement for this scenario. A test of the gravitational screen model would, therefore, be the detection of an actual deflecting object, which has, unfortunately, not been borne out in all the known superluminal sources.

There are certain features associated with the gravitational lensing effect which may even stand the scrutiny of future observations in the case of the superluminal sources. A notable feature of the gravitational bending of light is that the amount of deflection is independent of wavelength and we therefore expect the superluminal separation of components to be the same at all observing wavelengths. A definitive characteristic associated with the lensing phenomenon is the non-uniform amplification in directions perpendicular to the line of sight. Thus, the image of a linear trajectory would appear curved or bent, and it is only to be expected that the VLBI jets should be misaligned in relation to the extended structures. Such a misalignment property has, indeed, been noted in some of the quasars exhibiting superluminal motions. The superluminal acceleration or deceleration of the separating components is yet another consequence of the gravitational screen model, this could result from changes in the amplification of the light beams when the amount of relativistic bending varies with the density of the intervening matter. Furthermore, the local inhomogeneities in the deflecting object is also liable to produce short-term ($\delta \sim 1\,yr$) changes in the velocity;

the angular separation as a function of epoch is, therefore, unlikely to be a smooth curve, but rather should have a scatter around the linear trend. The apparent separation velocity observed in the source 3C345, for example, shows an increase from 7.5c to 12.2c (for $H_0 = 50\ km\, s^{-1}\ Mpc^{-1}$) which is a genuine case of superluminal acceleration.

A clear-cut prediction of the gravitational screen model would be the detection of superluminal separation of the VLBI components in the cores of AGNs and quasars which have been unambigously established gravitational lens systems. Thus, the twin quasar 0957 + 5c1 should show a magnificationi of velocity by a factor of 2-3 and consequently, should there be relativistically separating components in the source-quasars, we should see apparent superluminal motion, $v_{app} \leq 3c$. Likewise, the triple radio source 2016 + 112 should reveal an apparent superluminal speed, v_{app} exceeding 10c. Indeed, there are reported cases of highly magnified gravitational lens systems whose cores exhibit structures at submilliarcsecond scales. The VLBA features are detected in one of the images first, followed by their appearance in the second image of the lens system with a time-delay of several weeks to months. There is some observational evidence for the existence of such a superluminal motion in the lens system 1830-211 (Patnaik, Private Communication).

d) Relativistic beaming model was proposed by Rees ([19]) and later elaborated by Blandford and Königal ([20]). In this kinematic picture the superluminal motion is simulated by one or more blobs or plasma-components moving at a relativistic speed, v away from the core that is stationary in the rest frame of the observer. The transverse velocity of separation of the plasma blobs from the core is then given by

$$v_{app} = \frac{\beta \sin\theta}{1 - \beta\cos\theta}\, c, \qquad (\beta = v/c).$$

For the manifestation of apparent superluminal motion, the angle θ of the beaming plasmoid with the sightline has to be very small. The expression for the apparent velocity attains a maximum value, $v_{app}^{\max} = \sqrt{\gamma^2 - 1}\, c \simeq \gamma c$. This model makes an ingenious use of the kinematic effect, and was in fact, advanced even before the apparent faster-than-light phenomenon was discovered on the VLBI scale in the cores of AGNs and quasars.

The observed superluminal motions may be best interpreted in the framework of bulk relativistic motion beamed towards the observer. This is by far the most attractive model to explain the observed phenomena associated with the superluminal sources. However, the simple relativistic beaming model is not without its difficulties in accommodating various observational features.

Thus a successful model must be able to explain the emission characteristics of the superluminal sources such as the spectrum, polarization, flux variability and features like the curved trajectories of superluminal components, and their variable angular speeds, bent jets on the parsec scales and misalignment property of the extended structures. Should the relativistic beaming model be the correct description of superluminal sources, we would expect at least some quasars to show two-sided large-scale jets, unless the one-sided jet is an intrinsic property of quasars on both small and large scales. In any case the bright hot spots are expected to be on the jetside. Furthermore, because of the Doppler boosting the flux density of the approaching components is expected to exceed that of the receding (or stationary) component by several orders of magnitude, in conflict with the comparable flux densities of components.

The X-ray emission from superluminal radio sources is supposed to have provided a strong indication for the occurrence of relativistic beaming in their compact cores. For this purpose it is argued that the synchrotron radiation in a compact volume would produce X-ray flux, by an inverse Compton scattering of radio and infrared photons. The basic question to be addressed is whether the inverse Compton process is the underlying cause for the X-ray flux from superluminal sources, for which it is possible to constrain the physical parameters associated with the radio sources. It turns out that the observed X-ray emission is much smaller than what is expected from the parameters of the radio components. Essentially, the VLBI measurements determine v/c and the X-ray fluxes set limits on the Doppler factor, $\delta = 1/\gamma(1 - \beta\cos\theta)$, thus providing valuable constraints on the geometry and motion of the emitting components. It is usually argued (cf. [6]) that the observed superluminal motions, weak X-ray emission and variability of the sources are taken to provide strong evidence in support of the relativistic beaming model.

Marscher ([21]) has, however, pointed out certain difficulties encountered by the synchrotron-Compton emission process when applied to realistic radio sources. It turns out because of the complex nature of the compact sources, they are composed of a number of discrete components, and these could conspire to become self-absorbed at different frequencies to produce a remarkably flat composite spectrum. For the sake of simplicity, each component is assumed to have a spectrum of a uniform synchrotron source, but, then the resultant inverse-Compton X-ray flux density and the total energy requirement have a very strong dependence on the turnover frequency, and the angle of the bulk velocity vector with the line of sight. Based on the simplifying assumptions, the evidence for the inverse-Compton process generating the observed X-ray flux is favorable, though not overwhelming. But the discrepant time scales of variability in the wavebands ranging from millimeter ($\geq$ weeks) to X-rays ($\sim$ day) for the bright quasar 3C273 certainly casts a shadow on the tenability of the inverse-Compton hypothesis.

3. SPECULATIONS ON MICROLENSING AND SUPERLUMINAL MOTIONS

The phenomenon of gravitational lensing has been effectively used to gain valuable information about the masses and sizes of intervening deflectors. In most studies the lensing objects are generally assumed to be stationary, except in those cases where the effects of motion on the light curve have been important while crossing the caustics like in the microlensing events (cf. [22], [23], [24]). The usefulness of astronomic diagnostic properties of moving lenses was discussed by Chitre and Saslaw ([25]). It was demonstrated that with a suitable placement of the background source within the cone of inversion, the source velocities could conceivably be magnified by an order of magnitude or more and part of the image may even exhibit an apparent superluminal motion (cf. [26]; [27]).

A striking feature associated with moving lenses is the conversion of linear proper motion into rotational motion, since the lensing effect magnifies the velocities by different amounts in different directions. Consequently, we expect the conversion of uniform linear source motion to be accompanied by an apparent acceleration of the individual components in the source. Equally, the radial component of the source motion is also influenced by the moving lenses by converting it into a transverse component of the image motion.

One of the fascinating challenges in galactic astronomy is to surmise the presence of a putative massive black hole residing at the centre of our Galaxy. One obvious way to infer its existence and physical properties would be to search for its gravitational influence on the background sources such as maser complexes, relativistic jets of 'microquasars', lying on the far side of the galactic nucleus from us. Thus, it is tempting to imagine an individual relativistically moving source in a maser complex, or a relativistic beam of a microquasar located in the background to be lensed by the black hole in the galactic centre. This should almost certainly generate the resulting velocities which could apparently mimic superluminal motions. Such a suitable positioning of the background microquasar along the line of sight passing through the nuclear region should create an image morphology that could provide a valuable handle to infer the mass of the lensing black hole (cf. [25], [28]).

A possible velocity manifestation would consist of nearly circularly moving superluminal components, resulting from the lensing of a background relativistic jet by the massive black hole; the typical state of the velocity pattern would be of the order of several arcseconds. A definitive observation of superluminal motion in the direction of the galactic centre would provide further support to the existence of a massive black hole in the nucleus of our Galaxy.

A remarkable aspect of superluminality has been stressed by Gopal-Krishna and Subramaniam ([29]). This involves a superluminal microlensing scheme which combines beaming with the phenomenon of gravitational lensing. The microlensing of compact sources such as quasars by brown dwarfs has been invoked to account for their intensity variability on timescales of the order of several months to a few years. But some of the active quasars, in particular, blazars are known to show variability on a time-scale as short as hours in the optical waveband and days in the radio. The blazars have relativistically beaming jets composed of bright components which are known to make a small angle with the line of sight ([20]) and these knots are expected to exhibit apparent superluminal motions. Gopal Krishna and Subramanian ([30]) have invoked the superluminal microlensing of such ultra-rapidly moving components which causes an amplification of both the flux and velocity, over and above that resulting from the relativistic beaming or lensing phenomenon alone. Such a composite beaming-lensing scheme would also lead to the requisite short time-scale intensity variations. Furthermore, for the case of knots crossing a caustic this would lead to extraordinarily large apparent superluminal velocities exceeding 20-30c.

Thus, if the microlensing by a million solar mass black hole of a quasar or a relativistic jet were to happen, this will almost certainly lead to significant morphological distortions, variations in the flux ratios and velocities of the images over a very short time-scale ($\leq 1hr$). Clearly, the VLBA monitoring of the galactic nuclear region and of the cores of compact radio sources (e.g. AGNs, quasars, blazars) should reveal the existence of massive and supermassive black holes in the nuclei of galaxies. A definitive observation of superluminal motion in the direction of the Galactic centre would provide further support to the existence of a massive black hole in the nucleus of our Galaxy.

Acknowledgments

It was a pleasure to collaborate with Jayant Narlikar, some two decades ago, on problems relating to superluminal motions and physics of radio sources. Thanks are due to D. Narasimha for useful comments on the manuscript and valuable discussions.

References

[1] Gubbay, J., et al 1969, Nature, **224**, 1094

[2] Moffet, A.T. et al 1972 in IAU Symp. 44: External Galaxies and Quasi-Stellar Objects ed. D.S. Evans (Dordrecht: Reidel)

[3] Cohen, M.H. et al. 1971, Ap. J. **179**. 207

[4] Whitney, A.R. et al 1971, Science, **173**, 225

[5] Kellermann, K.I. 1978, Physica Scripta **17**, 257

[6] Cohen, M.H. & Unwin, S.C. 1984 in IAU Symp. 110: VLBI and Compect Radio Sources, ed. R. Fanti, K. Kellermann and G. Setti (Dordrecht: Reidel)

[7] Fanti, R., Kellermann, K. & Setti, G. 1984, IAU Symp. 110:VLBI and Compact Radio Sources (D. Reidel, Dordrecht)

[8] Vermeulen, R.C. & Cohen, M.H. 1994, Ap. J. **430**, 467

[9] Burbidge, G. & Burbidge, E.M. 1967, Quasistellar Objects (Freeman: San Francisco)

[10] Rees, M. J. 1966, Nature **211**, 468

[11] Burbidge, G. 1978, Physica Scripts, **17**, 281

[12] Arp H. 1983, in Liege Coll. 24: Quasars and Gravitational Lenses ed. J.-P. Swings (Univ. Liége)

[13] Dent, W.A. 1972, Science, **175**, 1105

[14] Cohen, M.H. et al 1977, Nature, **268**, 405

[15] Lynden-Bell, D. 1977, Nature, **270**, 396

[16] Chitre, S.M. & Narlikar, J. V. 1979, MNRAS, **189**, 655

[17] Chitre, S.M. & Narlikar, J.V. 1980, Ap. J. **235**, 335

[18] Walsh, D., Carswell, R.F. & Weymann, R.J. 1979 Nature, **279**, 381

[19] Rees, M. J. 1967, MNRAS, **135**, 345

[20] Blandford, R. & Königal, A. 1979, Ap. J. **232**, 34

[21] Marscher, A.P. 1987 in Superluminal Radio Sources, Ed. J.A. Zensun & T.J. Pearson (Cambridge University Press)

[22] Birkinshaw, M. & Gull, S.F. 1983, Nature **302**, 315

[23] Mitrofânov, J.G. 1981. Sovt.Astr. Lett. **7**, 39

[24] Gott, J.R. 1981, Ap. J. **243**, 140

[25] Chitre, S.M. & Saslaw, W.C. 1989, Nature, **341**, 38

[26] Liebes, S. 1964, Phys.Rev.B **133**, 835

[27] Saslaw, W.C., Narasimha, D. & Chitre, S.M. 1985, Ap. J. **292**, 348

[28] Narasimha, D. & Chitre, S.M. 1991 in Gravitational Lens, Hamburg, Ed., R. Kayser, T. Schramm, L. Nieser, p378

[29] Gopal-Krishna & Subramanian, K. 1991, Nature, **349**, 766

[30] Gopal-Krishna & Subramanian, K. 1996, Astron. Ap. **315**, 343

Chapter 8

DUAL SPACETIMES, MACH'S PRINCIPLE AND TOPOLOGICAL DEFECTS

Naresh Dadhich
Inter-University Centre for Astronomy and Astrophysics
Ganeshkhind, Pune 411 007, India

It is a matter of great pleasure and privilege to have known Jayant Narlikar and worked with him closely in setting up IUCAA. With deep affection and feeling I dedicate this work to him on his completing 60 years.

—

Abstract

By resolving the Riemann curvature relative to a unit timelike vector into electric and magnetic parts, we define a duality transformation which interchanges active and passive electric parts. It implies interchange of roles of Ricci and Einstein curvatures. Further by modifying the vacuum/flat equation we construct spacetimes dual to the Schwarzschild solution and flat spacetime. The dual spacetimes describe the original spacetimes with global monopole charge and global texture. The duality so defined is thus intimately related to the topological defects and also renders the Schwarzschild field asymptotically non-flat which augurs well with Mach's Principle.

1. INTRODUCTION

In analogy with the electromagnetic field, it is possible to resolve the gravitational field; i.e. Riemann curvature tensor into electric and mag-

N. Dadhich and A. Kembhavi (eds.), The Universe, 87–96.

netic parts relative to a unit timelike vector [1-2]. In general, a field is produced by charge (source) and its menifestation when charge is stationary is termed as electric and magnetic when it is moving. Electromagnetic field is the primary example of this general feature, which is true for any classical field. In gravitation, unlike other fields, charge is also of two kinds. In addition to the usual charge in terms of non-gravitational energy, gravitational field energy also has charge. Thus electric part would also be of two kinds corresponding to the two kinds of charge, which we term as active and passive.

The Einstein vacuum equation, written in terms of electric and magnetic parts is symmetric in active and passive electric parts. We define the duality relation as interchange of active and passive electric parts. Then it turns out that the Ricci and the Einstein tensors are dual of each-other. That is, the non-vacuum equation will in general distinguish between active and passive parts and we could have solutions that are dual of each-other [3]. In particular it follows that perfect fluid spacetimes with the equation of states, $\rho - 3p = 0$ and $\rho + p = 0$ are self dual $(\Lambda \rightarrow -\Lambda)$ while the stiff fluid is dual to dust.

The question is, can we obtain a dual to a vacuum soltuion? Since the equation is symmetric in active and passive parts, it would remain invariant under the duality transformation. However it turns out that in obtaining the well-known black hole solutions not all of the vacuum equations are used. In particular, for the Schwarzschild solution the equation $R_{00} = 0$ in the standard curvature coordinates is implied by the rest of the equations. If we tamper this equation, the Schwarzschild solution would remain undisturbed for the rest of the set will determine it completely. However this modification, which does not affect the vacuum solution, breaks the symmetry between active and passive electric parts leading to non-invariance of the modified equation under the duality transformation. Now we can have solution dual to vacuum which is different. This is precisely what happens for the Schwarzschild solution.

The Schwarzschild is the unique spherically symmetric vacuum solution, which means it characterizes vacuum for spherical symmetry. It is true that not all the equations are used in getting to the solution. In fact it turns out that ultimately the equations reduce to the Laplace equation and its first integral [4-5]. That means the Laplace equation becomes free as it would be implied by its first integral equation. Without disturbing the Schwarzschild solution we could introduce some energy density on the right which would be wiped out by the other equations. The modified equation would turn out to be not invariant under the duality transformation, yet however it admits the Schwarzschild solution as the unique solution. Now the dual set of equations also admits the unique solution, which could be interpreted as representing the Schwarzschild particle with global monopole charge [6]. The static black hole with and without global monople charge are hence dual of each-other.

Similarly it turns out that flat spacetime could as well be characterized by a duality non-invariant form of the equation. The static solution of the dual equation will represent massless global monopole (putting the Schwarzschild mass zero in the above solution) and the non-static homogeneous solution will give the FRW metric with the equation of state $\rho + 3p = 0$, which is the characterizing property of global texture [7-8]. The former could as well be looked upon as spacetime of uniform relativistic gravitational potential [4-5]. Global monopoles and textures are stable topological defects which are produced in phase transitions in the early universe when global symmetry is spontaneously broken [7-10]. In particular a global monopole is produced when the global $O(3)$ symmetry is broken into $U(1)$. A solution for a Schwarzschild particle with global monopole charge has been obtained by Barriola and Vilenkin [6]. It therefore follows that the Schwarzschild and the Barriola-Vilenkin solutions are related through the duality transformation. They are dual of each- other. Like the Schwarzschild solution, the global monopole solution is also unique. Applications to cosmology and properties of global monopoles [10-14] and of global textures [7-8,11,15-19] have been studied by several authors. What dual solution signifies is restoration of gauge freedom of choosing zero of relativistic potential which was not permitted by the vacuum equation that implied asymptotic flatness. This means that the dual solution breaks asymptotic flatness of the Schwazschild filed without altering its basic physical character. The relativistic potential is now given by $\phi = k - M/r$ instead of $\phi = -M/r$. This is precisely what is required to make the Schwarzschild field consistent with Mach's principle. The constant k brings in the information of the rest of the Universe, say for solar system moving towards the great attractor [20]. The important difference between the Newtonian and relativistic understanding of the problem is that constant k produces non-zero curvature and hence has non-trivial physical meaning. This is the most harmless way of making the field of an isolated body consistent with Mach's principle.
In sec. 2, we shall give the electromagnetic decomposition of the Riemann curvature, followed by the duality transformation and dual spacetimes in Section 3 and concluded with discussion in Section 4.

2. ELCTROMAGNETIC DECOMPOSITION

We resolve the Riemann curvature tensor relative to a unit timelike vector [1-2] as follows :

$$E_{ac} = R_{abcd}u^b u^d, \quad E_{ac} = *R*_{abcd}\, u^b u^d \tag{8.1}$$

$$H_{ac} = *R_{abcd}u^b u^d = H_{(ac)} - H_{[ac]} \tag{8.2}$$

where

$$H_{(ac)} = *C_{abcd}u^b u^d \tag{8.3}$$

$$H_{[ac]} = \frac{1}{2}\eta_{abce}R^e_d u^b u^d. \tag{8.4}$$

Here c_{abcd} is the Weyl conformal curvature, η_{abcd} is the 4-dimensional volume element. $E_{ab} = E_{ba}, \tilde{E}_{ab} = \tilde{E}_{ba}, (E_{ab}, \tilde{E}_{ab}, H_{ab})u^b = 0, \; H = H^a_a = 0$ and $u^a u_a = 1$. The Ricci tensor could then be written as

$$R_{ab} = E_{ab} + \tilde{E} - ab + (E + \tilde{E})u_a u_b - \tilde{E}g_{ab} + \frac{1}{2}H^{mn}u^c(\eta_{acmn}u_b + \eta_{bcmn}u_a) \tag{8.5}$$

where $E = E^a_a$ and $\tilde{E} = \tilde{E}^a_a$. It may be noted that $E = (\tilde{E} + \frac{1}{2}T)/2$ defines the gravitational charge density while $\tilde{E} = -T_{ab}u^a u^b$ defines the energy density relative to the unit timelike vector u^a.

3. DUALITY TRANSFORMATION AND DUAL SPACETIMES

The vacuum equation, $R_{ab} = 0$ is in general equivalent to

$$E \;or\; \tilde{E} = 0, \; H_{[ab]} = 0 = E_{ab} + \tilde{E}_{ab} \tag{8.6}$$

which is symmetric in E_{ab} and $\tilde{E}_{ab}$.

We define the duality transformation as

$$E_{ab} \longleftrightarrow \tilde{E}_{ab}, \; H_{[ab]} = H_{[ab]}. \tag{8.7}$$

Thus the vacuum eqaution (6) is invariant under the duality transformation (7). From eqn. (1) it is clear that the duality transformation would map the Ricci tensor into the Einstein tensor and vice-versa. This is because contraction of Riemann is Ricci while of its double dual is Einstein.

Consider the spherically symmetric metric,

$$ds^2 = c^2(r,t)dt^2 - a^2(r,t)dr^2 - r^2(d\theta^2 + sin^2\theta d\varphi^2). \tag{8.8}$$

The natural choice for the resolving vector in this case is of course it being hypersurface orthogonal, pointing along the t-line. From eqn. (6), $H_{[ab]} = 0$ and $E_2^2 + \tilde{E}_2^2 = 0$ lead to $ac = 1$ (for this, no boundary condition of asymptotic flatness need be used). Now $\tilde{E} = 0$ gives $a = (1 - 2M/r)^{-1/2}$, which is the Schwarzschild solution. Note that we did not need to use the remaining equation and $E_1^1 + \tilde{E}_1^1 = 0$, it is hence free and is implied by the rest. Without affecting the Schwarzschild solution, we can introduce some distribution in the 1-direction.

We hence write the alternate equation as

$$H_{[ab]} = 0 = \tilde{E}, \;\; E_{ab} + \tilde{E}_{ab} = \lambda w_a w_b \tag{8.9}$$

where λ is a scalar and w_a is a spacelike unit vector along 4-acceleration. It is clear that it will also admit the Schwarzschild solution as the general solution, and it determines $\lambda = 0$. That is for spherical symmetry the above alternate equation also characterizes vacuum, because the Schwarzschild solution is unique.

Let us now employ the duality transformation (7) to the above equation (9) to write

$$H_{[ab]} = 0 = E, \;\; E_{ab} + \tilde{E}_{ab} = \lambda w_a w_b. \tag{8.10}$$

Its general solution for the metric (8) is given by

$$c = a^{-1} = (1 - 2k - \frac{2M}{r})^{1/2}. \tag{8.11}$$

This is the Barriola-Vilenkin solution [6] for the Schwarzschild particle with global monopole charge, $\sqrt{2k}$. Again we shall have $ac = 1$ and $E = 0$ will then yield $c = (1 - 2k - 2M/r)^{1/2}$ and $\lambda = 2k/r^2$. This has non-zero stresses given by

$$T_0^0 = T_1^1 = \frac{2k}{r^2}. \tag{8.12}$$

A global monopole is described by a triplet scalar, $\psi^a(r) = \eta f(r) x^a/r, x^a x^a = r^2$, which through the usual Lagrangian generates energy-momentum distribution at large distance from the core precisely of the form given above in (12) [6]. Like the Schwarzschild solution the monopole solution

(11) is also the unique solution of eqn.(10).

If we translate eqns. (9) and (10) in terms of the familiar Ricci components, they would read as

$$R^0_0 = R^1_1 = \lambda, R^2_2 = 0 = R_{01} \tag{8.13}$$

and

$$R^0_0 = R^1_1 = 0 = R_{01}, R^2_2 = \lambda. \tag{8.14}$$

For the metric (8), we shall then have $ac = 1$ and $c^2 = f(r) = 1 + 2\phi$, say, and

$$R^0_0 = - \nabla^2 \phi \tag{8.15}$$

$$R^2_2 = -\frac{2}{r^2}(r\phi)' \tag{8.16}$$

Now the set (13) integrates to give $\phi = -M/r$ and $\lambda = 0$, which is the Schwarzschild solution while (14) will give $\phi = -k - M/r$ and $\lambda = 2k/r^2$, the Schwarzschild with global monopole charge. Thus global monopole owes its existence to the constant k, appearing in the solution of the usual Laplace equation implied by eqns. (14) and (15). It defines a pure gauge for the Newtonian theory, which could be chosen freely, while the Einstein vacuum equation determines it to be zero. For the dual-vacuum equation (14), it is free like the Newtonian case but it produces non-zero curvature and hence would represent non-trivial physical and dynamical effect (see $R^2_2 = -2k/r^2 \neq 0$ unless $k = 0$). This is the crucial difference between the Newtonian theory and GR in relation to this problem, that the latter determines the relativistic potential ϕ absolutely, vanishing only at infinity. This freedom is restored in the dual-vacuum equation, of course at the cost of introducing stresses that represent a global monopole charge. The uniform potential would hence represent a massless global monopole ($M = 0$ in the solution (11)), which is solely supported by the passive part of electric field. It has been argued and demonstrated [5] that it is the non-linear aspect of the field (which incorporates interaction of gravitational field energy density) that produces space-curvatures and consequently the passive electric part. It is important to note that the relativistic potential ϕ plays the dual role of the Newtonian potential as well as the non-Newtonian role of producing curvature in space. The latter aspect persisits even when potential is constant different from zero. It is the dual-vacuum equation that uncovers this aspect of the field.

On the other hand, flat spacetime could also in alternative form be characterized by

$$\tilde{E}_{ab} = 0 = H_{[ab]}, E_{ab} = \lambda w_a w_b \tag{8.17}$$

leading to $c = a = 1$, and implying $\lambda = 0$. Its dual will be

$$E_{ab} = 0 = H_{[ab]}, \tilde{E}_{ab} = \lambda w_a w_b \tag{8.18}$$

yielding the general solution,

$$c' = a' = 0 \Longrightarrow c = 1, a = const. = (1 - 2k)^{-1/2} \tag{8.19}$$

which is non-flat and represents a global monopole of zero mass, as it follows from the solution (11) when $M = 0$. This is also the uniform relativistic potential solution.

Further it is known that the equation of state $\rho + 3p = 0$ which means $E = 0$, characterizes global texture [7,19]. That is, the necessary condition for spacetimes of topological defects; global textures and monopoles is $E = 0$. Like the uniform potential spacetime, it can also be shown that the global texture spacetime is dual to flat spacetime. In the above eqns (13) and (14), replace $w_a w_b$ by the projection tensor $h_{ab} = g_{ab} - u_a u_b$. Then non-static homogeneous solution of the dual-flat equation (14) is the FRW metric with $\rho + 3p = 0$, which determines the scale factor $S(t) = \alpha t + \beta$, and $\rho = 3(\alpha^2 + \mathrm{k}) / (\alpha t + \beta)^2$, $\mathrm{k} = \pm 1, 0$. This is also the unique non-ststic homogeneous solution. The general solutions of the dual-flat equation are thus the massless global monopole (uniform potential) spacetime in the static case and the global texture spacetime in the non-static homogeneous case. Thus they are dual to flat spacetime.

It turns out that spacetimes with $E = 0$ can be generated by considering a hypersurface in 5-dimensional Minkowski space defined, for example, by

$$t^2 - x_1^2 - x_2^2 - x_3^2 - x_4^2 = k^2(t^2 - x_1^2 - x_2^2 - x_3^2) \tag{8.20}$$

which consequently leads to the metric

$$ds^2 = k^2 dT^2 - T^2[d\chi^2 + sinh^2\chi(d\theta^2 + sin^2\theta d\varphi^2)]. \tag{8.21}$$

Here $T^2 = t^2 - x_1^2 - x_2^2 - x_3^2$ and $\rho = 3(1-k^2)/k^2T^2$. The above construction will generate spacetimes of global monopole, cosmic strings (and their homogeneous versions as well), and global texture-like depending upon the dimension and character of the hypersurface. Of course, $E = 0$ always; i.e. zero gravitational mass [11]. The trace of active part measures the gravitational charge density, responsible for focussing of congruence in the Raychaudhuri equation [21]. The topological defects are thus characterized by vanishing of focussing density (tracelessness of active part).

Application of the duality transformation, apart from vacuum/flat case considered here, has been considered for fluid spacetime [3]. The duality transformation could similarly be considered for eletrovac equation including the Λ-term. Here the analogue of the master equation (10) is

$$H_{[ab]} = 0, E = \Lambda - \frac{Q^2}{2r^4}, \; E^b_a + \tilde{E}^b_a = (-\frac{Q^2}{r^4} + \lambda) w_a w^b \tag{8.22}$$

which has the general solution $c^2 = a^{-2} = (1 - 2k - 2M/r + Q^2/2r^2 - \Lambda r^2/3)$ and $\lambda = 2k/r^2$. The analogue of eqn. (6) will have $\tilde{E} = -\Lambda - Q^2/2r^4$ instead of E in (20). Thus the duality transformation works in general for a charged particle in the de Sitter universe. Similarly spacetime dual to the NUT solution has been obtained [22]. In the case of the Kerr solution it turns out, in contrast to others, that dual solution is not unique. The dual equation admits two distinct solutions which include the original Kerr solution [23].

4. DISCUSSION

First of all let us try to get some physical feel of active, passive and magnetic parts. For a canonical resolution relative to a hypersurface orthogonal unit timelike vector, it follows that E_{ab} would refer to the curvature components R_{0a0a}, $\tilde{E}_{ab}$ to R_{abab} and H_{ab} to R_{0aab}. With reference to the spherically symmetric metric (8), it can be easily seen that the active part is crucially anchored onto the Newtonian potential appearing in $g_{00} = 1 + 2\phi$, while the passive part to the relativistic potential, $g_{11} = -(1 + 2\phi)^{-1}$. Note that in obtaining the Schwarzschild solution we ultimately solve the Laplace equation, which does not take into account contribution of gravitational filed energy as source. It can be shown that contribution of gravitational field energy goes into curving the space through $g_{11} \neq 1$ leaving the Laplace equation undisturbed [4-5]. Thus passive part is created by the field energy while the active by non-gravitational energy distribution. The magnetic part would as expected be due to motion of sources.

Under the duality transformation, the vacuum equation remains invariant leading to to the same solutions, but the Weyl tensor changes sign which would mean $GM \to -GM$. This happens because active part is produced by positive non-gravitational energy while passive part by negative field energy, and the interchange of active and passive would therefore imply interchange of positive energy and negative field energy [2].

Consider the Maxwell like duality $E \to H, H \to -E$ as given by

$$E_{ab} \to H_{ab},\ H_{ab} \to -\tilde{E}_{ab}, \qquad \tilde{E}_{ab} \to -E_{ab} \tag{8.23}$$

This implies $E = 0, H_{[ab]} = 0, E_{ab} + \tilde{E}_{ab} = 0$ which is the vacuum equation (6) and keeps the Einsten action invariant because $R = 2(E - \tilde{E})$. This is a remarkable result indicating that vacuum equation is implied by the duality symmetry of the action [2]. Note also that duality transformation of the action does not permit the cosmological constant which could however be brought in as matter with the specific equation of state. This result is similar to the well-known property of GR that equation of motion for free particle is contained in the field equation.

The duality transformation (7) introduces in most harmless manner a global monopole in the Schwarzschild black hole which amounts to breaking the asymptotic flatness. The latter is a necessary requirement for the field to be consistent with Mach's principle at the very elementary level. In essence, it is obtained by simply retaining the constant of integration in the solution of the Laplace equation. Thus it makes no difference at the Newtonian level and hence its contribution is purely relativistic.

The most general duality-invariant expression consisting of the Ricci and the metric is $R^a_b - (\frac{R}{4} - \Lambda)g^a_b$. This, without Λ equal to zero would be the equation for gravitational instanton, which follows from the R^2-action. The instanton action and the field equation are duality-invariant. They are also conformally invariant as well. As a matter of fact conformal invariance singles out the R^2-instanton action. That means the conformal invariance includes the duality invariance, while the duality invariance of the Palatini action with the condition that the resulting equation be valid for all values of R would lead to the conformal invariance [24-25]. The simplest and well-known instanton solution is the de Sitter spacetime. Here the duality only leads to the anti-de Sitter.

Acknowledgments

I have pleasure in thanking Jose Senovilla, and LK Patel and Ramesh Tikekar for useful discussions.

References

[1] M.A.G. Bonilla and J.M.M. Senovilla, 1997, *Gen. Rel.Grav.*29,91.
[2] N. Dadhich, 1999, *Mod. Phys. Letts. A*14, 337, 759.
[3] N. Dadhich, L.K. Patel and R. Tikekar (1998) Class. Quantum Grav. **15**, L27.
[4] N. Dadhich, 1995, GR-14 Abstracts, A.98.
[5] N.. Dadhich, 1997, On the Schwarzschild field, gr-qc/9704068.
[6] M. Barriola and A. Vilenkin, 1989, *Phys. Rev. Lett.* **63**,341.
[7] R.L. Davis, 1987, *Phys. Rev. D* **35**, 3705.
[8] N. Turok, 1989, *Phys. Rev. Lett.* **63**, 2625.
[9] T.W.B. Kibble, 1979, *J. Phys. A*9, 1347.
[10] A. Vilenkin, 1985, *Phys. Rep.* **121**,263.
[11] N. Dadhich & K. Narayan, 1998, *Gen. Rel.Grav.*30, L1133.
[12] D. Harari & C. Lousto, 1990, *Phys. Rev. D* **42**,2626.
[13] G.W. Gibbons, M.E. Oritz, F. Ruiz Ruiz a& T.M. Samols, 1992, *Nucl. Phys. B*385, 127.
[14] N. Dadhich, K. Narayan & U.A. Yajnik, 1998, *Pramana* **50**, 307.
[15] R.J. Gott III and M.J. Rees (1987) *Mon. Not. Roy. astr. Soc.* **227**, 453.
[16] E.W. Kobl, 1989, *Astrophys. J.* **344**, 543.
[17] M Kamionkowski & N. Toumbas, 1996, *Phys. Rev. Lett.* **77**, 587.
[18] V. Sahni, 1991, *Phys. Rev. D* **43**, R301.
[19] D. Notzold, 1991, *Phys. Rev. D* **43**, R961.
[20] D. V. Ahluwalia, 1998, *Mod. Phys. Letts. A*13, 1397.
[21] A.K. Raychaudhuri, 1955, *Phys.Rev.*90, 1123.
[22] M. Nouri-Zonoz, N. Dadhich & D. Lynden-Bell, 1999, *Class. Quant. Gra.*16, 1.
[23] N. Dadhich and L. K. Patel (1999) *Gravo-electric dual of the Kerr solution*, submitted.
[24] K.B. Marathe and G. Martucci (1998) Nuovo Cim. **B113**, 1175.
[25] N. Dadhich and K.B. Marathe (1998) Electromagnetic resolution of curvature and gravitational instantons, submitted to Nuovo Cim.

Chapter 9

NONCOSMOLOGICAL REDSHIFTS OF QUASARS

Prashanta Kumar Das
Indian Institute of Astrophysics,
Bangalore 560034, India

Abstract This article very briefly reviews the evidence for an against the cosmological hypothesis (CH) viz. the redshifts of all extragalactic objects are due to the expansion of the Universe. In the latter part various theoretical noncosmological alternatives are discussed with a special emphasis on quasars.

1. INTRODUCTION

The discovery of the velocity-distance relation by Hubble established cosmology as an observationally testable and hence falsifiable subject. The theoretical basis for the observed Hubble relation was provided by the 'expanding Universe' models of Friedmann as well as Robertson and Walker. The Hubble's law is best stated in the form of the 'Cosmological Hypothesis' (CH): 'The redshift of an extragalactic object is (almost) entirely due to the expansion of the Universe'. The term 'almost' allows for a small noncosmological component. Thus if z and z_C are the total and cosmological redshifts of an extragalactic object and z_{NC} stands for redshift(s) due to Doppler, gravitational or any other effects or a combination of them then the validity of CH demands $z \approx z_C$ and $| z_{NC} | \ll 1$.

Ever since its early success CH has remained the most favoured hypothesis for extragalactic redshifts and the entire edifice of modern cosmology critically depends upon it. Hence it is imperative that its validity is assessed in the light of modern observational data. We proceed to do so in the next section.

N. Dadhich and A. Kembhavi (eds.), The Universe, 97–103.

2. OBSERVATIONAL EVIDENCE

Following the approach of Burbidge (1973) and Narlikar (1989) we segregate the observations as (i) evidence consistent with the CH (ii) neutral evidence (iii) discordant evidence.

2.1 CONSISTENT EVIDENCE

Observations which have a natural interpretation in terms of the CH fall in this category. The following may be included in this group.

A reasonably scatter-free magnitude-redshift (m-z) relation, consistent with the CH, is obtained for galaxies - albeit for carefully chosen samples. Similarly the number-magnitude [N(m)] and number-flux density [N(s)] observations can be accommodated in the CH. Also the observations of absorption line systems in QSO spectra and gravitational lensing can be fairly plausibly explained if the CH is valid. An evolutionary connection between the QSOs and active galactic nuclei (AGNs) can be made on the basis of the Hubble law. The data on QSO-galaxy (Q-G) associations provide, with some reservations, evidence for the cosmological nature of QSOs with moderate redshifts.

2.2 NEUTRAL EVIDENCE

In this we group the observations which do not suggest the CH directly but can be made compatible with it with the help of suitable epicycles.

It is well known that the m-z plot for QSOs is a complete scatter diagram. Unlike in the case of galaxies there is a clear lack of any correlation between the magnitudes and the redshifts in the case of quasars. This does not disprove the CH but, at the same time, it does not give any support to it either. Similarly the angular size-redshift $[\theta(z)]$ observations for galaxies, radio sources and QSOs do not provide any conclusive evidence for the CH. The observed superluminal motions in QSOs can be made consistent with the CH in the somewhat contrived relativistic beaming models. The energy problems of QSOs, which largely disappear if the QSOs are local, can be somehow reconciled with the CH. Lastly the absence of significant absorption in the continuum blueward of Lyman - α in the spectra of the QSOs with $z \geq 2$ does not have a very satisfactory explanation in the CH.

2.3 DISCORDANT EVIDENCE

In this section we present evidences which, if real, imply that some objects atleast possess substantial noncosmological redshifts and thus seriously question the validity of the CH.

Nonlinear Hubble Relation. Segal (1976, 1980) has argued for a quadratic velocity-distance relation (v $\approx D^2$) instead of the linear Hubble law for nearby galaxies on the basis of his chronometric cosmology.

Periodicities in Redshift Distribution. Periodicities and peaks in the redshift distributions (redshift quantization) have been claimed both for nearby galaxies (Tifft, 1997 and references therein) and QSOs (Duari, 1997 and references therein), observations which go against the Cosmological Principle underlying the CH.

Galaxy-Galaxy Associations. Arp (1987 and references therein) has reported several cases where the compact companion galaxies have excess redshifts compared to the bright main galaxy in a group, with an apparent luminous connection joining the two in some cases.

QSO-Galaxy Associations. There seems to be a strong evidence that normal galaxies and QSOs tend to cluster together irrespective of their redshifts (Burbidge 1981, Arp 1981; Burbidge et al. 1990) implying physical associations and a substantial noncosmological (anomalous) redshift component in the QSO if its redshift is large.

Alignments and Redshift Bunching. There are several examples of remarkable alignments between QSOs around a central galaxy (in some cases without a galaxy). Also in many cases the QSO redshifts are bunched around in relatively small intervals (Arp 1987, 1997a, 1997b) which are difficult to explain on the basis of the CH.

3. NONCOSMOLOGICAL OPTIONS

We feel that an unbiased assessment of the observational data shows that, though the CH may be applicable to ordinary galaxies, a substantial noncosmological (anomalous, discordant) component of redshifts may be present in the QSOs and to a lesser extent in compact, companion galaxies. Thus we may have $z_{NC} \approx z_C$ or even $z_{NC} > z_C$. We now discuss various theroetical alternatives for z_{NC}.

3.1 THE DOPPLER EFFECT

The Doppler shift, which occurs when there is relative motion between the source and the observer, was first considered as a mechanism for quasar redshifts by Terrell (1964) and Hoyle and Burbidge (1966). The Doppler model is attractive because the problems of energy generation and superluminal motion in quasars largely disappear, the scatter in the Hubble diagram for quasars ceases to be a matter of concern and it offers a natural explanation to the phenomenon of alignments. The main problem with the Doppler effect is the absence of blueshifts which should strongly predominate over redshifts in a normal situation. However, this can be overcome in a scenario suggested by Hoyle (1980) in which the QSO radiates within a backward cone. Doppler models based on this hypothesis have been considered by Narlikar and Edmunds (1981), Narlikar and Subramanian (1982, 1983) to explain quasar alignments such as the Arp-Hazard triplets (Arp and Hazard, 1980). However, these models still predict a small number of blueshifted QSOs and the actual mechanism of ejection of quasars from galaxies still needs to be investigated.

3.2 THE GRAVITATIONAL REDSHIFT

The gravitational redshift, predicted by Einstein's General Theory of Relativity, was considered as a possible explanation for quasar redshifts soon after the discovery of QSOs. But there were two problems with it. The first was based on the observations of the very first two QSOs to be discovered, viz. 3C48 and 3C273. (Greenstein and Schmidt 1964). The second, of a theoretical nature, was due to Bondi (1964) who showed that with physically plausible equations of state, the surface gravitational redshift from a spherical body could not exceed a value ~ 0.62.

However, both these could be overcome in an ingenious model proposed by Hoyle and Fowler (1967), who visualized the QSO as a composite object in which the observed emission is from a central emitting region, which gets largely redshifted by the gravitational potential provided by the largely transparent envelope composed of highly collapsed compact objects.

Das and Narlikar (1975), Das (1975, 1976, 1979; 1984) have developed detailed Hoyle-Fowler type core-envelope models for QSOs. From their work it seems possible to have bound, stable, massive (M$\sim 10^{10} M_\odot$) objects with realistic equations of state and non-negative distribution functions, capable of central redshifts upto ~ 1.5. On the whole, these models could be satisfactory for isolated QSOs with $z \sim 1.5$ but cannot offer a suitable explanation for alignments and associations.

3.3 THE SPECTRAL COHERENCE EFFECT

The spectral coherence effect (Wolf 1986, 1987), which can give rise to both blueshifts and redshifts, has been suggested as an explanation for QSOs with $z_{NC} > 0$. But the Wolf effect is yet to be studied in detail in the astrophysical scenario.

3.4 CHRONOMETRIC COSMOLOGY

Segal (1976) has suggested the chronometric cosmology as an alternative to the expanding Universe which gives a quadratic redshift-distance relation for nearby objects and has also claimed a better agreement with $m(z), N(m)$ and $\theta(z)$ observations.

3.5 THE TIRED LIGHT THEORY

Pecker (1976) has developed the idea that the photon may have a small but nonzero rest mass and travelling through space would lose energy progressively (tired light) through interaction with a specific form of matter (ϕ -matter) and be thus redshifted. The anomalous redshifts are attributed to local inhomogeneities in the ϕ - bath. Again the applicability of this theory remains to be critically assessed.

3.6 THE VARIABLE MASS HYPOTHESIS

Based on the Variable Mass Hypothesis (VMH) of the Hoyle and Narlikar (HN) theory of conformal gravitation (Hoyle and Narlikar 1974, Narlikar 1977) Narlikar and Das (1980) developed a model for the anomalous redshift QSOs. In this scenario the redshifts are due to variable particle masses and kinks (inhomogeneities) in the zero mass hypersurface give rise to the anomalous redshifts. It is hypothesised that quasars are born in and ejected from the nuclei of parent galaxies as massless objects and the masses in them systematically increase with epoch. Analysis of the dynamics of such a quasar-galaxy (Q-G) pair shows that, depending upon the initial conditions, Q may escape the gravitational influence of G to emerge as a field quasar or may form a bound system with G undergoing damped oscillations of decreasing periods. The Narlikar-Das model is fairly successful in explaining the observed features of Q-G associations, alignments and redshift bunching as well as the luminous connections observed between objects with vastly dissimilar redshifts [Das 1993] The VMH can also, in principle, offer an explanation to redshift quantization where a discrete mass spectrum would lead to discrete values of z.

4. CONCLUDING REMARKS

We have presented a very short overview of the present state of the observational evidence for an against the Hubble law and also discussed briefly the noncosmological alternatives. While no definite conclusions about the validity of the CH can be drawn (nor was it hoped) we feel that the discordant evidence cited above point to the fact that the noncosmological, 'antiestablishment' alternatives merit a far more serious consideration than they have been accorded till now. While some of them (3.1. - 3.3.) fall within 'conventional' physics, the rest (3.4. - 3.6.) involve 'unconventional' ideas. Perhaps before invoking such 'radical' ideas the former should be investigated in more detail.

Acknowledgments

I would like to thank Arun Thampan for his help. It gives me great pleasure to thank Ajit Kembhavi and Naresh Dadhich for their invitation to write this article on the occasion of Jayant Narlikar's 60th Birthday. It has been a sheer delight to know Jayant Narlikar first as a teacher and then as an esteemed colleague and friend over the past twenty five years.

References

[1] Arp, H.: 1981, *Astrophys. J.* **250**, 31.

[2] Arp, H.: 1987, *Quasars, Redshifts and Controvrsies*, Interstellar Media, Berkeley.

[3] Arp, H.: 1997a, *Cosmological Parameters and Evolution of the Universe, IAU Symp.***183**.

[4] Arp, H. :1997b, *Journ. Astrophys. Astron.*, 393.

[5] Arp, H. and Hazard, C.: 1980, *Astrophys. J.* **240**, 726.

[6] Bondi, H.: 1964, *Proc. Roy. Soc. Lond. A* **282**, 303.

[7] Burbidge, G.R.: 1973, *Nature* **246**, 17.

[8] Burbidge, G.R.: 1981, in R.Ramaty and F.C.Jones (eds), Tenth Texas Symposium on Relativistic Astrophysics, New York Academy of Sciences, P.123.

[9] Burbidge, G.R., Hewitt, A., Narlikar, J.V., and Das Gupta, P.: 1990, *Astrophys. J. Suppl.* **74**, 675.

[10] Das, P.K.: 1975, *Mon. Not. Roy. astr. Soc.* **172**, 623.

[11] Das, P.K.: 1976, *Mon. Not. Roy. astr. Soc.* **177**, 391.

[12] Das, P.K.: 1979, *Mon. Not. Roy. astr. Soc.* **186**, 1.

[13] Das, P.K.: 1984, *Mon. Not. Roy. astr. Soc.* **210**, 753.

[14] Das, P.K.: 1993, Contributed paper at 6th Asian Pacific Regional Meeting on Astronomy, Pune, India.

[15] Das, P.K. and Narlikar, J.V.: 1975, *Mon. Not. Roy. astr. Soc.* **171**, 87.

[16] Duari, D: 1997, *Journ. Astrophys. Astron.*, 441.

[17] Greenstein, J.L. and Schmidt, M.: 1964, *Astrophys. J.* **140**, 1.

[18] Hoyle, F.: 1980 'Astrophysics and Relativity', Preprint No.63, University College, Cardiff.

[19] Hoyle, F.and Burbidge, G.R.: 1966, *Astrophys. J.* **144**, 534.

[20] Hoyle, F. and Fowler, W.A.: 1967, *Nature* **213**, 373.

[21] Hoyle, F. and Narlikar, J.V.: 1974, *Action at a Distance in Physics and Cosmology*, W.H. Freeman and Co., New York.

[22] Narlikar, J.V.: 1977, *Ann. Phys.* **107**, 325.

[23] Narlikar, J.V.: 1989, Space Science Reviews, **50**, 523.

[24] Narlikar, J.V. and Das, P.K.: 1980, *Astrophys. J.* **240**, 401.

[25] Narlikar, J.V. and Edmonds, M.G: 1981, *Journ. Astrophys. Astron.*, 289.

[26] Narlikar, J.V.and Subramanian, K.: 1982, *Astrophys. J.* **260**, 469.

[27] Narlikar, J.V. and Subramanian, K.: 1983, *Astrophys. J.* **273**, 44.

[28] Pecker, J.-C: 1976, IAU/CNRS Colloquium (Paris: (CNRS), p.451.

[29] Segal, I.E.: 1976, *Mathematical Cosmology and Extragalactic Astronomy*, Academic Press, New York.

[30] Segal, I.E.: 1980, *Mon. Not. Roy. astr. Soc.* **192**, 755.

[31] Terrell, J.: 1964, *Science* **145**, 918.

[32] Tifft W.G.: 1997, *Journ. Astrophys. Astron.*, 415.

[33] Wolf, E.: 1986, *Phys. Rev. Lett.* **56**, 1370.

[34] Wolf, E.: 1987, *Nature* **326**, 363.

Chapter 10

EXTRAGALACTIC FIRE-WORKS IN GAMMARAYS

Patrick Das Gupta

Department of Physics

University of Delhi, Delhi 110 007, India

The enigma of gamma-ray flashes from random locations in space releasing bursts of photons with energy mostly above ~ 0.1 MeV, on time-scales ranging from about 30 milli-seconds to $\sim 10^3$ seconds with diverse time-profiles, gives rise to one of the most challenging problems in astronomy that cajoles many theorists to create exotic models to understand the nature of gamma-ray bursts. Even professor Jayant Narlikar was not spared from the charm of gamma-ray flashes, having been provoked into working out a whitehole description of gamma ray bursters (GRBs) in late seventies! As a tribute to Prof. Narlikar, I will presently discuss some of the observational aspects of GRBs, and conclude with a brief description of our recent findings pertaining to temporal profiles of short-duration bursts.

1. FLASHES AND AFTERGLOWS

Three decades have gone by since the serendipitous discovery of 16 GRBs by US watchdog satellites Vela 5 and Vela 6, that was reported in scientific literature later in 1973 [1] However, it appears now that the first detected GRB may have been an earlier event recorded by the Vela 4a satellite on July 2, 1967 [2]. About a quarter of a century later, with the launching of the Compton Gamma-Ray Observatory (CGRO) by NASA, there was a major jump in the understanding of GRBs. BATSE on board the CGRO started detecting GRBs at a rate of ~ 1 per day, and further, led to the discovery that GRBs are distributed isotropically on the celestial sphere [3], ruling out those models that involved neutron stars in the Galactic disc and suggesting, rather, that GRBs lie at cosmological distances. BATSE detected gamma-ray events with intensity as high as 10^{-3} erg/sec/cm^2.

N. Dadhich and A. Kembhavi (eds.), The Universe, 105–109.

The next phase transition in the understanding of GRBs started on February 28, 1997, when with the help of an Wide Field Camera on board the Italian-Dutch satellite BeppoSAX , astronomers zeroed in to the location of gamma-ray burst event GRB 970228, and detected transient afterglows from this object both in X-rays as well as in optical band [4]. Since then, there has been a string of detection of afterglows in the lower energy bands corresponding to a handful of GRBs [5,10], with an added bonus - distance information for two GRBs. Observed absorption lines suggests that the redshift of GRB 970228 is larger than 0.83 [6], while GRB 971214 appears to be on top of a galaxy with a high star-formation rate at a redshift of 3.42 [8]. If redshift estimates are correct then the debate between local versus extragalactic origin of GRBs gets settled in favour of the latter. GRBs with observed afterglows like GRB 970508, GRB 971214, GRB 980326 and GRB 980329 appear to be on top of different host galaxies, strengthening the case for their extragalactic origin.

The afterglow flux in X-rays and in optical band for these GRBs appears to decrease with time steadily as $\frac{1}{t}$, in agreement with the prediction of Fire-ball models [11,15]. Most fire-ball models start with the assumption that energy $\sim 10^{53}$ ergs is very quickly released in a region of size $\sim 100km$ leading to a rapid expansion of this hot ball of photons, neutrinos and electron-positron pairs, and subsequent conversion of thermal energy into relativistic bulk motion of the outer shell [11,12]. Whether the observed gamma-ray burst take place when the expanding fire-ball becomes optically thin or when the shocked outer shell moving relativistically interacts with external matter is still a subject of intense activity [16,19]. However, the afterglows in lower energy photons is expected to be due to the interaction of the expanding ball with the ambient matter [13,15].

GRB 971214, observed a year back, poses a brain teaser. The estimated energy released above 20 keV from GRB 971214 is found to be $\sim 10^{53}$ ergs (provided emission is isotropic, and provided its redshift is indeed ~ 3.4) embarrasingly large for models that invoke merging of binary neutron-stars to trigger a fire-ball [8]. Furthermore, merging scenario faces problem from different quarters - numerical calculations suggest that it is unlikely that neutrino annihilation could produce required photons and electron-positron pairs to initiate a powerful fire-ball, and to make matter worse, both the coalescing neutron stars seem to collapse to form blackholes before the final merging [20]. The key question at the moment is : what gives rise to an expanding fire-ball? A total energy release of $\sim 10^{53}$ ergs to about $\sim 10^{55}$ ergs in a region of size ~ 100 km appears to be the requirement, indicating the basic source of energy to be of gravitational binding energy origin either from binary systems of compact objects or from cores of massive stars. It has been suggested in the literature that GRBs may be associated with cataclysmic end of massive stars, linking the rate of gamma-ray burst events with the massive star-formation rate [20,22]. However, it is fair to say that the current situation is open to new ideas.

2. LONG AND SHORT

In gamma-rays, GRBs display a wide variety in their temporal profiles - single pulse events, smooth well-defined multiple peaks, chaotic events or distinct peaks with long gaps in between [23]. During the burst, flux typically varies on time-scales of few milli-seconds, although a case with sub-millisecond structure has been reported [24]. The peaks are normally asymmetric with shorter leading edges and longer trailing edges, suggesting an explosive origin [23]. If the individual peaks were due to sweeping beams as in the case of pulsars, one would expect symmetric pulses on an average.

Duration of a burst is normally characterised by T_{90}, defined as the time-interval in which the gamma-ray fluence increases from 5 to 95 percent of its total gamma-ray fluence. It has been pointed out that although for single pulse events the duration is a measure of pulse-width, in the case of bursts with multiple, narrow peaks as well as those with long gaps sandwiched between peaks, the duration characterises pulse-separation time, and therefore ought to be distinguished from the pulse width, as the respective origins could be due to distinct physical processes.

The histogram of T_{90} exhibits a bimodality with a distinct dip in number of GRBs, around $T_{90} \sim 2$ seconds [25]. The short-duration bursts (defined to be those with duration less than 2 seconds) are found to be roughly one-fourth of the total number of observed GRBs. It appears that bursts with shorter duration tend to be brighter and harder [23] (i.e. fraction of number of photons detected in higher energy channels is larger). Time-profiles of a given GRB in different energy channels show that duration in high energy channels tend to be shorter while the corresponding sub-pulses are sharper. Those bursts that display sharp rise followed by a long decay period exhibit hard to soft evolution with time [23]. One may also look for gravitationally lensed GRBs by analysing two or more time-profiles with apparent similarity [26,27].

Recently, we have studied 65 short duration bursts belonging to the 3B catalogue [28]. In this sample, time-profiles with single and double peaks are 23 and 16 in number, respectively, forming the majority, while triple peaked bursts are 12 in number. Bursts with larger number of peaks are relatively fewer, culminating with a solitary case of eight-peaked time-profile. Since, peaks in the time-profiles display temporal asymmetry, individual peaks have been fitted with log-normal functions. We have characterised the rise-time and decay-time of individual peaks by the time taken by the observed photon counts to increase from 5 to 95 percent of the peak height and the corresponding decrease from 95 to 5 percent, respectively.

Since the distances of the bursts are unknown, we have studied r_{rd}, the ratio of rise to decay time, against other parameters. Taking the ratio has a merit in the sense that stretching of time-intervals due to cosmological expansion gets cancelled. We find that single-peaked bursts tend to be highly asymmetric with an average value for r_{rd} to be ~ 0.3 (average has been taken over bursts with only one peak). Considering the sub-sample of bursts with two or more peaks, one finds a systematic increase in the average value of r_{rd} as one moves from first to second peak and so on. We also find a strong positive correlation between rise-time and decay-time. Scatter-diagram of energy evolution parameter versus burst duration, corresponding to this sample of short-duration bursts, appears to suggest that there is a greater lag between hard and soft photons for longer bursts in this sample. Details of these analysis will appear later [28].

3. TAIL END

Both supernovae as well as GRBs involve roughly similar energy scales suggesting that latter may be associated with the end product of either single stellar or binary evolution. With $\sim 10^{11}$ stars in individual galaxies, and 'seeing distance' in gamma-rays extending upto few Gpc, the high rate of GRB detection may not pose serious threat to such scenarios. GRBs are also expected to be strong sources of gravitational radiation, and their high event-rate heralds a promising future for LIGOs.

However, mechanisms that lead to the fire-ball and the observed multiple peaks as well as the bimodal distribution of duration, still remain a puzzle. In the post-BeppoSAX times, the extragalactic origin along with the observed high degree of isotropy of GRB distribution may be used as a reverse-argument to claim that universe is isotropic even in the gamma-ray regime!

Acknowledgments

It is a pleasure to thank Prof.P.N.Bhat and Ms.Varsha Gupta for stimulating discussions.

References

[1] Klebesadel, R.W. et al, 1973. *Ap. J.* **182**,L85.

[2] Bonnell, J.T. & Klebesadel,R.W., 1995. In *Gamma-ray Bursts*, AIP Conf. Proc. **384**, Eds. Kouveliotou C. Briggs, M. F. & Fishman, G.J., p977.

[3] Meegan, C.A. et al 1992. *Nature* **355**,143.

[4] van Paradijs, J. et al 1997. *Nature* **386**,686.

[5] Djorgovski, S.G. et al 1997. *Nature* **387**,876.

[6] Metzer, M.R., et al, 1997. *Nature* **387**,878.

[7] Frail, D.A., et al 1997. *Nature* **389**,261.

[8] Kulkarni, S.R. et al 1998. *Nature* **393**,35.

[9] Halper, J.P., et al 1998. *Nature* **393**,41.

[10] Ramaprakash, A.N. 1998. et al,*Nature* **393**,43.

[11] Cavallo, G. & Rees 1987. M.J. *MNRAS* **183**,359(1987).

[12] Shemi, A. & Piran,T. 1990. *Ap. J.* **365**,L55.

[13] Meszaros,P. & Rees 1997. M.J., *Ap. J.* **476**,232.

[14] Wijers,R.A.M.J., et al 1997. *MNRAS* **288**,L51.

[15] Waxman,E., 1997. *Ap. J.* **489**,L33.

[16] Meszaros,P.& Rees,M.J.1992. *MNRAS* **258**,41P.

[17] Narayan,R.,Paczynski,B.,Piran,T., 1992. *Ap. J.* **395**,L83.

[18] Rees,M.J. & Meszaros,P., 1994. *Ap. J.* **430**,L93.

[19] Sari,R. & Piran,T., 1997. *MNRAS* **287**,110.

[20] Woosley,S.E., 1995. In *Gamma-ray Bursts*, AIP Conf. Proc. **384**, Eds. Kouveliotou C. Briggs, M. F. & Fishman, G.J., p977., and the references therein.

[21] Totani,T., 1997. *Ap. J.* **486**,L71.

[22] Paczynski,B., 1997. & Kouveliotou,C.,*Nature* **389**, 548.

[23] Fishman,G.J. & Meegan,C.A., 1995. *Ann. Rev. Astron. Astrophys.* **33**, 415, and the references therein.

[24] Bhat,P.N., et al 1992. *Nature* **359**,217.

[25] Kouveliotou,C., et al 1993. *Ap. J.* **413**,L101.

[26] Wambsganss,J., 1993. *Ap. J.* **406**,29.

[27] Das Gupta,P. & Ramaprakash,A.N., 1995. *Suppl.J.Astrophys.Astr.* **16**,141.

[28] Gupta,V., Das Gupta,P. & Bhat,P.N., 1999. In preparation.

Chapter 11

INSTABILITIES IN OPTICAL CAVITIES OF LASER INTERFEROMETRIC GRAVITATIONAL WAVE DETECTORS

S. V. Dhurandhar
Inter University Centre for Astronomy & Astrophysics, Ganeshkhind, Pune 411 007, India
and
Department of Physics and Astronomy
University of Wales, Cardiff CF2 3YB, UK

Dedicated to Professor Jayant Narlikar.

Abstract

The large scale interferometric gravitational wave detectors consist of Fabry-Perot cavities operating at very high powers ranging from tens of kW to MW. The high powers may result in several nonlinear effects which would affect the performance of the detector. In this article I will consider two such major effects which could result in degrading the performance of the detector. The first is the thermal distortion of the mirrors due to temperature gradients and the second is effect of radiation pressure which can displace the freely hanging mirrors. Both these effects tend to drive the cavity out of resonance degrading the optimal performance of the detector. These effects are likely to be important in the optimal functioning of the full-scale interferometers such as the VIRGO and LIGO.

N. Dadhich and A. Kembhavi (eds.), The Universe, 111–121.

1. INTRODUCTION

This article is in honour Prof. Jayant Narlikar who has been a wonderful teacher and a source of great inspiration to me. He is responsible for the crucial 'phase transition' which launched me in my research career.

The direct detection of gravitational radiation is one of the major challenges in these millenary years. The existence of gravitational waves (GW) was predicted by Einstein as early as 1916. It was not until forty years later, that relativists proved rigourously that gravitational radiation was in fact a physically observable phenomenon and that GW carry away energy. In some ways, in the general theory of relativity, GW are similar to electromagnetic waves, in that they travel in vacuum with the universal speed $c \sim 3 \times 10^8$ metres per second and have two polarisations. But in many crucial ways they differ from electromagnetic waves so that they can bring to us information about the universe which is complementary, in fact, almost orthogonal, to that of electromagnetic waves. While electromagnetic waves are generated by matter on the atomic scale, GW are generated by bulk motions of matter. The crucial point is that since gravity couples very weakly to matter, GW are not easily scattered by intervening matter, unlike electromagnetic waves, and thus carry high fidelity information about the source. Astrophysically powerful sources of GW must be compact and relativistic. Compact objects possess high potential energies which can give rise to relativistic velocities in surrounding matter, thus producing powerful GW. Such sources are normally shrouded by dust or plasma, the fact that GW are not easily scattered, as opposed to electromagnetic waves, it makes them ideal probes of such objects. However, the other side of the coin is that this very weak coupling makes them hard to detect. So much so that, physicists have not seriously considered them for experimental observation or detection until recently.

But thanks to the enormous strides technology has taken in the past few decades and simultaneously the efforts by astronomers that it has become viable to seriously consider the observation of GW. The advent of radio astronomy established that the universe exhibits violent phenonmena such as radio jets, quasars etc. Technology at the same time made it possible to make high precision measurements and produce instruments of unprecedented sensitivities which could in principle detect GW from the violent phenomena in the universe. At first the sensitivities required to detect GW were beleived to be naively optimistic. But subsequent negative results obtained by experimentalists, coupled with the up to date and careful estimates of the strengths of the sources obtained from astrophysics and highly directed and focussed R and D for better detection techniques, led to the construction of three large scale and two medium scale laser interferometric detectors. The three large scale detectors comprise of two detectors of the US LIGO [1] project with arm lengths of 4 km. and the one detector of the French/Italian VIRGO [2] project with an arm length of 3 km. In the medium scale there are the German-British project GEO600 [3] with an arm length of 600 metres and the Japanese TAMA300 [4] of 300 metres arm length. Also initial funding has been obtained for the Australian AIGO500 [5] project. There are also separate proposals for space-based detectors which could be operational twenty-five years from now (e.g., LISA: the Laser Interferometer Space Antenna, a cornerstone project of the European Space Agency) [6]. The ground based interferometers will use Fabry-Perot cavities in their arms and arm lengths of a few kilometers.

2. NONLINEAR EFFECTS IN HIGH POWERED CAVITIES

There are several noise sources which plague the detector. Amongst them, the photon shot noise is dominant at high frequencies. It is reduced by increasing the amount of power of the laser source, as the noise scales inversely as the square root of the power. Therefore to attain the desired sensitivities, the cavities envisaged will operate at very high powers – tens of kW for initial detectors and perhaps MW in advanced detectors.

However, the high power stored in the cavities can generate a number of nonlinear effects which would adversely affect the operation of the optical cavity. The most evident effect is that of the absorption of the light power in the substrates of the mirrors resulting in temperature gradients across the mirror. The temparature gradients can cause thermal lensing finally leading to loss of power in the cavity. Moreover, the temperature changes deform the mirror, detuning the cavity in the process. The other major effect is that of the radiation pressure exerted on the mirror surface. Since in the detector the mirrors are 'freely' hanging, the radiation pressure can change the position of the mirror, driving the cavity out of resonance and thus degrading the sensitivity of the detector. Therefore, it is essential that experimentalists have a quantitative idea about the magnitude of these effects and when these effects must be seriously combated.

In this article I will restrict myself to two of these effects, (a) the thermo-elastic deformation of the mirrors, (b) radiation pressure. The

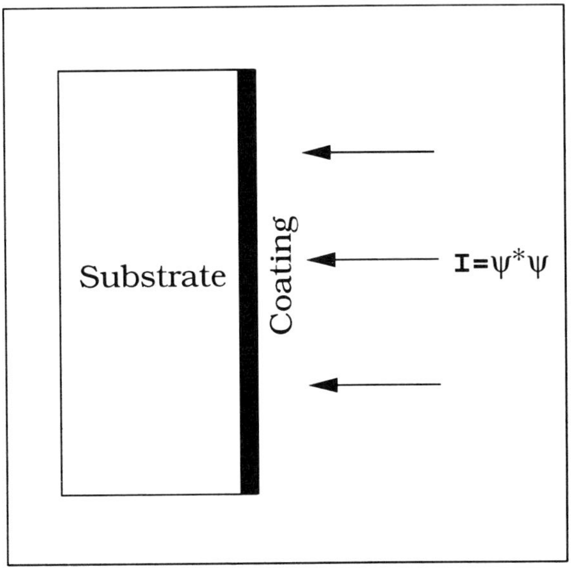

Figure 11.1 The mirror comprising of substrate and coating heated by the laser beam

analysis is particularly important within the bandwidth of the detector between few Hz to a few kHz where the servo is in effect inoperative. The emphasis therefore lies in this regime, where the analysis is necessarily dynamical.

3. THERMO-ELASTIC DEFORMATION OF THE MIRRORS

In order to study the dynamics we must first obtain the *time dependent* transfer function which connects the power in the cavity to the deformation of the mirror. Then secondly we must obtain the change in power due to the deformation. We then get a coupled system which we solve self-consistently [7, 8]. We proceed in three steps:

1. Obtain the time dependent temperature profile inside the mirror substrate.

2. Solve the thermo-elastic problem to obtain the deformation for the temperature profile.

3. Evaluate the change in power due to the deformation of the mirror.

The mirror is in the shape of a cylinder of radius a and thickness h and consists of a substrate, usually silica, and a high-quality reflective coating. In gravitational wave interferometers typically, $a \sim 0.1$ m and $h \sim 0.1$ m (for the input mirrors). The axis of the mirror is the z-axis and it is coated on the face at $z = 0$ (see Fig.11.1). The cavity lies along the positive z-axis with the other mirror (suitably curved) at $z = L$. The intra-cavity laser light is incident on the mirror at $z = 0$ which gets heated due to absorption in the coating. We neglect absorption in the substrate of the mirror. The mirror loses heat to its surroundings by radiation. We consider a time varying intensity profile $I(r)e^{-i\Omega t}$ with a single Fourier component at Ω. $I(r)$ is the modulus square of the TEM00 mode of the electric field and has a Gaussian profile. The temperature T obeys the diffusion equation,

$$(i\Omega C\rho + K\nabla^2)T = 0, \tag{11.1}$$

where (for pure silica) ρ is the mass density (2202 kg m^{-3}), C is the specific heat capacity (745 J kg^{-1} K^{-1}) and K is the thermal conductivity (1.38 W m^{-1} K^{-1}). We solve the equation with radiative boundary conditions. An approximate but adequately accurate, axially symmetric, solution is the temperature profile given by,

$$T(t, r, z) = \frac{\epsilon I(r)\delta}{\sqrt{2}K} \exp\left[\frac{z}{\delta} - i\left(\frac{z}{\delta} - \frac{\pi}{4} + \Omega t\right)\right]. \tag{11.2}$$

where, δ is the 'skin depth' defined by,

$$\delta = \sqrt{\frac{2K}{\Omega C\rho}}. \tag{11.3}$$

For the VIRGO parameters the skin depth is a fraction of a millimetre.

The time varying temperature causes time varying deformation near the lit surface of the mirror producing acoustic waves. To quantify this we solve the thermoelastic equations for the temperature profile above. We do not write down the equations here but just mention that the one must obtain the displacement vector field $\mathbf{u}$ in the mirror by solving the elastic equations in which the forcing term arises from the temperature. However what is important to our analysis is the z component of the displacement field, u_z, at the lit surface of the mirror. It is in fact u_z averaged over the Gaussian beam profile which determines the detuning of the cavity. It is given by,

$$\langle u_z \rangle (\Omega) = \int_{\mathcal{A}} u_z |\Phi_{00}|^2 dS, \tag{11.4}$$

where $\mathcal{A}$ is the area of the mirror and Φ_{00} is the TEM00 mode. For 1 Watt of absorbed power and for the VIRGO parameters, $\langle u_z \rangle (\Omega) \sim 0.72i \times 10^{-9}\ \Omega^{-1}$ m.

The phase ψ corresponding to this deformation is obtained by multiplying this average displacement by twice the wave number $k = 2\pi/\lambda$ of the laser light. Thus we have,

$$\psi(\Omega) = 2k \langle u_z \rangle (\Omega). \tag{11.5}$$

This is the detuning phase which must be substituted in the expression for the intra-cavity power. Note that this information in ψ is incomplete since the above expression is valid only at frequencies much greater than the diffusion time-scale $\sim$ 3 hours for the VIRGO mirror. We include the static part as given by [9, 10] in our analysis in a phenomenological manner and obtain $\psi(\Omega)$ for $\Omega \sim 0$. We then have a transfer function connecting $\psi(\Omega)$ to the power $P(\Omega)$. By taking inverse Fourier transforms we obtain a differential equation governing $\psi(t)$, namely,

$$\frac{d\psi}{d(\Omega_0 t)} = -\psi + \frac{a}{b - \cos(\psi + \phi_0)}, \tag{11.6}$$

where, $a = \frac{\alpha(0) t_1^2 \epsilon P_{in}}{2R\Omega_0}$, $b = \frac{1+R^2}{2R}$, $R = r_1 r_2$, the product of the reflectivities of the mirrors, t_1 is the transmission coefficient of the mirror M_1, P_{in} is the input power and ϵ is the absorption coefficient. $\alpha(0)$ is the detuning phase per Watt of absorbed power in the static case and α is the corresponding quantity in the dynamic case. The ratio of α to $\alpha(0)$ is Ω_0. Note that since $R \leq 1$, $b \geq 1$. For $b > 1$ we obtain stable solutions. For small values of ψ and $\phi_0 = 0$ we can integrate the equation to yield,

$$\psi \sim \frac{a}{b-1}\left(1 - e^{-\Omega_0 t}\right). \tag{11.7}$$

This is inherently a stable solution. It is to be noted however that eq. 11.6 is suspect in the regime when the variations in ψ occur near the thermal diffusion time scale $\sim$ 3 hours. Our interest however lies in the bandwidth of the detector which is above a few Hz where the solution is certainly valid. In any case, variations below the bandwidth will be removed by the servo-control.

4. DYNAMICS OF RADIATION PRESSURE EFFECTS

The other important effect is that of radiation pressure. The intra-cavity power P will produce a radiation pressure force on the mirrors $\sim 2P/c$, where c is the speed of light. Even in initial detectors, the intra-cavity power will be of the order of tens of kW, which will produce a force of the order of 10^{-4} Newtons. This force is sufficient to displace the mirrors by order of the wavelength $\lambda \sim 10^{-6}$m of the laser (Nd-Yag) light, thus driving the cavity out of resonance [11, 12]. In fact, the situation is even worse because, there is the so called 'delay effect' [13, 14] which leads to a continuous gain in energy, if the mirrors are left 'free' meaning that no servo-control is used. However, in actual detectors a servo will be used, and even then the delay effect cannot be ignored in the action of the servo. Therefore in this section we will first consider the case for the free mirrors and then just describe the results for mirrors with servo control or the 'locked cavity'.

4.1 FREE MIRRORS

The only forces acting on the mirrors are the radiation pressure forces and gravity which manifests itself as the restoring force of the pendulum.

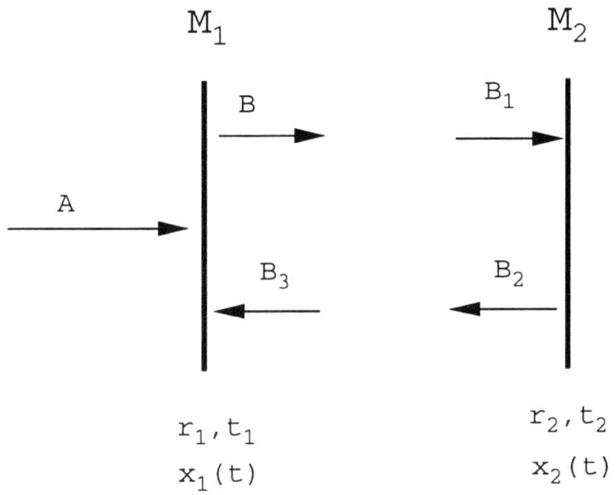

Figure 11.2 Schematic diagram of the cavity and the intra-cavity fields

We consider a single cavity with mirrors M1 and M2 which are suspended as shown in fig. 11.2. The input beam A enters the cavity from mirror M1 and bounces back and forth between the mirrors. The field builds up inside the cavity and this magnitude depends on the finesse which is dependent on the reflectivities of the two mirrors and the detuning of the cavity. The field or the power produces radiation pressure force which pushes on the mirrors driving them apart, thus changing the original distance between them. This in turn changes the power inside the cavity. For instance if the mirrors were hanging in a position of resonance, the radiation pressure force drives the cavity out of resonance, reducing the radiation pressure force. The mirrors start swinging with radiation pressure force adjusting to the continuously varying length of the cavity. The crucial point is that the radiation pressure force does not adjust instantaneously to the new length but *lags* behind the expected static force (the force if the cavity had this fixed length) by an amount comparable to the storage time of the cavity.

We will consider two situations:

1. The cavity is in resonance and the laser is switched on.

2. The mirrors are hanging in an equilibrium state with the radiation pressure forces balancing the restoring forces of the suspension.

Since it is the distance between the mirrors which determines resonance, we will consider the differential mode $\psi = k(x_2 - x_1)$, where x_1, x_2 are the positions of the mirrors. The appropriate equation of motion is:

$$\ddot{\psi} + \omega^2 \psi = F(t), \tag{11.8}$$

where $F(t)$ is the total radiation pressure force acting on the two mirrors. We have ignored damping because the delay effects occur on much smaller time scales than the damping time-scale of $\sim 10^6$ sec (VIRGO). Denoting by $F_s(\psi)$ the force when the mirrors are stationary, we find that for low mirror velocities $\sim 1\mu$m / sec the force profile 'follows' the mechanical motion retarded by an effective delay τ_{lag}, or,

$$F(t) \simeq F_s(\psi(t - \tau_{lag})) \simeq F_s(\psi) - \tau_{lag} \frac{dF_s}{d\psi} \dot{\psi}. \tag{11.9}$$

The equation of motion becomes,

$$\ddot{\psi} + \tau_{lag} \frac{dF_s}{d\psi} \dot{\psi} + \omega^2 \psi = F_s(\psi) \tag{11.10}$$

The $\dot{\psi}$ term is the damping/anti-damping term and depending on its sign the system gains or loses energy. We can write an expression for the energy gain as,

$$\Delta E = - \int_{t_1}^{t_2} \tau_{lag}(t) \frac{dF_s}{d\psi} \dot{\psi}^2 dt \tag{11.11}$$

where, ΔE is the energy gain/loss in the time interval between t_1 and t_2. For the VIRGO finesse the effective delay τ_{lag} is around 16 to 30 times the round trip time τ of the cavity, near the resonance peak.

We can integrate the above equations numerically. We find that the net effect of the $\dot{\psi}$ term is that of *anti-damping* and energy is *gained* as the system completes an oscillation.

The results are similar in the case when the mirror hangs in equilibrium between gravitational and suspension forces. The perturbation $\delta\psi$ about the equilibrium point ψ_{eq} satisfies the equation,

$$\delta\ddot{\psi} - \frac{2}{\tau_{eq}}\delta\dot{\psi} + \Omega_{eq}^2\delta\psi = 0, \tag{11.12}$$

where τ_{eq} and Ω_{eq} depend on the finesse, power, ψ_{eq} and τ. Since $\tau_{eq} > 0$, the system is unstable. Figure (3) displays the phase space trajectory of the mirror.

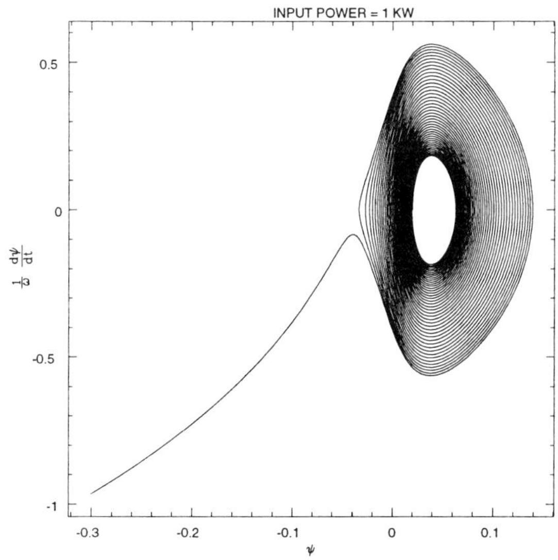

Figure 11.3 The phase space trajectory for 1 kW of input power.

4.2 LOCKED CAVITY

We now include the effect of the servo-system since the mirrors in the actual cavities will be locked by the servo. We have used the servo-control transfer function given by Caron et al.[15] in incorporating the effect of the servo. We assume the displacement to be small enough so that the radiation pressure force $F_s(\psi)$ is linear in the displacement. Then it is possible to use Laplace transform methods to analyse the system. In effect, we obtain a feedback loop and thus a characteristic equation for the mirror displacement. The roots of this equation determine the stability of the system and they essentially depend on the phase offset δ of the operating point, the finesse of the cavity and the input power. We find that for $\delta > 0$ and for typical parameters, the system is stable. For $\delta < 0$, but chosen within the linewidth, for a given finesse and above a certain critical power the cavity becomes unstable.

Given the finesse, the mass and the servo transfer function parameters, we can write, the critical input power P_{crit}, above which instability occurs, as,

$$P_{crit} = P_{char}\frac{(1+\alpha^2)^3\Theta^2}{\alpha(1+\alpha^2-3\Theta)(1+1.5\Theta)}, \tag{11.13}$$

where $\Theta = 1 - R$ and α is defined through $\delta = \alpha\Theta$. For VIRGO, $P_{char} \simeq$ 340 kW and the critical power can range from few kW to hundreds of kW depending on the detuning.

5. CONCLUSION

Since as yet, there are no optical cavities operating at such high powers, it is all the more important to have simulated results in the absence of any experiments. We draw the following conclusions from our analysis:

Thermoelastic deformation of the mirrors will not cause power variations within the bandwidth of the detector and therefore this is not a cause for worry to GW experiments. On the other hand, radiation pressure makes the GW interferometer without servo control, intrinsically unstable at all powers. A servo will combat the instability only below a certain critical power P_{crit} given above. But since the critical power is high, the initial detectors operating at relatively low powers will not encounter the instability. However, for advanced detectors operating at high powers, one must seriously consider the implications.

Acknowledgments

A substantial part of the work described in this article has been done under the IFCPAR project no. 1010-1. I thank the PPARC for financial support.

References

[1] A.Abramowici et al., Science.,256-325 (1992).

[2] C.Bradaschia et al., Nucl.Instrum. Methods Phys.Res., Sect.**A**, 518 (1990).

[3] K. Danzmann et al., Proc. 1st Amaldi conference, Frascati (1994); also J. Hough et al., Proc. MG7, Stanford (1994).

[4] K. Tsubono et al., in Proceedings of the TAMA Workshop (1997), to appear.

[5] R.J.Sandeman et al.,A.I..G.O. Prospectus (1997), unpublished.

[6] P. Bender et. al., *LISA, Laser interferometer space antenna for gravitational wave measurements: ESA Assessment Study Report. R. Reinhard, ESTEC* (1990)

[7] S. V. Dhurandhar, P. Hello, B. S. Satyaprakash and J-Y Vinet, *Stability of giant Fabry-Perot cavities of interferometric gravitational-wave detectors*, Applied Optics, Vol.36, No.22, 5325-5334.

[8] W. Winkler, K. Danzmann, A. Rudiger, and R. Schilling, *Heating by optical absorption and the performance of interferometric gravitational-wave detectors*, Phys. Rev. A **44**, 7022-7036 (1991).

[9] P. Hello and J-Y. Vinet, *Analytical models of thermal aberations in massive mirrors heated by high power laser beams*, J. Phys. (Paris)**51**, 1267-1282(1990).

[10] P. Hello and J.-Y. Vinet, *Analytical models of transient thermoelastic deformations of mirrors heated by high power cw laser beams*, J. Phys. (Paris) **51**, 2243-2261 (1990).

[11] V. Chickarmane, S. V. Dhurandhar, R. Barillet, P. Hello and J.-Y. Vinet, *Radiation pressure and stability of interferometric gravitational-wave detectors*, Applied Optics, Vol.37, 3236-3245.

[12] A. Pai, S. V. Dhurandhar, P. Hello and J-Y Vinet, *Radiation Pressure Induced Instabilities in Laser Interferometric Detectors of Gravitational Waves* (1998), to appear.

[13] N. Deruelle and P. Tourrence, *The problem of the optical stability of a pendular Febry-Perot*, Gravitation, Geometry and Relativistic Physics, (Springer-Verlag, Berlin, 1984).

[14] B. J. Meers and N. McDonald, *Potential radiation pressure induced instabilities in cavity interferometers*, Phys. Rev.A **40**, 3754-AAAA(1989).

[15] B. Caron et.al *A preliminary study of the locking of an interferometric for gravitational wave detection*, Astropart. Phys **6**, 245-256 (1997).

Chapter 12

THE EPISTEMOLOGY OF COSMOLOGY

George F. R. Ellis
Mathematic Department, University of Cape Town
Cape Town 7700, South Africa

Fundamental issues in cosmology have always been of major concern to Jayant Narlikar. It is thus a pleasure to dedicate this reflection on the nature of epistemology in cosmology to him, on the occasion of his 60th birthday.

—

Abstract This article reviews epistemological issues that arise in cosmology, which is different from other subjects particularly because the universe is unique. A series of philosophical assumptions underlie our present-day spatially homogeneous and isotropic world models, whose assumed geometry is not directly testable because of limitations on what can be measured; alternative models are also viable. Nevertheless the standard model has strong support from evidence and can with full justification be adopted as a solid basis for cosmological investigation. However physical cosmology rests on, and is unable to investigate in terms of its own methods, a further series of metaphysical issues to do with the existence and nature of physical laws. Examination of these issues has of necessity to rest on appropriate philosophical and metaphysical approaches.

1. INTRODUCTION

Scientific Cosmology is the study of the large scale structure of the physical Universe (by definition, that single physical entity consisting of

N. Dadhich and A. Kembhavi (eds.), The Universe, 123–140.

all events that are causally connected to each other and that includes our world), and in particular examines the expansion of the universe from a Hot Big Bang and the subsequent formation of structure (including galaxies and galaxy clusters). It has unique features that make consideration of its epistemology[1] a central concern in any mature approach to the subject (see [2, 13] for useful discussions).

2. THE CONTEXT

Three somewhat mundane features are crucial in terms of their effect on the epistemology of cosmology. These restrictions cannot be evaded.

The first and most fundamental is that there is only one physical universe. While there is a vast variety of conceivable or possible universes, there is only one that actually exists and that we have access to (we live in it, and can observe it and experimentally interact with some parts of it). I discount here as a serious part of classical physics[2], loose talk about 'many universes' - if they are directly or indirectly physically connected to us, they are part of our one universe, and the terminology is seriously misleading; if they are not, we cannot interact with them or observe them, so we can say anything we like about them without fear of disproof; thus any statements we make about them have no scientific status.

The implication is that we cannot compare the universe with any similar object that we know exists. We can compare it with hypothetical objects, considered as possible under various kinds of hypotheses, for example possible universes as legislated by Einstein's theory - but this is quite different to comparing it with real objects, demonstrating the existence in Nature of some kind of behaviour; they are possible, but not actual[3]. Nor can we scientifically establish 'laws of the universe' that might apply to the class of all such objects - for there is no way we can test any such proposed law (we cannot re-run the universe, nor obtain statistical properties of a class of physically existing universes). The concept of a 'law' becomes doubtful when there is only one object to which it applies.

The second is that we can only view the universe, considered on a cosmological scale, from one spacetime event ('here and now'), because of its vast scale. If we were to move away from this spatial position at almost the speed of light for say 10,000 years, we would not succeed in leaving our own galaxy, much less in reaching another one; and if we were to start a long term astronomical experiment that would store data for say 20,000 years and then analyze it, the time at which we observe the universe would be essentially unchanged (because its age is of the order of 10^{10} years).

The third is that the region of the universe we can see from this vantage point is restricted, because a finite time has elapsed since the universe became transparent to radiation, and light can only have traveled a finite distance in that time. As no signal can travel to us faster than light, we cannot receive any information from galaxies more distant than our visual horizon - essentially the distance light can have traveled since the decoupling of matter and radiation as the hot early universe cooled down [30, 5].[4]. There are many galaxies - perhaps an infinite number - at a greater distance, that we cannot observe by any type of radiation. The exception to this is if we live in a small universe we have already seen around (because it has closed spatial sections whose size is smaller than the Hubble scale). This is a possibility, and is in principle testable [7]; however there is no solid evidence to show that the real universe is like this.

3. OBSERVATIONS

As in other sciences, the epistemology of cosmology is based on the types of observation we can make. Null cone observations of sources and background radiation are obtained from telescopes operating at all wavelengths (optical, infrared, ultraviolet, radio, X-ray), giving detailed observations (including visual pictures, spectral information, and polarization measurements) of the matter this side of the visual horizon. We can also aspire to use neutrino and gravitational wave telescopes to obtain information on matter lying between the visual horizon and the particle horizon[5]. However distant sources appear very faint, both because of their distance, and because their light is highly redshifted (due to the expansion of the universe). Additionally, absorption of intervening matter can interfere with light from distant objects. The further back we look, the worse these problems become; thus our reliable knowledge of the universe decreases rapidly with distance (although the situation has improved greatly owing to the new generation of telescopes and detectors, particularly the Hubble Space Telescope and the COBE satellite).

Three interrelated problems occur in interpreting these observations. The first is that (because we can only view the universe from one point) we only obtain a 2-dimensional projection on the sky of the 3-dimensional distribution of matter in the universe. To reconstruct the real distribution, we need reliable distance measurements to the objects we see. However because of variation in the properties of sources, we lack reliable standard candles or standard size objects to use in calibrating distances, and have to study statistical properties of classes of sources. Second, because the radiation by which we obtain information travels to us at the speed of light[6], we see distant sources at an earlier epoch when their properties may have been quite different. The inevitable lookback-time involved in our observations means we need to understand evolution effects which can cause systematic changes in the properties of sources we observe; but we do not have good theories of source evolution. Thirdly, a variety of selection effects interfere with observations, because some sources are easy to detect but others are not. Most notably, some kinds of matter emit very little radiation and are not easy to detect by absorption, hence the dark matter problem: we do not know the amount of matter in the universe to within an order of magnitude.

Another source of cosmological information is data of a broadly geological nature; that is, the present day status of rocks, planets, star clusters, galaxies, and so on contains much information on the past history of the matter comprising those objects. Thus we can obtain detailed information on conditions near our spatial position[7] at very early times if we can interpret this data reliably, for example by relating theories of structure formation to statistical studies of source properties. Particularly useful are measurements of the abundances of elements which resulted from nucleosynthesis in the Hot Big Bang, and age estimates of the objects we observe. If we can obtain adequate quality data of this kind at high redshifts, we can use this type of argument to probe conditions very early on at some distance from our past worldline [9].

A final - much more controversial - source of data about the universe is the nature of local physical laws. For example it might be that the local inertial properties of matter are related to the distribution of matter in the distant universe, or that the existence of the local arrow of time is related to boundary conditions in the distant past and future. The problem is that in the cosmological context, because of the uniqueness of the universe, it is difficult to distinguish between laws of nature and boundary conditions governing solutions to those laws; and any proposal in this regard is untestable. Thus while there may be invaluable information hidden here, it is difficult to decode it uniquely.

4. GEOMETRY

One of the prime aims of cosmology is to determine the spacetime geometry of the universe. The standard models of cosmology are the Friedmann-Lemaître (FL) family of universe models expanding from a Hot Big Bang, based on the Robertson-Walker (RW) geometries, that is, spacetimes that are exactly spatially homogeneous and isotropic everywhere. We want to know to what degree observational data supports these universe models, and what parameters are appropriate for a 'best-fit' FL model.

Observational support for the basic idea of expansion from a hot big bang is very strong [10]: the linear magnitude-redshift relation demonstrates expansion, with the blackbody Cosmic Background Radiation (CBR) being strong evidence that there was indeed evolution from a hot early stage[8], and agreement between measured light element abundances and the theory of nucleosynthesis in the early universe confirming this interpretation. The question is how good are the RW geometries as models of the real universe.

The first issue that arises is that of averaging scales and scale of description. The real universe is obviously neither spatially homogeneous nor isotropic. Thus this idealized model is meant to represent the universe in some smoothed out or averaged sense. It is difficult to define such an averaging procedure in an adequate way within the context of the curved spacetimes of general relativity theory (the theory of gravity in the majority of cosmological models). Thus underlying our models is an ill-defined averaging process that is rarely examined [12, 13]; indeed it is seldom one even sees an explicit statement as to what averaging scale is understood when the RW geometry is used, but this is an important parameter in cosmology, characterizing the minimum scale at which the universe can be validly described as homogeneous and isotropic.

Assuming one is talking about a 'large enough scale', the second issue is, why should we believe that the universe is both spherically symmetric and spatially homogeneous? What is important here is the relation between spatial homogeneity[9] and isotropy[10]. If (i) a universe is spatially homogeneous, and is also isotropic about one point, then clearly it is isotropic about every point. If (ii) a universe is isotropic about at least 3 points at any time, and in particular if it is isotropic about every point, then it is necessarily spatially homogeneous. If either of these relations hold everywhere at some initial time, then[11] they will hold at all later times, and the universe has a RW geometry. It will then be characterized by an isotropic background radiation spectrum, and by isotropic source observations (magnitude-redshift relation, angular size-distance relation, number counts) of a specific FL form, seen equally by all observers, with the shape of the curves depending on the deceleration parameter q_0 and density parameter Ω_0.

Considered on a large enough angular scale, astronomical observations are isotropic about us, both as regards source observations and background radiation; indeed the latter is spectacularly isotropic, better than one part in 10^5 after a dipole anisotropy, understood as resulting from our motion relative to the rest frame of the universe[12], has been removed. Because this isotropy applies to all observations (not just the background radiation, which by itself cannot establish the required result), this establishes that in the observable region of the universe, both the space-time structure and the matter distribution are isotropic about us. If we could additionally show that the source observations had the unique FL form as a function of distance, this would additionally establish spatial homogeneity, and hence a FL geometry. However the observational problems mentioned earlier - specifically, unknown source evolution - prevent us from carrying this through. Indeed the actual situation is the inverse: taking number-count data at its face value, without allowing for source evolution, contradicts a RW geometry. The usual procedure is to assume spatial homogeneity is known some other way, and deduce the evolution required to make the observations compatible with this assumption (it is always possible to find a source evolution that will achieve this [14]). Thus attempts to observationally prove spatial homogeneity this way fail [15, 16].

What about an alternative route? If we could show isotropy about more than two observers, we would prove spatial homogeneity. Now the crucial point has already been made: we cannot observe the universe from any other point, so we cannot observationally establish this requirement. Hence the standard argument is to assume a Copernican Principle: that we are not privileged observers. This is plausible in that all observable regions of the universe look alike: we see no major changes in conditions anywhere we look. Combined with the isotropy we see about ourselves, this implies that all observers see an isotropic universe, and hence by (ii) establishes the RW geometry. The result holds if we assume isotropy of all observations; a powerful enhancement was proved by Ehlers, Geren, and Sachs [17], who showed that it follows if one assumes simply isotropy of freely-propagating radiation about each observer: using the Einstein and Liouville equations, exact isotropy of the CBR at each point implies an exact RW geometry.

This is currently the most persuasive observationally- based argument we have for spatial homogeneity[13]. A problem is that it is an exact result, assuming exact isotropy of the CBR; is the result stable? Recent work has shown that indeed it is: almost-isotropy of the CBR everywhere in some region proves the universe geometry is almost-RW in that region [19]. Thus the result applies to the real universe - provided we make the Copernican assumption that all other observers, like us, see almost isotropic CBR. And that is the best we can do. The observational situation is clear: the result follows not directly from astronomical data, but because we add to the observations a philosophical principle that is plausible but untested. It may or may not be true[14].

What is the alternative? It is that we live in a spherically symmetric inhomogeneous universe [22, 23], where we are situated somewhere near the centre (as otherwise our observations would not be almost isotropic), with the cosmological redshift being partly gravitational[15]. Most people regard this proposal as very unappealing - but that does not show it is incorrect. One can claim that physical processes such as inflation make existence of almost-RW regions highly likely, indeed much more probable than spherically symmetric inhomogeneous regions. This is a viable argument, but we must be clear what is happening here - we are replacing an observational test by a theoretical argument based on a physical process that may or may not have happened. It will be strongly bolstered if current predictions for the detailed pattern of CBR anisotropy on small scales, based on the inflationary universe theory, are confirmed; but that argument will only become rigorous if it is shown that spherically symmetric inhomogeneous models (with or without inflation) cannot produce similar patterns of anisotropy[16].

The purpose of the above analysis is not to seriously support the view that the universe is inhomogeneous, but rather to show clearly the nature of the best observationally-based argument by which we can (quite reasonably) justify the assumption of spatial homogeneity. Accepting this argument, the third question is, in which spacetime regions does it establish a RW-like geometry? We consider separately when the result may be supposed to hold, and where it is established.

The CBR we detect probes the state of the universe from the time of decoupling of matter and radiation (at a redshift of about 1100) to the present day. The argument from CBR isotropy can legitimately be applied for that epoch. However it does not necessarily imply isotropy of the universe at much earlier or much later times, because there are spatially homogeneous anisotropic perturbation modes that are unstable in both directions of time; and they will occur in a generic situation. Indeed, if one examines the Bianchi (spatially homogeneous but anisotropic) universes, using the powerful tools of dynamical systems theory, one can show that intermediate isotropisation can occur [24, 25]: despite being highly anisotropic at very early and very late times, such models can mimic a RW geometry arbitrarily closely for an arbitrarily long time, and hence can reproduce within the errors any set of RW- like observations. We can obtain strong limits on the present-day strengths of these anisotropic modes from CBR anisotropy measurements and from data on element abundances, the latter being a powerful probe because (being of the 'geological' kind) it can test conditions at the time of element formation, long before decoupling. But however low these observational limits, anisotropic modes can dominate at even earlier times as well as at late times (long after the present). If inflation took place, this conclusion is reinforced: it washes out any information about very early universe anisotropies and inhomogeneities in a very efficient way.

As well as this time limitation on when we can regard isotropy as established, there are major spatial limitations. The above argument for homogeneity does not apply to domains beyond the visual horizon - for we have no evidence at all as to what conditions are like there; observers there may or may not see near isotropy. Indeed in the currently popular chaotic inflationary models [26]it is a definite prediction that the universe will not be like a RW geometry on a very large scale - rather it will consist of many RW-like domains, each with different parameter values, separated from each other by highly inhomogeneous regions lying outside our visual horizon[17]. This prediction is as untestable as the previously prevalent assumption[18] that the universe is everywhere RW-like. The point here is that the verification status of the spacetime regions inside and outside our past light cone are totally different, see [30][19]. For example, it is commonly stated that if the density parameter is less than unity and the cosmological constant vanishes, then the universe has infinite spatial sections. However this deduction only applies if the RW-like nature of the universe within the past light cone continues to be true indefinitely far outside it - and there is no way we can obtain observational evidence that this is the case.

The final issue regarding the best-fit FL model for the observed region of the universe is, what are the values of the parameters characterizing such a model? Establishing the Hubble constant H_0, deceleration parameter q_0, and density parameter Ω_0 has been the subject of intensive work for the past 30 years. However there is still major uncertainty about their values[20], essentially because of the observational problems discussed in section 2. Particularly important are estimates of the age of the universe (dependent on the Hubble constant and the density parameter), as compared to the age of objects in the universe; this is the one area where the standard models are presently vulnerable to being shown to be inconsistent, hence the vital need to establish reliable distance scales, basic to estimates of both H_0 and the ages of stars.

Two further points here are significant from the viewpoint of epistemology. Firstly, because of our lack of adequate theories for the objects we observe[21], there are a variety of conflicting estimates for these cosmological parameters, based on different lines of argument; particularly, many of the methods of estimating Ω_0 depend on studying the growth and nature of inhomogeneities in the universe; this makes them rather model-dependent, and introduces a further set of parameters (describing the statistical properties of the matter distribution) to be determined by observation. To obtain believable answers one has to use informed judgment to decide which methods are more reliable, and give them more weight.

Secondly, determination of the values of cosmological parameters is an issue that must be ultimately decided by observations. One has to specifically state this because there has been a tendency by some to claim that the inflationary models make such a definite prediction that the density parameter Ω_0 must be unity, that observational data is irrelevant[22]. More recently a variety of inflationary models have arisen that do not predict $\Omega_0 = 1$ (the initial ones, e.g. [35], having being ignored). This may be connected with a growing perception that after all this prediction is not true. The viewpoint of this article is that rather one should insist on a methodology that respects the basic canons of science.

5. THE DIFFERENT APPROACHES TO MODELING

The underlying tension here is that between theory and observation. In essence, three approaches[23] to relating models to observations have been used [36].

The standard approach, implicitly assumed above, is to (1) choose a family of spacetime geometries and use them to obtain universe models dependent on a few parameters; (2) determine observational relations in such universes as a function of these parameters; and then (3) determine the best-fit values for the parameters by fitting these theoretical curves to astronomical observations. This approach is essentially theory based, for it allows one to use physically based models, such as an inflationary universe, to make observational predictions; when these are found to be true, theorists are justly satisfied. The model chosen in most cases is a FL model based on the RW geometry; the more adventurous use Bianchi spatially homogeneous models, or Tolman-Bondi spherically symmetric models. The approach is popular because of its high explanatory power - geometric and physical features are related in a satisfactory way, for example in the case of nucleosynthesis (where the expansion timescale determined by the spacetime geometry together with physical reaction rates lead to good predictions of light element abundances).

The problem is that there are many possibilities; there is no guarantee we have chosen the best model to describe the real universe. The retort that the choice made is justified because we get a good fit to the observations can be regarded with a bit of skepticism, because this is only true once one has added extra freedom to the model (if we use FL models we have to introduce source evolution functions to make number counts consistent - which will then allow any spherical geometry to fit the observations [14]).

The inverse approach works differently. Here we make no a priori assumptions about the matter distribution and space-time geometry, but rather try to determine them directly from observations [37, 38] on the basis of assumed knowledge about properties of the distant sources we observe - supernovae, for example. This approach is observationally based rather than theory based[24]. It has no explanatory power, and for that reason is not popular. However it has high descriptive power, and without it we would never have discovered features such as structure in the large-scale distribution of matter - superclusters, voids, walls. Using the standard approach, we can only discover what is already built into our models.

Thus both approaches have elements we need. The third approach, not yet fully developed, combines features of the other two. It is based on an optimal fitting procedure for a chosen model, which aims not just to determine the parameters of the model, but to obtain a detailed fitting of the model to the real universe, enabling a pointwise characterization of the deviation of the universe from a FR geometry [40]. This then allows one to quantify goodness of fit of the model, and hence criteria that a model be acceptable as a good description of the real universe. This process underlies approaches where density inhomogeneities are mapped in detail from large scale velocity flows[25], and can be the basis for a series of successive approximations to the real universe, based on stepwise refinement of an initial idealized model.

Reflection on these different approaches to modeling in cosmology may assist in developing the best way to integrate explanatory theory with detailed observational data; the first two approaches have tended to occur rather independently of each other as almost independent strands. Something like the third way may be optimal.

6. THE UNIVERSE AND PHYSICS

On the standard view, local physical behaviour (with given initial conditions) determines the geometry of the universe, which then in turn serves as the background for local physics. Operating in this context, physical laws lead inter alia to nucleosynthesis, creation of structure, and the existence of life.

The further feature mentioned earlier is that it is possible the cosmos influences the nature of local physical laws - for example leading to the arrow of time[26] or a time variation in physical constants[27]. This link should enable us to determine features of the cosmos by carefully examining features of local physics: for example perhaps deducing the expansion of the universe from the fact that the night sky is dark [28, 42]. A recent argument of this kind is Penrose's suggestion that the existence of the arrow of time undermines standard inflationary universe models [43, 44]. However as has already been discussed, such proposals are intrinsically untestable, and so are unlikely to gain consensus.

7. PHYSICS AND THE UNIVERSE

The underlying program of the standard approach is to use only known local physics, pushed as far as far as possible, to explain the structure of the Universe, giving a solely physical explanation of what we see. The relevant local physics is General Relativity (the classical theory of gravity) plus a suitable matter description, possibly including some approach to quantum cosmology at very early times. Two problems arise here[28].

The first is our inability to test the physics that applies in the early universe. The highest energies we can attain in accelerators on Earth cannot reach those relevant to the very early universe, hence our understanding of physics at that time has to be based on extrapolation of known physics way beyond the circumstances in which it can be tested. The result is we cannot be confident of the validity of the physics we use, and this becomes particularly so in the presumed quantum gravity era. We end up rather testing theoretical proposals for this physics by exploring their implications in the early universe (which is the only 'laboratory' where we can test some of our ideas regarding fundamental physics). The problem is we cannot simultaneously do this and also carry out the aim of the program stated above: if we don't know the relevant physics, we can't use it to predict anything. Guessing this physics and then confirming our guesses only by their implications for the early universe gives support to a particular proposal for the physics only insofar as no other proposal can give similar cosmological outcomes. A particular example is the inflationary universe proposal: the supposed inflaton field underlying an inflationary era of rapid expansion in the early universe has not even been identified, much less shown to exist by any laboratory experiment, or demonstrated to have the properties required in order that inflation took place as proposed. The hypothesis that no inflation took place is as viable (although not as satisfying from an explanatory viewpoint).

Second, this verification problem occurs a fortiori in considering the creation of the universe itself, and the associated problem of what determines initial conditions for the universe. No physical experiment at all can help here because of the uniqueness of the universe, and the feature that no spacetime exists before such a beginning; so brave attempts to define a 'physics of creation' stretch the meaning of 'physics'[29]. The Hartle-Hawking 'no-boundary' proposal [45] gets round the issue of a time of creation in an ingenious way, but cannot get around the basic problem: a purely scientific approach (as usually understood) cannot succeed in explaining why the universe has one specific form rather than another, when other forms seem perfectly possible. A choice between different contingent possibilities has somehow occurred; but no experimental test can determine the nature of any mechanisms that may be in operation in the relevant circumstances, when even the concepts of cause and effect are suspect[30]. Unavoidably, whatever approach one may take, metaphysical issues inevitably arise.

8. METAPHYSICS

A series of profound questions lie at the base of cosmology, whose nature is metaphysical[31] rather than physical: their status is philosophical rather than scientific, for they are issues that cannot be resolved purely scientifically. These include the more profound forms of the Anthropic question:

- why does the universe allow the existence of life?

[46], because they rest on the basic cosmological questions of this kind:

- why does the universe exist ?
- why do the laws of physics exist ?
- why do they have the form they do ?
- why do boundary conditions have the form they do ?

At this point the issue becomes, what is scope of cosmology? This is a choice one has to make.

These further questions need further assumptions if answers are to be given; standard cosmology cannot answer them without supplement. One option is to decide to treat cosmology as far as possible in a strictly scientific way; one ends up with a solid technical subject that by definition excludes such fundamental philosophical questions[32], because they cannot be solved scientifically. This is a consistent and logically viable option. One should note here that in any case there will be philosophical assumptions underlying the practice of cosmology even if carried out as a purely technical exercise, but those assumptions will not enter this kind of arena.

The second option is to decide that these kinds of philosophical questions are of such interest and importance that one will tackle them, even if that leads one outside the strictly scientific arena. This is also a legitimate exercise, investigating the various options available here, provided one follows three basic guidelines. First, one must avoid the claim that scientific methods can resolve these questions: it is essential to respect the limits of what the scientific method can achieve[33], and acknowledge clearly when arguments and conclusions are based on some metaphysical philosophical stance rather than purely on scientific argument. If we acknowledge this and make that stance explicit, then the bases for different viewpoints are clear, and alternatives can be argued rationally.

Second, in undertaking this task, one must be aware of the limitations of the models of reality we use as our basis for understanding. They are necessarily partial and incomplete reflections of the true nature of reality, helpful in many ways but also inevitably misleading in others. No model (literary, intuitive, or scientific) can give a perfect reflection of reality; so they must not be confused with reality[34].

Finally, if one wants to seriously tackle issues in the relation of cosmology to humanity, one must include in one's analysis data of a broad enough scope to reflect fully the nature of human beings.

As well as taking into account that we are complex structures based on the physics and chemistry of organic molecules who have evolved by natural processes in the context of the expanding universe, such attempts must acknowledge our truly human attributes and experience - consciousness and emotion, love and pain, free will and ethical choice [49, 50]. Only if we add to the cosmological data considered above the much broader range of data of this kind can we hope to obtain a world view of adequate scope to be a worthy theory of humanity and cosmology - that is, of Cosmology in the broad sense that relates fully to philosophy and the humanities as well as to science[35].

This can be undertaken as a perfectly rational project; it is a question of choice as to whether one wants to embark on a study of this broader scope, or to restrict one's consideration to the physical aspects of cosmology. Confusion will be avoided if one makes quite clear at the outset what is the scope of the theory one wishes to consider.

Acknowledgments

I thank John Norton for helpful comments. This article is a modified version of a paper that appeared in *La Recherche* [51].

Notes

1. ' epistemology' = 'the philosophical theory of knowledge which seeks to define it, distinguish its principle varieties, identify its sources, and establish its limits' [1].

2. The situation in quantum cosmology is not included in the scope of discussion in this paragraph.

3. Here one must try to make sense of notions of possibility and necessity, and to distinguish between what might happen in some possible worlds and what must happen in all possible worlds, as has been investigated in David Lewis' theory of counterfactuals. I am indebted to John Norton for emphasizing this.

4. One should realize here that we can in principle feel the gravitational effect of matter beyond the horizon; however we cannot uniquely decode that signal to determine what matter distribution caused it, see [6].

5. The furthest matter with which we can have had any causal connection, see [8] (Despite its name, this paper actually deals with causal horizons).

6. Hence on light rays lying in our past light cone.

7. More accurately, near our past world-line in spacetime.

8. Particularly important are measurements of the CBR temperature at high redshift, confirming the standard interpretation of this radiation, see [11]

9. All physical and geometrical quantities are the same at each point of space (i.e. at a constant time).

10. All observations are the same in all directions about the point of observation.

11. Provided the matter content is a perfect fluid, as usually assumed.

12. An alternative interpretation would be that this is evidence of spatial inhomogeneity.

13. Another proposal is to use the uniformity in the nature of the objects we see to deduce they must have all undergone essentially the same thermal history, and then to prove from this uniformity of thermal histories that the universe must be spatially homogeneous; however this effort has not succeeded so far, see [18]. Nevertheless observations of element abundances at high z are very useful in constraining inhomogeneity.

14. Weak tests of the isotropy of the CBR at other spacetime points come from the Sunyaev-Zeldovich effect [20], and from CMB polarization measures [21], but not enough to give good limits on spatial inhomogeneity through this line of argument.

15. And conceivably a contribution to the CBR dipole from this inhomogeneity (if we are a bit off-centre).

16. They may be able to do so, because the source of the expected 'Doppler Peaks' in the CBR spectrum is pressure- generated waves in the matter-radiation mixture before decoupling, rather than any specific feature of the RW geometry.

17. See also [27] for arguments on large-scale inhomogeneity.

18. Formalised as the Cosmological Principle, see for example [28, 29].

19. The following analogy is used there: consider an ant surveying the world from the top of a sand dune in the Sahara desert. Her world model will be a universe composed only of sand dunes - despite the existence of cities, oceans, forests, tundra, mountains, etc. beyond her horizon.

20. For overviews of current estimates, see [31], [32], [33].

21. Cepheids, supernovae, galaxies, galaxy clusters for example.

22. For example, after I wrote a paper with Peter Coles summarizing the data and deriving a value for Ω_0 of between 0.2 and 0.3 [34], Coles was told by a senior cosmologist that he should not question the theoretically preferred value of unity; if he continued to do so, he would be considered a crank.

23. Associated with different criteria for satisfactoriness of a cosmological model, see Section 1.3 of Coles and Ellis [32].

24. The work of Edwin Hubble is a classic example. His 200 page book *The Realm of the Nebulae* [39] relegates theory to 4 pages at the end; rather than using the FL models, he simply fitted curves to the data.

25. Definition of both the flows and the inhomogeneities are based on such a fitting procedure, which defines a specific perturbation gauge.

26. The fundamental physical laws by themselves being time symmetric, and so unable to explain this feature.

27. This is to some degree open to observational test, see e.g. [41].

28. Apart from the averaging issue: some averaging of descriptions on different scales is involved here, because the scale on which General relativity is tested (the solar system scale) is quite different than the cosmological scale on which we apply it; but the dynamical equations do not commute with this averaging process. Extra 'polarization' type terms result in the field equations, which may be significant under some circumstances, but handling these adequately in the context of curved spacetimes is difficult.

29. These usually rely on an array of properties of quantum field theory and of fields that seem to hold sway independent of the existence of the universe and of space and time (for the universe itself is to arise out of their validity). The locus of their existence or other grounds for their validity in this context are unclear.

30. As are the concepts of 'occurred', 'circumstances' and even 'when' - for we are talking inter alia about the existence of spacetime. Our language can hardly deal with this.

31. i.e. beyond or behind physics.

32. They are sometimes labeled as meaningless; but this is true only if one chooses to restrict one's method of investigation to the purely scientific.

33. An example where this is not the case is Frank Tipler's book *The Physics of Immortality: Modern Cosmology, Physics, and the Resurrection of the Dead* [47]; see also [48].

34. An example where such confusion takes place is Tipler's book.

35. If we propose a 'thin' theory that does not reflect human experience adequately, the broader public and our academic colleagues in other disciplines will rightly dismiss it as simplistic and inadequate as a full view of the nature of the universe. The full range of human experience is indeed evidence about the universe both because we exist in the universe, and because we have arisen from it.

References

[1] A Bullock, O Stallybrass, and S Trombley (Eds). *The Fontana Dictionary of Modern Thought*, (Fontana, 1988).

[2] G F R Ellis. "Major Themes in the relation between Philosophy and Cosmology". *Mem Ital Ast Soc* **62**, 553-605 (1991).

[3] W R Stoeger (Ed). *Theory and Observational Limits in Cosmology.* (Vatican Observatory, Castel Gandolfo, 1987).

[4] G F R Ellis. "Cosmology and Verifiability". *Qu Journ Roy Ast Soc* **16**, 245-264 (1975).

[5] G F R Ellis and W R Stoeger. "Horizons in inflationary universes". *Class Qu Grav* **5**, 207 (1988).

[6] G F R Ellis and D W Sciama. "Global and non-global problems in cosmology", in *General Relativity*, ed. L. O'Raifeartaigh (Oxford University Press, 1972), 35-59.

[7] G F R Ellis and G Schreiber. "Observational and dynamic properties of small universes". *Phys Lett* **A115**, 97-107 (1986).

[8] W Rindler. "Visual horizons in world models". *Mon Not Roy Ast Soc* **116**, 662 (1956).

[9] G F R Ellis. "Observations and cosmological models". In *Galaxies and the Young Universe*, Ed. H Hippelein, K Meisenheimer and H-J Roser (Springer,1995), 51-65.

[10] P J E Peebles D N Schramm E L Turner and R G Kron. "The case for the Relativistic hot big bang cosmology". *Nature* **352**, 769-776 (1991).

[11] D M Meyer. "A distant space thermometer". *Nature*: **371**, 13 (1994).

[12] G F R Ellis. "Relativistic cosmology: its nature, aims and problems". In *General Relativity and Gravitation*, Ed B Bertotti et al (Reidel, 1984), 215-288.

[13] W R Stoeger, G F R Ellis, and C Hellaby. "The relationship between continuum homogeneity and statistical homogeneity in cosmology". *Mon Not Roy Ast Soc* **226**: 373 (1987).

[14] N Mustapha, C Hellaby, G F R Ellis. "Large scale inhomogeneity versus source evolution: can we distinguish them observationally?" *Mon Not Roy Ast Soc* **292**: 817-830 (1998).

[15] G F R Ellis. "Limits to verification in cosmology". *Ann New York Acad Sci* **336**: 130-160 (1980).

[16] M H Partovi and B Mashhoon. "Toward verification of large-scale homogeneity in cosmology". *Astrophys Journ* **276**: 4 (1984).

[17] J Ehlers, P Geren and R K Sachs. "Isotropic solutions of the Einstein-Liouville equations". *J Math Phys* **9**, 1344- 1349 (1968).

[18] W B Bonnor and G F R Ellis. "Observational homogeneity of the universe". *Mon Not Roy Ast Soc* **218**, 605-614 (1986).

[19] W Stoeger, R Maartens and G F R Ellis. "Proving almost- homogeneity of the universe: an almost-Ehlers, Geren and Sachs theorem". *Astrophys Journ* **443**, 1-5 (1995).

[20] J Goodman. "Geocentrism reexamined". *Phys Rev* **D52**:1821 (1995).

[21] M Kamionkowski and A Loeb. "Getting around cosmic variance", to appear, *Phys Rev* **D**, (1997).

[22] G F R Ellis, R Maartens, and S D Nel. "The expansion of the universe". *Mon Not Roy Ast Soc* **184**, 439-465 (1978).

[23] G F R Ellis. "Alternatives to the Big Bang". *Ann Rev Astron Astrophys* **22**, 157-184 (1984).

[24] J Wainwright and G F R Ellis (Eds). *The dynamical systems approach to cosmology.* (Cambridge University Press, 1996).

[25] J Wainright, G F R Ellis and M Hancock. "On the Isotropy of the Universe: Do Bianchi VIIh universes isotropize?" *Class Qu Grav* **15** 331-350 (1998).

[26] A D Linde. *Particle Physics and Inflationary Cosmology.* (Harwood Academic, 1990).

[27] G F R Ellis. "The homogeneity of the universe". *Gen Rel Grav* **11**: 281-289 (1979).

[28] H Bondi. *Cosmology* (Cambridge University Press,1960).

[29] S W Weinberg. *Gravitation and Cosmology* (Wiley, 1972).

[30] G F R Ellis. "Cosmology and Verifiability". *Qu Journ Roy Ast Soc* **16**, 245-264 (1975).

[31] Particle Review Group. *Reviews of Modern Physics* **68**: 708-722 (1996).

[32] P Coles and G F R Ellis. *Is the Universe Open or Closed: The Density of Matter in the Universe* (Cambridge University Press, 1997).

[33] G Boerner and S Gottlober (Eds). *The Evolution of the Universe*, Dahlem Workshop Report (Wiley, 1997).

[34] P Coles and G F R Ellis. "The case for an open universe". *Nature* **370**, 609-615 (1994).

[35] G F R Ellis, D H Lyth, and M B Mijic. "Inflationary Models with Ω not equal to 1". *Phys Lett* B271, 52 (1991).

[36] D R Matravers, G F R Ellis, and W R Stoeger. "Complementary approaches to cosmology: Relating theory and observations". *Qu J Roy Ast Soc* **36**, 29-45 (1995).

[37] J R Kristian and R K Sachs. "Observations in cosmology". *Astrophys Journ* **143**: 379-399 (1966).

[38] G F R Ellis, S D Nel, W Stoeger, R Maartens, and A P Whitman. ' 'Ideal Observational Cosmology". *Phys Reports* **124**: 315-417 (1985).

[39] E Hubble. *The Realm of the Nebulae.* (Yale University Press, 1936, 1982).

[40] G F R Ellis and W R Stoeger. "The Fitting Problem in Cosmology". *Class Qu Grav* **4**: 1679-1690 (1987).

[41] L Cowie and A Songaila. "Astrophysical limits on the evolution of dimensionless physical constants over cosmological time", *Astrophys Journ* **453**: 596 (1995).

[42] E R Harrison. *Cosmology: The Science of the Universe.* (Cambridge University Press, 1981).

[43] R Penrose. "Difficulties with inflationary cosmology". Proc 14th Texas Symposium on Relativistic Astrophysics (Ed. E Fenves), *Ann New York Academy of Science* (1989).

[44] R Penrose. *The Emperor's New Mind.* (Oxford University Press, 1989), Chapter 7.

[45] S W Hawking. *Hawking on the Big Bang and Black Holes.* (World Scientific, 1993).

[46] J Barrow and F J Tipler. *The Anthropic Cosmological Principle.* (Oxford University Press, 1986).

[47] F J Tipler. *The Physics of Immortality: Modern Cosmology, Physics, and the Resurrection of the Dead.* (MacMillan, 1995).

[48] P W Atkins. "The limitless power of science" . In *Nature's Imagination: The Frontiers of Scientific Vision.* Ed J Cornwell (Oxford University Press, 1995).

[49] G F R Ellis. *Before the Beginning.* (Bowerdean Press/Marion Boyars, 1993).

[50] G F R Ellis. "Modern Cosmology and the Limits of Science". *Trans Roy Soc S Africa*, **50**:1-25 (1995).

[51] G F R Ellis. "Les Limites De L'Enterprise Cosmologique". *La Recherche* (April 1998), 114-120.

Chapter 13

MATHEMATICS AND SCIENCE

Fred Hoyle
102, Admiral's Walk
West Cliff Road, West Cliff
Bournemouth, Dosset BH2 5HF United Kingdom

The simplest Friedmann models, about which astronomers were still arguing in the late nineteen fifties and sixties are dominated by the equation

$$\frac{\dot{S}^2 + kc^2}{S^2} = \frac{A}{S^3} \tag{13.1}$$

in which S is the scale factor of the universe, a function of the time, A is a positive constant and k is a topological factor which can be 0 or ± 1. No zero of $\dot{S}$ exists for $S < S_0$ where S_0 is the present day value of S, and this result is not affected by adding positive terms to the right hand side of (1), as for instance a term B/S^4 due to relativistically-moving particles, e.g. photons or neutrinos. Thus if the simple Friedmann models were correct, a spaceti me singularity, $S \longrightarrow 0$, would inevitably occur in a time-reversed form of the models. This was the Big-Bang, which has been well-known to astronomers since Hubble and Humason had discovered the expansion of the universe in about 1930.

However, it was implicit in equation (1) that the universe is homogeneous and isotropic. Otherwise the spacetime structure cannot be described by the single function $S(t)$. Several functions become involved and they depend on several of the coordinate s of spacetime. When Jayant Narlikar began research in 1960, doubts were felt among astronomers that the universe could have originated in a singularity and ways to avoid the implications of (1) were being sought. In particular by E.M. Lifshitz in the Sov iet Union and by O. Heckmann in Germany. Both relied on complicating the spacetime structure, Heckmann through rotation and Lifshitz by inhomogeneity in a very difficult paper of about fifty pages. Such matters were extensively discussed at international c onferences in those days.

N. Dadhich and A. Kembhavi (eds.), The Universe, 141–144.

This was not a problem as I would not normally have asked a research student to tackle - it was much too difficult. But Jayant had distinguished himself by an outstanding performance in the Cambridge Mathematical Tripos, and then he was in the summer o f 1960 asking for a problem in relativity and cosmology, so I suggested that he looked into these claims by Heckmann and Lifshitz. It seemed reasonable to simplify things a bit. In Britian at the time there had been a strong emphasis, first from Milne, th en from McCrea and then Bondi, on the fact that equation same as (1) can be obtained by Newtonian methods as well as from general relativity. The likelihood, therefore, was that the same would apply more generally, when isotropy and homogeneity were aban doned. And subject to the resulting simplification Jayant was able to show that the claims which had been advanced with great confidence in Germany and the Soviet Union were wrong. Deviations from isotropy and homogeneity made no difference to the conclus ion that in a time-reversed model the Universe plunges into a singularity, and in essentially the same time interval, about 10^{10} years, as in the simple Friedmann models. This I think was Jayant's first paper, published in the Monthly Notices of the R.A.S.

Some years later, the same result was obtained from general relativity, by Penrose and Hawking, confirming our expectation that Newtonian cosmology was just as reliable in this more complicated case as it had been in the simple models. It was found nec essary by Penrose and Hawking to assume that the tt-component of the energy momentum tensor is not negative, which prevents negative terms from appearing on the right-hand side of (1). The form of the steady-state theory in use in the 1960's had such a ne gative term, of the form

$$\frac{\dot{S}^2 + kc^2}{S^2} = \frac{A}{S^3} - \frac{B}{S^6}, \tag{13.2}$$

where A and B are both positive constants. Now a time-reversed solution leads to $\dot{S} = 0$ and contraction of the Universe switches to expansion. Indeed in the case of $k = 1$ there can be two values of S at which $\dot{S} = 0$, with the Universe oscillating between them. What makes for such a negative term in (2) is that creation of matter occurs with *conservation of energy*, a negative term being required to balance the positive energy of matter.

I had not intended to involve Jayant with anything as controversial as the steady-state theory, but in 1961 Martin Ryle announced the first result of the 4C survey, in which he claimed that the number of radio sources continued to increase at a super-Euclidean rate down to significantly lower flux values than the 9Jy of the 3C survey. This as it has turned out was on exaggeration of a situation in which a super-Euclidean count occurs over a much more restricted flux range and is appreciably less in amount than those first claims. The super-Euclidean behaviour arises because the observer lies in a partial void with respect to radio sources, a void with a radius of about $0.1cH_0^{-1}$. In 1961-62 we published two papers on this problem, in the seco nd of which we adopted this inhomogeneity solution. But the thought that the universe might be irregular on such a scale did not recommend itself to astronomers of the day, although it turned out eventually to be so.

During the source-counting episode we made a trip to Jodrell Bank to see Robert Hanbury Brown, who had expressed doubts about the uniformity of the radio sources that were being counted in the Cambridge surverys. It was during this visit that we were s hown Henry Palmer's first results from a wide interferometric survey of sources then being carried out between Jodrell Bank and a mobile field station. About a dozen sources were of too small an angular diameter to have been resolved at that time. As we w ere shown the list, we did not dream that here wan an astonishing new class of objects, later to become known as quasi-stellar objects (QSOs).

The history of the discovery of QSO's in early 1963 is complicated. The decisive step was the identification of the Balmer series at a redshift of 0.158 in the spectrum of the source 3C 273, $H\beta, \gamma$ and δ being detected photographically by Maarten Schmidt and H_α detected in the infrared by J.B. Oke. Since the visual magnitude of 3C 273 was about +13 this meant that according to the usual Hubble relation 3C 273 was intrinsically brighter than major galaxies by a large factor (~ 100), suggesting that all of cosmology would soon be revolutionised by the adoption of QSO's as standard candles instead of galaxies. But cosmologists were due for a surprise. For when by 1966 some 50 QSO's had been detected and these redshifts and magnitudes had been measured, it was found that they did not fit a Hubble diagram. Those who continued to believe that QSO redshifts arose like the redshifts of galaxies from the expansion of the universe, probably hoped that things would eventually settle down into a Hubble diagram as soon as the number of QSO's increased sufficiently. But this has not happened, even with more than 10,000 QSO's now available. What has indeed emerged is a relation of magnitude to $(1 + Z)^{-2}$. This simply reflects the fact that objects lose energy from the redshift and from the counting effect of the redshifts. In an expanding universe the appeared luminosity l of an object of intrinsic luminosity L situated at a radial coordinate r with a redshift Z is given by

$$l = \frac{L}{4\pi r^2(1+Z)^2}. \tag{13.3}$$

Galaxies satisfy this relation, but QSO's do not show the effect of the radial coordinate r, although they do show the $(1+Z)^{-2}$ factor.

And in the late 1960's too many cases of QSO's near bright galaxies were found for their jaxtapositions to be chance projections on the sky. But all such evidence was ignored by most of the astronomical community. It seemed more and more determined, the stronger the evidence becomes, to force the world to conform with their own wishes.

An investigation carried out in 1989 by Jayant in collaboration with Geoffrey Burbidge, A. Hewitt and P.DasGupta seemed to me to settle the matter beyond all doubt. A computed search of a catalogue of about 7000 QSO's and of the Revised New Cat alogue of Non-Stellar Astronomical Objects was made with a view to determining all cases in which a QSO and a galaxy lay within an angular separation $\theta \leq 600''$ of one other. About 400 cases were found, in most of which the galaxy in question had a measured redshift value, which was considered to give the distance d of the galaxy. Then θ was plotted against d, with the result $\theta d \simeq$ constant. Since this relation was maintained over a range of about 500 in d, the evident implication was that the QSO was in physical association with the galaxies. Therefore the redshifts of the QSO's, which were mostly much larger than those of the galaxies, must come mainly from a source other than the expansion of the universe. What this so urce may be remains a problem, but the circumstance that we may not understand a phenomenon is no reason to ignore it. Science would have made little, if any, progress if this had been the dominant attitude throughout history. It is covered today by too much money, especially in the United States, by too many people, and by a weakness in the training of scientists that was first pointed out to me by Jayant Narlikar.

Whereas in mathematics one is constantly seeking to learn new tricks for solving problems, in laboratory work in sciences the student only obtains known answers to known situations. The one produces a willingness to consider new possibilities, while th e other leads to a rigidity that takes for granted more than it should. The mathematician likes to work with a clean sheet, whereas the scientist likes to work along fixed tracks.

Chapter 14

RADIATION REACTION IN ELECTRODYNAMICS AND GENERAL RELATIVITY

Bala R. Iyer
Raman Research Institute,
Bangalore 560 080, India.

1. PROLOGUE

It is a privilege and pleasure to be invited to contribute an article to the JVN Fest. When I received this invitation, I tried to go back along my world-line and look for intersections with Jayant. A popular article by Jayant Narlikar entitled 'The Arrow of Time' [1] mystified and fascinated me. It roused an almost romantic longing and an urge to appreciate, if not investigate, such basic problems. Probably it was these subconscious fantasies that propelled me towards physics and eventually, general relativity. I still remember the first time I heard a public talk by Narlikar on Cosmology after his return to India. It was at the Homi Bhabha auditorium of TIFR in 1972. The hall was overflowing and I heard his (favorite?) joke on the mathematician, physicist and astronomer for the first time. I heard it again this year in his talk at the Academy and was impressed by his un-apologetic use of it to make his point! I met Jayant Narlikar at the Einstein centenary symposium in Ahmedabad in 1979 and his interests then included scale invariant cosmology (with Ajit Kembhavi) and black holes as tachyon detectors (with Sanjeev Dhurandhar). He carried his fame lightly, was unassuming and though he was not very talkative, he felt very approachable. When I finished my Ph.D. with Arvind Kumar at the Bombay University, I could not get a post doc at TIFR or work with Jayant, since he was away that particular year. Over the last sixteen years, I have had much overlap with Jayant in the organization of Relativity related activities in India. There is much to admire in Jayant and emulate. His time management, missionary zeal to the popularization of science, vision and hard work, pedagogic skills, fervor for the non-standard and ability to play devil's

N. Dadhich and A. Kembhavi (eds.) The Universe, 145-160.

advocate in his research almost as a point of faith. In addition to the above, personally, I also admire him for his ability to take criticism and his democratic mode of functioning.

I am always impressed by Jayant's ability to start a lecture on fairly profound, subtle and technical themes like action at a distance in physics and cosmology or Mach's principle from a very elementary basic discussion. In every lecture of his that I have heard he covers a fair amount of ground starting from the very beginning and leading to what he is currently researching on. He reminds me of a capable, composed and competent guide taking a group of motley tourists up a mountain, leading everyone to the heights their capability can reach. Everyone gets a view, maybe a different glimpse, but everyone is happy to have participated in the trek and adventure that Jayant leads them on. No wonder he is a populariser par excellence and probably holds a record for such lectures and writing at least in India.

I have heard that Jayant has a soft corner for his work related to electrodynamics and action-at-a-distance [2]; he considers it to be one of the important topics in his research career. Recent progress in theoretical gravitational radiation research is very reminiscent of this research and as a tribute to Jayant, I shall try to imitate him and without getting lost in technical details compare these developments in general relativity to those in electrodynamics.

2. GRAVITATION AND ELECTROMAGNETISM

The similarity of gravitation and electromagnetism does not escape any thoughtful student of an elementary physics course [3]. Both Newton's law of gravitation and Coulomb's law of electrostatics are inverse square laws. They are proportional to their respective charges: gravitational mass and electric charge. The gravitational charge is of only one kind, while there are two kinds of electric charges, conventionally denoted as positive and negative. In electrostatics, like charges repel, while unlike charges attract. Gravitation on the other hand is always attractive and in gravitation, like charges attract! Though functionally similar, the numerical strengths of these forces is very different. The gravitational force is about 10^{39} times weaker than the electrical force and this has experimental implications, as we shall see later. Unlike electromagnetic forces, gravitation cannot be screened out. Moreover, matter in the universe is predominantly neutral. This is why, in spite of its enormous weakness, gravitation determines the large scale structure of the universe.

Both Newton's law of gravitation and Coulomb's law of electrostatics assume instantaneous action-at-a-distance. Thus they cannot be consistent with the principle of special relativity. Coulomb's law is not adequate to describe moving charges. Electromagnetic phenomena are more simply described by field equations and a moving charge produces both an electric field and a magnetic field. The laws of electromagnetism are summarized by Maxwell's equations and Lorentz equation of motion. These equations are relativistically invariant. However, in Newtonian gravitation, there is no analogue of the magnetic field; a moving mass produces the same field, as a mass at rest, if the mass distributions are identical. The situation is different in Einstein's general theory of relativity and closer to electromagnetism. Here the gravitational field produced by a body depends not only on the distribution of matter but also the state of its motion. Mathematically, the source of the gravitational field is the energy momentum tensor whose components include mass, motion and stresses. The gravitational analogue of the magnetic force is called gravimagnetism and like the Lorentz force in electrodynamics, depends on the test particle velocity. It has physical consequences like the dragging of inertial frames, Lense Thirring effect or precession of gyroscopes. Usual tests of general relativity normally involve only the gravielectric component. Like the magnetic force, the gravimagnetic component is usually smaller by a factor of v/c relative to the gravielectric part and experiments are under way to verify it directly. One can set up a detailed analogy between rotation in general relativity and magnetism. In electromagnetism, there has long been a conjecture about the possible existence of magnetic monopoles. Given the detailed similarity between rotation and magnetic fields, one can ask, if there is such a thing as the gravimagnetic monopole. The answer is in the affirmative. The famous NUT solution is the gravimagnetic monopole [4]. Of course, the Schwarzschild solution the gravielectric monopole.

3. ELECTROMAGNETIC WAVES AND GRAVITATIONAL WAVES

As mentioned earlier, the laws of electromagnetism are summarized by Maxwell's equations. Maxwell's equations admit wave like solutions and these are electromagnetic (EM) waves. EM waves are produced by accelerated electric charges. The dominant radiation is dipole radiation and is caused by the time varying dipole moment of the charge distribution. The EM field is of spin one (a vector field) and has a conserved quantity associated with it: charge. Consequently there is no monopole EM radiation. EM waves propagate at speed of light c, they are transverse and have two independent states of linear polarization corresponding to oscillations of the electric field in two perpendicular directions. The effect of an EM wave can be seen by its action on a test particle. If a sinusoidally varying EM wave is incident on a test particle, it impresses on it this sinusoidal motion. Thus, by studying the motion of a test particle, we can infer the passage of a EM wave. EM is a strong force. Consequently by the oscillation of charges and currents we can produce EM waves at one end of the laboratory and detect it at the other end: the famous Hertz experiment.

Similarly, the best description of gravitation is via Einstein's equations. These equations also admit wave like solutions. Gravitational waves are not mere artefacts of our choice of coordinates, but indeed physical, in that they carry energy. For a fascinating historical account of these debates, see Kennefick [5]. Gravitational waves are produced by accelerated motions of masses. The dominant radiation is quadrupolar and caused by the second time variation of the quadrupole moment of the mass energy distribution. The gravitational field is of spin two (a tensor field) and has conserved quantities associated with it corresponding to mass, linear momentum and angular momentum. Consequently, there is no monopole or dipole radiation. Gravitational waves also propagate with speed c, are transverse and have two independent states of linear polarization. The effect of a gravitational wave *cannot* be seen by its action on a single test particle. Gravity obeys the equivalence principle and consequently a uniform gravitational field can be transformed away by going to an accelerated frame. Tidal fields cannot be so transformed and provide a true measure of gravitational fields. Gravitational waves induce a weak time-dependent tidal field and thus, a gravitational wave can be detected by letting it impinge on a circular ring of *particles.* Due to the tidal field, the ring is squeezed in one direction and elongated along the perpendicular direction. Since the tidal field oscillates in time, the ring will go through a pattern of shapes, characteristic of the tidal field. Starting out as a ring of particles, after a quarter of a period the ring elongates into a ellipse, say along the x axis, back to a circle, then an ellipse elongated along y axis and back again to a circular shape. This pattern repeats thereafter and is characteristic of spin two. This is referred to as plus polarization. The other independent mode of polarization yields an ellipse rotated by 45^0 and is referred to as the cross polarisation. Gravitational wave detectors differ in the way they measure this minute tidal effect. Broadly we can classify them as bars (spheres), interferometers on earth and interferometers in space.

Unlike EM, gravitation is a very weak force. Consequently, the oscillation of masses in the laboratory cannot produce gravitational waves of measurable strength. The detection by any suitable method is equally difficult for the same reason. A Hertz type experiment is not possible in this case and one is forced to appeal to astronomy, to provide sources that will radiate in this bandwidth.

4. INSPIRALING COMPACT BINARIES AND GW PHASING

The Binary pulsars 1913+16 and 1534+12 establish the reality of gravitational radiation [6]. They provide proof of the validity of Einstein's general relativity in the strong field regime. More importantly, they are prototypes of inspiralling compact binaries, which are strong sources of gravitational waves for ground based laser interferometric detectors like LIGO and VIRGO [7]. The phenomenal success of the high-precision radio wave observation of the binary pulsar makes crucial use of an accurate relativistic 'Pulsar timing formula' [8, 9]. Similarly, precise gravitational-wave observation of inspiraling compact binaries would require an equivalent accurate 'Phasing formula' [7, 10] i.e. an accurate mathematical model of the continuous evolution of the gravitational wave phase. The lowest order gravitational wave radiation reaction is sufficient to treat pulsar timing. Gravitational wave phasing, on the other hand, requires higher post-Newtonian order gravitational radiation reaction, since in the final stages the systems are highly relativistic.

At this point, it is worth comparing the situation here in general relativity (GR) to that in electrodynamics (ED) to illustrate the issues. For instance, in ED we have the following categories of problems: (a) Given the charge and current distribution, compute the electromagnetic field; e.g. evaluate fields in wave-guides. (b) Given the external electromagnetic field, compute the effect on charges and currents; e.g. energy losses of charged particles moving past a nucleus. (c) Given the energy loss by say the Larmor formula, compute the reaction on the motion; e.g. Abraham-Lorentz, Planck. The corresponding situation in GR, in the inspiraling binary problem, is the following: (i) *Generation Problem:* Given the compact binary and its orbital motion, compute the gravitational field in this situation. (ii) Given the gravitational field, compute the far-zone energy and angular momentum fluxes. (iii) *Radiation Reaction problem:* Given the far zone fluxes of energy and angular momentum, compute the reaction on the near zone motion, assuming energy (angular momentum) balance. Or compute it directly, by a higher iteration of the equations of motion.

In what follows, we will discuss briefly aspects of motion, generation and radiation reaction and draw parallels to the EM case, where possible.

5. MOTION

It may be worth mentioning that unlike linear EM, non-linear GR has the feature, that its field equations contain the equations of motion. For discussions on the relation between the above feature, non-linearity and tensor nature of the field, see the review article by Havas [11]. The N-body problem as in Newtonian gravity is decomposed into an external problem and an internal problem. The former refers to the problem of defining and determining the motion of the center of mass and the latter to motion of each body around the center of mass. The effacement of internal structure in the external problem and effacement of external structure on the internal problem involves subtle issues in the problem of motion and we cannot do better than refer the reader to the beautiful review by Damour [12].

The topic of EOM for compact binary systems received careful scrutiny in the years following the discovery of the binary pulsar. There have been three different approaches to the complete kinematical description of a two body system upto the level where radiation damping first occurs (2.5PN). Damour's method explicitly discusses the external motion of two condensed bodies without ambiguities, using harmonic coordinates, in which all metric deviation components satisfy hyperbolic (wave) equations. The method employs the best techniques to treat various subproblems. (a) A Post-Minkowskian approximation to obtain the gravitational field outside the bodies incorporating a natural 'no incoming-radiation condition' whose validity is not restricted to only the near-zone. (b) A matched asymptotic expansion scheme to prove effacement and uniquely determine the gravitational field exterior to the condensed bodies. (c) An Einstein Infeld Hoffmann Kerr type approach to compute equations of orbital motion from knowledge of the external field only. The nth approximate EOM is obtained from the integrability condition on the $(n+1)^{th}$ approximated vacuum field equations. (d) Use of Riesz's analytic continuation technique to evaluate surface integrals. The final EOM at 2.5PN level are expressed only in terms of instantaneous positions, velocities and spins in a given harmonic coordinate system and given explicitly in Ref.[12]. The two mass parameters in these formulas are the Schwarzschild masses of the two condensed bodies.

The conservative part of the EOM upto 2PN (excluding the secular 2.5PN terms) are not deducible from an conventional Lagrangian (function of positions and velocities) in harmonic coordinates, but only from a generalized Lagrangian (depending on accelerations). This is consistent with the result in classical field theory that in Lorentz-covariant field theories there exists no (ordinary) Lagrangian description at $O(c^{-4})$ [13]. This Lagrangian is invariant under the Poincare group and thus allows one to construct ten Noetherian quantities that would be conserved during the motion. These include the 'Energy', 'Angular Momentum', 'Center of Mass' and thus a solution to the problem of 'motion' provides the Energy that enters into the phasing formula. The EOM for the general case is given in [12] and crucially used in the following studies of generation and radiation reaction. All the above has detailed parallels in the electromagnetic case and the relevant Lagrangian and associated subtleties are discussed in the Les Houches lecture by Damour [9].

Schafer's [14] approach, on the other hand, is based on the Hamiltonian approach to the interaction of spinless point particles with the gravitational wave field. The Hamiltonian formulation is best done in the Arnowitt-Deser-Misner (ADM) coordinates, in which two metric coefficients satisfy hyperbolic equations (evolution) while the remaining eight are of elliptic type (constraints). It uses a different gauge that allows an elegant separation of conservative and damping effects. One recovers the damping force acting on the Hamiltonian subsystem of instantaneously interacting particles coming from its interaction with the dynamical degrees of freedom of the gravitational field. In this approach, point masses are used as sources and regularisation uses Hadamard's 'partie finie' based on Laurent's series expansion regularisation.

The last approach due to Grischuk and Kopejkin [15] on the other hand is based on (a) Post-Newtonian approximation scheme (b) assumption that bodies are non-rotating 'spherically-symmetric' fluid balls. The symmetry is in the coordinate sense. The EOM of the center of mass of each body are obtained by integration of the local PN EOM. These are explicitly calculated retaining all higher derivatives that appear. One then reduces the higher derivatives by EOM and obtains the final results. Formally collecting the various relativistic corrections into a 'effective mass', one can have a PN proof of effacement of internal structure and provide a plausibility argument for validity of 'weak field formulas' for compact objects.

The fact that three independent methods give formally identical equations of motion at 2PN order is a strong confirmation of the validity of the numerical coefficients in the EOM. This work provides the basis for the timing formula mentioned earlier. The damping terms can be considered as perturbation to a Lagrangian system which is multi-periodic – a radial period and a angular period corresponding to periastron precession – and leads to the observed secular acceleration effect in the binary pulsar. No balance argument is involved at any stage.

The situation is now under investigation at the 3PN level. The work on 3PN generation crucially requires the EOM at 3PN accuracy and work is in progress to obtain the 3PN contributions by different techniques. These include the MPM method supplemented by Hadamard 'partie-finie' [16], the Epstein Wagoner Will Wiseman method [17] as also the Hamiltonian formalism [18]. As mentioned above, upto 2.5PN, three distinct computational techniques led to a unique EOM. Preliminary investigations have even raised questions about whether this sort of uniqueness will persist at 3PN.

It is interesting to note that both the Riesz regularisation and the Hadamard finite part averaged over all directions of approach to the singularity are techniques employed in the discussions of EOM in EM [19]. Both continuous source distributions and point sources (delta functions) have also been used in these computations. However, the situation in EM is much better than in gravitation because all the divergent terms can be renormalized into the mass after regularization. In gravitation, these offensive terms have a more complicated structure and we do not renormalize and simply throw away these divergent terms. The procedure in EM is also different since it is Lorentz invariant. In gravitation on the other hand we work in a particular frame and *hope* that in the end the EOM is nevertheless Lorentz invariant. Of course, if they are, it is a very powerful check that all is well with the computation [20]!

6. GENERATION

There are two approaches to calculate gravitational wave generation to higher orders, philosophically following the approaches of Fock and Landau-Lifshitz; the Blanchet-Damour-Iyer (BDI) [21] approach and the Epstein-Wagoner-Thorne-Will-Wiseman (EWTWW) [22, 23] approach respectively. Blanchet, Damour and Iyer build on a Fock type derivation using the double-expansion method of Bonnor. This approach makes a clean separation of the near-zone and the wave zone effects. It is mathematically well defined, algorithmic and provides corrections to the quadrupolar formalism in the form of compact support integrals or more generally well defined analytically continued integrals. The BDI scheme has a modular structure: the final results are obtained by combining an 'external zone module' with a 'radiative zone module' and a 'near zone module'. For dealing with strongly self-gravitating material sources like neutron stars or black holes one needs to use a 'compact body module' together with an 'equation of motion module'. It correctly takes into account all the nonlinear effects. It should be noted that, in generation problems, as one goes to higher orders of approximation, two independent complications arise. Though algebraically involved in principle, the first is simpler: contributions from higher multipoles. The second complication is not only algebraically tedious but technically more involved: contributions from higher nonlinearities e.g for 2PN generation cubic nonlinearities need to be handled.

The Epstein and Wagoner (EW) approach, also starts by rewriting the Einstein equations in a "relaxed" form. As in electromagnetism, one can write down a *single* formal solution valid everywhere in spacetime based on the flat-spacetime retarded Green function. The retarded integral equation for $h^{\alpha\beta}$, can then be iterated in a slow-motion ($v/c < 1$), weak-field ($||h^{\alpha\beta}|| < 1$) approximation as shown by Thorne [22]. Unlike in the electromagnetic case, however, the non-linear field contributions make the integrand of this retarded integral non-compact. The EW formalism leads to integrals that are not well defined, or worse, are divergent. Though at the first few PN orders, different arguments were given to ignore these issues, they provide no justification that the divergences do not become fatal at higher orders. Consequently, the EW formalism did not appear to be a reliable route to discuss higher PN approximations. Recently, Will and Wiseman have critically examined the EW formalism and provided a solution to the problem of its divergences by taking literally the statement that the solution is a *retarded* integral, *i.e.* an integral over the *entire* past null cone of the field point. The new EW method proposed by Will and Wiseman can be carried to higher orders in a straightforward, albeit very tedious manner and the result is a manifestly finite, well-defined procedure for calculating gravitational radiation to high PN orders.

The end result of the computations are expressions for the radiative mass and current multipole moments characterizing the source distribution. Once they are on hand, one can proceed to compute the associated gravitational waveform. From the waveform, the far zone energy flux may be computed by time differentiation (this is why one needs the EOM) and integration over all directions. The energy flux can also be computed directly from the moments and this provides a simple check on the algebraic correctness of the long computations. The angular momentum flux can also be computed for non-circular orbits. At the 2PN level this program is complete not only for circular, but also general orbits [24]. The extension to spinning bodies is also available [25]. The extension of these results to 3PN accuracy is an algebraically heavy and conceptually involved exercise, under investigation since 1996, using the multipolar post-Minkowskian approach [26]. The Hadamard regularization, based on the Hadamard partie finie, used in the computation of motion is also used in generation and provides consistent results. Though the known test particle limits are recovered, the finite mass correction introduces a plethora of new contributions. Hopefully in the near future the EW and ADM formalisms [17, 18] should provide a check on these results.

The solution to the generation problem thus provides the second input for phasing once we make the assumption of energy balance.

7. RADIATION REACTION IN ELECTRODYNAMICS

The idea of a damping force associated with an interaction that propagates with a finite velocity was first discussed in the context of electromagnetism by Lorentz. He obtained it by a direct calculation of the total force acting on a small extended particle due to its 'self-field'. The answer was incorrect by a numerical factor and the correct result was first obtained by Planck using a 'heuristic' argument based on energy balance which prompted Lorentz to re-examine his calculations and confirm Planck's result, $F^i = \frac{2}{3}\frac{e^2}{c^3}\ddot{v}^i$, where v_i is the velocity of the particle. The relativistic generalization of the radiation reaction by Abraham based on arguments of energy and linear momentum balance preceded by a few years the direct relativistic self-field calculation by Schott and illustrates the utility of this heuristic, albeit less rigorous, approach [9]. The argument based on energy balance proceeds thus: A non-accelerated particle does not radiate and satisfies Newton's (conservative) equation of motion. If it is accelerated, it radiates, loses energy and this implies damping terms in the equation of motion. Equating the work done by the reactive force on the particle in a unit time interval, to the negative of the energy radiated by the accelerated particle in that interval (Larmor's formula) the reactive acceleration is determined and one is led to the Abraham-Lorentz equation of motion for the charged particle. Lorentz's direct method of obtaining radiation damping, on the other hand, is based on the evaluation of the retarded action of each piece of the charge on the other parts. Starting with the momentum conservation law for the electromagnetic fields, one rewrites this as Newton's equation of motion, by decomposing the electromagnetic fields into an 'external field' and a 'self-field'. Expanding the self-field in terms of potentials, solving for them in terms of retarded fields and finally making a retardation expansion, one obtains the required equation of motion, when one goes to the point particle limit. For a historical summary of classical theories of radiation reaction see Erber's account [27].

There have been two broad approaches to radiation reaction later: The **field theory** one originally due to Dirac [28], that considers the *total* field at all points in space to be a fundamental physical quantity and point charges as singularities of the field; **the action-at-a-distance** one originally due to Wheeler and Feynman [29], that considers only forces exerted on the charge by *other* charges as physically meaningful. Each approach strictly goes beyond Maxwell's equations and uses an additional assumption: the conservation law for the EM energy momentum tensor in field theory and the relation between Lorentz force and momentum of the particle in action-at-a-distance theory. Though the plausibility of the physical idea of reducing everything to interaction of particles is the fascinating advantage of action-at-a-distance theories, none of the viewpoints appears preferable to the other from considerations of simplicity. Hoyle and Narlikar [3] have assessed the status of action-at-a-distance theories both in classical and the quantum electrodynamics. As there are no fields, the usual problems of divergences are absent in this treatment. When considered within cosmological models, these theories place stringent requirements on the future and past null cones of the universe. The theories will not work in Friedman cosmologies but do in steady state or quasi-steady state models. Issues related to the use of advanced fields in the Dirac derivation, were clarified later [30] and an approach to radiation reaction without advanced fields was presented by properly taking into account the retarded self-field of the point charge as required by the idea of energy-momentum localization. Since the retarded field diverges on the world line of the particle and the 'limit' depends on the direction of approach, one defines the field at the singularity as the average value over all possible directions [19]. A recent novel approach to radiation reaction is due to Gupta and Padmanabhan [31]. They show that fields of charged particles moving on arbitrary trajectories in an inertial frame can be related in a simple manner to the fields of a uniformly accelerated charged particle in its proper rest frame. Since the latter field is static and easily calculable, the former field is obtained by a coordinate transformation. It also allows them to compute the self force on the charged particle and recover the Dirac result.

8. RADIATION REACTION IN GR

As in electromagnetism, radiation reaction forces arise in gravitation from the use of retarded potentials satisfying time asymmetric boundary conditions like no-incoming boundary condition at past null infinity.

The problem is more complicated because of the nonlinearity of general relativity.

The approach to gravitational radiation damping has been based on the balance methods, the reaction potential or a full iteration of Einstein's equation. The first computation in general relativity was by Einstein who derived the loss in energy of a spinning rod by a far-zone energy flux computation. The same was derived by Eddington by a direct near-zone radiation damping approach. He also pointed out that the physical mechanism causing damping was the effect discussed by Laplace, that if gravity was not propagated instantaneously, reactive forces could result. An useful development was the introduction of the radiation reaction potential by Burke and Thorne [32] using the method of matched asymptotic expansions. In this approach, one derives the equation of motion by constructing an outgoing wave solution of Einstein's equation in some convenient gauge and then matching it to the near-zone solution. Restricting attention only to lowest order Newtonian terms and terms sensitive to the outgoing (in-going) boundary conditions and neglecting all other terms, one obtains the required result. The first complete direct calculation à la Lorentz of the gravitational radiation reaction force was by Chandrasekhar and Esposito. Chandrasekhar and collaborators [33] developed a systematic post-Newtonian expansion for extended perfect fluid systems and put together correctly the necessary elements like the Landau-Lifshitz pseudo-tensor, the retarded potentials and the near-zone expansion. These works established the balance equations to Newtonian order, albeit for weakly self-gravitating fluid systems. The revival of interest in these issues following the discovery of the binary pulsar and the applicability of these very equations to binary systems of compact objects follows from the works of Damour [9] and Damour and Deruelle [8] discussed earlier.

Many other approaches to radiation reaction problems have emerged in the last five years. For instance, given the formulas for the far-zone energy and angular momentum fluxes to a particular PN accuracy, to what extent can one infer the radiation reaction acceleration in the (local) EOM? Given the algebraic complexity of various computations and subtle evaluations of various small coefficients, it is worthwhile to check the obvious consistency requirement on the far-zone fluxes. To this end, Iyer and Will (IW) [34] proposed a refinement of the text-book treatment of the energy balance method used to discuss radiation damping. This generalization uses both energy and angular momentum balance to deduce the radiation reaction force for a binary system made of non-spinning structureless particles moving on general orbits. Starting from the 1PN conserved dynamics of the two-body system, and the radiated energy and angular momentum in the gravitational waves, and taking into account the arbitrariness of the 'balance' upto total time derivatives, they determined the 2.5PN and 3.5PN terms in the equations of motion of the binary system. The part not fixed by the balance equations was identified with the freedom still residing in the choice of the coordinate system at that order. The explicit gauge transformations they correspond to has also been constructed. Blanchet [35], on the other hand, obtained the post-Newtonian corrections to the radiation reaction force from first principles using a combination of post-Minkowskian, multipolar and post-Newtonian schemes together with techniques of analytic continuation and asymptotic matching. By looking at "antisymmetric" waves – a solution of the d'Alembertian equation composed of retarded wave minus advanced wave, regular all over the source, including the origin – and matching, one obtains a radiation reaction tensor potential that generalizes the Burke-Thorne reaction potential, in terms of explicit integrals over matter fields in the source. The *validity* of the balance equations upto 1.5PN is also proved. By specializing this potential to two-body sytems, Iyer and Will [34] checked that this solution indeed corresponds to a unique and consistent choice of coordinate system. This provides a delicate and non-trivial check on the validity of the 1PN reaction potentials and the overall consistency of the direct methods based on iteration of the near-field equations and indirect methods based on energy and angular momentum balance. It should be noted that the 'balance method' by itself cannot fix the particular expression for the reactive force in a given coordinate system. In order to solve a practical problem (in which we erect a particular coordinate system), the method is in principle insufficient by itself, but it provides an extremely powerful check of other methods based on first principles. Gopakumar, Iyer and Iyer [36] have applied the refined balance method to obtain the 2PN radiation reaction – 4.5PN terms in the equation of motion. Different facets of the IW choice like the functional form of the reactive acceleration have been systematically and critically explored and a better understanding of the origin of redundant equations is provided by studying variants obtained by modifying the functional forms of the ambiguities in energy and angular momentum. These reactive solutions are general enough to treat as particular cases any reactive acceleration obtained from first principles in the future.

Within the ADM approach, the radiative 3.5PN terms in the ADM Hamiltonian has been obtained by Jaranowski and Schafer [37]. Work is in progress to check that this leads to expressions for 3.5PN acceleration that is a particular case of the general IW solution. In the test particle case, work on radiation reaction has focussed on understanding the evolution of Carter constant in Kerr geometry by a variety of methods. Issues related to radiative versus retarded fields, adaptation of Dewitt-Brehme and asymptotic matching methods, axiomatic treatments as well as extension to spinning particles have also been investigated in the last three years [38].

9. CONCLUSION

It is amazing that in the macroscopic world, the computations of small higher order corrections so reminiscent of Lamb shift corrections in quantum electrodynamics (microscopic world) are in-expendable to extract the best from the LIGO and VIRGO facilities that will be able to look for gravitational wave signals by 2001. General relativity, far from being an esoteric and abstruse theory driven by aesthetic considerations is in a situation where experiments are driving the theory. We are on the threshold of opening another window to this marvelous universe and gravitational wave astronomy could well be the new astronomy of the 21st century. With the inauguration of the Gravitational Wave Astronomy, more than ever before, General Relativity will have found its true home.

References

[1] J. V. Narlikar, Science Today, p. 11, (June 1969).

[2] F. Hoyle and J. V. Narlikar *Action at a distance in physics and cosmology*, (W. H. Freeman, San Francisco 1974).

[3] F. Hoyle and J. V. Narlikar, Rev. Mod. Phys., **67**, 113 (1995).

[4] J. Samuel and B. R. Iyer, Current Science, **55**, 818 (1986).

[5] D. Kennefick, grqc-9704002.

[6] J. H. Taylor, A. Wolszczan, T. Damour, and J. M. Weisberg, Nature, **355**, 132 (1992).

[7] Kip Thorne in *Compact Stars in Binaries*, Eds. J. Van Paradijs, E. Van den Heuvel and E. Kuulkers, (Kluwer, Dordrecht 1995).

[8] T. Damour and N. Deruelle, C. R. Acad. Sci. Paris **293**, 537 (1981); **293**, 877 (1981); Phys. Lett., 87A, 1981, 81.

[9] T. Damour, *Gravitational Radiation*, Eds. N. Deruelle and T. Piran (North Holland, Amsterdam, 1983), p.59 and references therein.

[10] T. Damour, B. R. Iyer and B. S. Sathyaprakash, Phys. Rev. D **57**, 885 (1998).

[11] P. Havas, in *Isolated gravitating systems in general relativity*, Ed. J. Ehlers, (1979), p.74 and references therein.

[12] T. Damour, in *300 Years of Gravitation*, edited by S. W. Hawking and W. Israel (Cambridge University Press, London, 1987), p. 128.

[13] J. Martin and J. L. Sanz, J. Math. Phys. **20**, 25 (1979).

[14] G. Schafer, Ann. Phys. (N.Y.) **B161**, 81 (1985); *Gravitational wave detection*, Ed. A. Krolak, (Banach center publications, Warzawa, 1997), p. 43.

[15] L. P. Grishchuk and S. M. Kopejkin, in *Relativity in Celestial Mechanics and Astrometry*, edited by J. Kovalevsky and V. A. Brumberg (Reidel, 1986), p. 19.

[16] L. Blanchet, G. Faye and B. Ponsot, grqc-9804079 L. Blanchet and G. Faye, work in progress.

[17] C. M. Will and M. E. Pati, work in progress;

[18] G. Schäfer and P. Jaranowski, grqc-9712075 (1997), grqc-9802030 (1998).

[19] C. Teitelboim, Phys. Rev. **D1**, 1572 (1970); **2**, 1763 (E)(1970); **3**, 297 (1971); **4**, 345 (1971).

[20] Luc Blanchet, private communication.

[21] Luc Blanchet, in *Relativistic gravitation and gravitational radiation*, Eds. J.- P. Lasota and J.- A. Marck, Cambridge Univ. Press, Cambridge (1997), p.33.

[22] K. Thorne, Rev. Mod. Phys. **52**, 299 (1980).

[23] C. M. Will and A. G. Wiseman, Phys. Rev. D **54**, 4813 (1996).

[24] L. Blanchet, T. Damour, B.R. Iyer, C. M. Will and A. G. Wiseman, Phys. Rev. Lett. **74**, 3515 (1995); L. Blanchet, T. Damour, and B. R. Iyer, Phys. Rev. D **51**, 5360 (1995); C. M. Will and A. G. Wiseman, Phys. Rev. D **54**, 4813 (1996); L. Blanchet, B. R. Iyer, C. M. Will and A.G. Wiseman, CQG **13**, 575, (1996); A. Gopakumar and B. R. Iyer, Phys. Rev. D **56**, 7708 (1997); in preparation (1998).

[25] L. E. Kidder, Phys. Rev. D **52**, 821 ;1995); B. J. Owen, H. Tagoshi and A. Ohashi, Phys. Rev. D **57**, 6168 (1998).

[26] L. Blanchet, CQG **15**, 89, 113, 1971 (1998); L. Blanchet, B. R. Iyer and B. Joguet, paper in preparation;

[27] T. Erber, Fortschritte der Physik, **9**, 343 (1961).

[28] P. A. M. Dirac, Proc. R. Soc Lon., A **167**, 148 (1938).

[29] J. A. Wheeler and R. P. Feynman Rev. Mod. Phys., **17**, 157 (1945); **21**, 425 (1949).

[30] P. Havas, Phys. Rev. **74**, 456 (1948).

[31] A. Gupta and T. Padmanabhan, Phys. Rev. D **57**, 7241 (1998).

[32] W. L. Burke, J. Math. Phys. **12**, 401 (1971); K. S. Thorne, Astrophys. J. **158**, 997 (1969).

[33] S. Chandrasekhar, Astrophys. J. **158**, 45 (1969); S. Chandrasekhar and Y. Nutku, Astrophys. J. **158**, 55 (1969); S. Chandrasekhar and F. P. Esposito, Astrophys. J. **160**, 153 (1970).

[34] B. R. Iyer and C. M. Will, Phys. Rev. Lett. **70**, 113 (1993); Phys. Rev. D **52**, 6882 (1995).

[35] L. Blanchet, Phys. Rev. D **47**, 4392 (1993); **55**, 714 (1997).

[36] A. Gopakumar, B. R. Iyer and Sai Iyer, Phys. Rev. D **55**, 6030 (1997).

[37] P. Jaranowski and G. Schafer, Phys. Rev. D **55**, 4712 (1997).

[38] For references see e.g. B. R. Iyer, in *Black holes, gravitational radiation and the universe*, Eds. B. R. Iyer and B. Bhawal, Kluwer, Dordrecht (1998).

Chapter 15

GRAVITATIONAL COLLAPSE: THE STORY SO FAR

Pankaj S. Joshi
Tata Institute of Fundamental Research
Homi Bhabha Road, Bombay 400005, India.

Abstract We discuss here some recent developments in the theory of gravitational collapse, examining the issue of the final fate of continual collapse of a matter cloud. It is pointed out that it is basically the nature of the regular initial data that decides whether the collapse ends in a black hole or a naked singularity. We outline here some problems which remain as yet unresolved regarding the naked singularities and cosmic censorship.

A central issue in the gravitation theory today is that of cosmic censorship and asymptotic predictability [1]. Eventhough the singularity theorems predict the existence of singularities under fairly general conditions on a spacetime, they are silent on the nature of these singularities. It is thus necessary to understand the nature of singularities arising as end state of collapse. The important assumption fundamental to black hole physics is that the singularities forming at the end point of gravitational collapse of a massive object will necessarily be covered by the event horizons of gravity. Such a cosmic censorship hypothesis remains fundamental to the theoretical foundations of black hole physics. On the other hand, existence of visible or naked singularities would offer a new approach on these issues, offering the possibilities of their observational effects.

We review here some recent developments in this direction, examining the possible final fate of gravitational collapse in general relativity. Dynamical collapse scenarios have been examined in the past decade or so for cases such as clouds composed of dust, radiation, perfect fluids, and also of matter compositions consisting of type I general matter fields.

N. Dadhich and A. Kembhavi (eds.), The Universe, 161–168.

We discuss these conclusions and some open problems in the field are pointed out.

Consider, for example, the collapse of a dust cloud. It is well known that this will result into a black hole in the case of homogeneous dust collapse. What will be the situation when the perturbations over a homogeneous density profile are taken into account? It is important to include effects of inhomogeneities because typically a realistic collapse would start from an inhomogeneous initial data with a centrally peaked density profile. This problem can be investigated using the Tolman–Bondi-Lemaitre models [2]. This is an infinite dimensional family of asymptotically flat solutions of Einstein's equations, which is matched to the Schwarzschild space-time outside the boundary of the collapsing star. The Oppenheimer and Snyder [3] homogeneous dust ball collapse is a special case of this class of solutions. This question has now been investigated in detail (see e.g. [4], and references therein), and it is seen that the introduction of inhomogeneities leads to a qualitatively different picture of gravitational collapse. The metric for spherically symmetric collapse of inhomogeneous dust, in comoving coordinates (t, r, θ, ϕ), is given by,

$$ds^2 = -dt^2 + \frac{R'^2}{1+f}dr^2 + R^2(d\theta^2 + sin^2\theta\, d\phi^2) \tag{1}$$

$$T^{ij} = \epsilon\delta^i_t\delta^j_t, \quad \epsilon = \epsilon(t,r) = \frac{F'}{R^2R'} \tag{2}$$

where T^{ij} is the stress-energy tensor, ϵ is the energy density, and R is a function of both t and r given by

$$\dot{R}^2 = \frac{F}{R} + f \tag{3}$$

Here the dot and prime denote partial derivatives with respect to the parameters t and r respectively. As we are considering collapse, we require $\dot{R}(t,r) < 0$. The quantities F and f are arbitrary functions of r and $4\pi R^2(t,r)$ is the proper area of the mass shells. The area of such a shell at $r =$ const. goes to zero when $R(t,r) = 0$. For gravitational collapse situation, we take ϵ to have compact support on an initial spacelike hypersurface and the space-time can be matched at some $r =$ const. $= r_c$ to the exterior Schwarzschild field with total Schwarzschild mass $m(r_c) = M$ enclosed within the dust ball of coordinate radius of $r = r_c$. The apparent horizon in the interior dust ball lies at $R = F(r)$.

With the integration of equation for $\dot{R}$ above we have in all three arbitrary functions of r, namely $f(r)$, $F(r)$, and $t_0(r)$ where the last indicates the time along the singularity curve. One could use the coordinate freedom left in the choice of scaling of r to reduce the number of arbitrary functions to two. Rescaling R such that $R(0, r) = r$ leaves us with only two free functions f and F. The time $t = t_0(r)$ corresponds to $R = 0$ where the area of the shell of matter at a constant value of the coordinate r vanishes. The singularity curve $t = t_0(r)$ corresponds to the time when the matter shells meet the physical singularity. Thus, the coordinates are given by $0 \leq r < \infty$, $-\infty < t < t_0(r)$. It follows that unlike the collapsing Friedmann case, or the homogeneous dust case, where the physical singularity occurs at a constant epoch of time (say, at $t = 0$), the singular epoch is a function of r as a result of inhomogeneity in the matter distribution. One could recover the Friedmann case if we set $t_0(r) = t_0'(r) = 0$. The function $f(r)$ classifies the space-time as bound, marginally bound, or unbound depending on the range of its values which are $f(r) < 0, f(r) = 0, f(r) > 0$, respectively. The function $F(r)$ is interpreted as the weighted mass within the dust ball of coordinate radius r. For physical reasonableness the weak energy condition is assumed, that is, $T_{ij}V^iV^j \geq 0$ for all non-spacelike vectors V^i. This implies that the energy density ϵ is everywhere positive, ($\epsilon \geq 0$) including the region near $r = 0$. ¿From the scaling above, the energy density ϵ on the hypersurface $t = 0$ is written as $\epsilon = F'/r^2$. Then the weak energy condition implies that $F' \geq 0$ throughout the space-time.

Within this framework, the nature of the shell-focussing singularity at $R = 0$ can be examined. In particular, the problem of nakedness or otherwise of the singularity can be reduced to the existence of real, positive roots of an algebraic equation, constructed out of the free functions F and f and their derivatives [5], which constitute the initial data of this problem. It is then seen that for a wide variety of physically reasonable regular initial data, the singularity can be naked.

We call the singularity to be a *central singularity* if it occurs at $r = 0$. Partial derivatives R' and $\dot{R}'$ can be written as,

$$\left(\frac{\partial R(t,r)}{\partial r}\right)_{t=\text{const.}} = R' = (\eta - \beta)P - \left[\frac{1+\beta-\eta}{\sqrt{\lambda+f}} + (\eta - \tfrac{3}{2}\beta)\tfrac{t}{r}\right]\dot{R} \quad (4)$$

$$\left(\frac{\partial R'(t,r)}{\partial t}\right)_{r=\text{const.}} = \frac{\beta}{2r}\dot{R} + \frac{\lambda}{2rP^2}\left[\frac{1+\beta-\eta}{\sqrt{\lambda+f}} + (\eta - \tfrac{3}{2}\beta)\tfrac{t}{r}\right] \quad (5)$$

where we have used the notation,

$$R(t,r) = rP(t,r), \quad \eta = \eta(r) = r\frac{F'}{F}, \quad \beta = \beta(r) = r\frac{f'}{f}, \quad F(r) = r\lambda(r) \quad (6)$$

To focus the discussion, we restrict to functions $f(r)$ and $\lambda(r)$ which are analytic at $r = 0$ such that $\lambda(0) \neq 0$. The tangents $K^r = dr/dk$ and $K^t = dt/dk$ to the outgoing radial null geodesics, with k as the affine parameter, satisfy

$$\frac{dK^t}{dk} + \frac{\dot{R}'}{\sqrt{1+f}} = 0, \quad K^r K^t = 0, \quad \frac{dt}{dr} = \frac{K^t}{K^r} = \frac{R'}{\sqrt{1+f}} \tag{7}$$

Our purpose is to find whether these geodesics terminate in the past at the central singularity $r = 0, t = t_0(0)$. The exact nature of this singularity $t = 0, r = 0$ could be analyzed by the limiting value of $X \equiv t/r$ at $t = 0, r = 0$. If the geodesics meet the singularity with a definite value of the tangent then using l'Hospital rule we get

$$X_0 = \lim_{t\to 0, r\to 0} \frac{t}{r} = \lim_{t\to 0, r\to 0} \frac{dt}{dr} = \lim_{t=0, r=0} \frac{R'}{\sqrt{1+f}} \tag{8}$$

where the notation is, $\lambda_0 = \lambda(0), \beta_0 = \beta(0), f_0 = f(0)$ and $Q = Q(X) = P(X, 0)$. Using the expression for R' earlier, the above can be written as $V(X_0) = 0$, where

$$V(X) \equiv (1-\beta_0)Q + \left(\frac{\beta_0}{\sqrt{\lambda_0 + f_0}} + (1 - \frac{3}{2}\beta_0)X\right)\sqrt{\frac{\lambda_0}{Q} + f_0} - X\sqrt{1+f_0} \tag{9}$$

Hence if the equation $V(X) = 0$ has a real positive root, the singularity could be naked. In order to be the end point of null geodesics at least one real positive value of X_0 should satisfy the above. Clearly, if no real positive root of the above is found, the singularity $t = 0, r = 0$ is not naked. It should be noted that many real positive roots of the above equation may exist which give the possible values of tangents to the singular null geodesics terminating at the singularity. However, such integral curves may or may not realize a particular value X_0 at the singularity. Suppose now $X = X_0$ is a simple root to $V(X) = 0$. To determine whether X_0 is realized as a tangent along any outgoing singular geodesics to give a naked singularity, one can integrate the equation of the radial null geodesics in the form $r = r(X)$ and it is seen that there is always atleast one null geodesic terminating at the singularity $t = 0, r = 0$, with $X = X_0$. In addition there would be infinitely many integral curves as well, depending on the values of the parameters involved, that terminate at the singularity. It is thus seen that the existence of a positive real root of the equation $V(X) = 0$ is a necessary and sufficient condition for the singularity to be naked. Finally, to determine the curvature strength of the naked singularity at $t = 0$, $r = 0$, one may analyze the quantity $k^2 R_{ab} K^a K^b$ near the singularity. Standard analysis shows that the strong curvature condition is satisfied, in that the above quantity remains finite in the limit of approach to the singularity.

The assumption of vanishing pressures here may be considered a limitation of dust models. It is argued sometimes that in the final stages of collapse, the dust equation of state could be relevant and at higher densities the matter may behave more and more like dust. Further, if there are no large negative pressures (as implied by the energy conditions), the pressure also might contribute gravitationally in a positive manner to the effect of dust and may not alter the conclusions.

In any case it is important to consider collapse situations with matter with non-zero pressures and reasonable equations of state. It is possible that pressures may play an important role for the later stages of collapse and one must investigate the possibility if pressure gradients could prevent the occurrence of naked singularity. This issue has been examined in a number of papers [6], for both self-similar as well as non-self-similar collapse models. In particular, for self-similar models the results are, if in a self-similar collapse a single null radial geodesic escapes the singularity, then an entire family of non-spacelike geodesics would also escape provided the positivity of energy density is satisfied as above.

The results on matter forms such as directed radiation, dust, perfect fluids etc imply some general pattern emerging about the final outcome of gravitational collapse. Hence one could ask the question whether the final fate of collapse would be independent of the form of matter under consideration. An answer to this is important because it has often been conjectured that once a suitable form of matter with an appropriate equation of state, and satisfying energy conditions, is considered then there may not be naked singularities. After all, there is always a possibility that during the final stages of collapse matter may not have any of the forms considered above, because such relativistic fluids are phenomenological and one must treat matter in terms of fundamental fields, such as for example, a massless scalar field.

Recent efforts in this direction are worth mentioning where the above results on perfect fluid were generalized to matter forms without any restriction on the form of T_{ij}, which was supposed to satisfy the weak energy condition only [7]. The main argument of the results such as these is along the following lines. In the discussion above it was pointed out that naked singularities could form in the gravitational collapse from a regular initial data, from which non-zero measure families of non-spacelike trajectories come out. The criterion for the existence of such singularities was characterized in terms of the existence of real positive roots of an algebraic equation constructed out of the field variables. A similar procedure was developed now for general form of matter. In comoving coordinates, the general matter can be described by three functions, namely the energy density and the radial and tangential pressures. The existence of naked singularity is again characterized in terms of the real positive roots of an equation, constructed from the equations of non-spacelike geodesics which involve the three metric functions. The field equations then relate these metric functions to the matter variables and it is seen that for a non-zero measure subspace of this free initial data in terms of matter variables, the above equation will have real positive roots, producing a naked singularity in the space-time.

It is thus seen that the occurrence of naked singularity is basically related to the choice of initial data to the Einstein field equations, and would therefore occur from regular initial data within the general context considered, subject to the matter satisfying weak energy condition. The condition on initial data which leads to the formation of black holes is also similarly characterized. It would then appear that the occurrence of naked singularity or a black hole is more a problem of choice of the initial data for field equations rather than that of the form of matter or the equation of state. This has important implications for the cosmic censorship in that in order to preserve the same one has to avoid all such regular initial data causing naked singularity, and hence a deeper understanding of the initial data space is required in order to determine such initial data and the kind of physical parameters they would specify. This would classify the range of physical parameters to be avoided for a particular form of matter. More importantly, it would also pave the way for the black hole physics to use only those ranges of allowed parameter values which would produce black holes, thus putting black hole physics on a more firm footing.

Much attention has been devoted in past years to analyze the collapse of a scalar field, both analytically as well as numerically [8]. This is a model problem of a single massless scalar field which is minimally coupled to gravitational field and it provides possibly one of the simplest scenarios to investigate the nonlinearity effects of general relativity. On the analytic side, the results by Christodoulou show that when the scalar field is sufficiently weak, there exists a regular solution, or global evolution for an arbitrary long time of the coupled Einstein and scalar field equations. For strong enough fields, the collapse is expected to result into a black hole. Such an approach helps study the cosmic censorship problem as the evolution problem in the sense of examining the global Cauchy development of a self-gravitating system outside an event horizon. The problem of scalar field collapse has been numerically studied by Choptuik and others, where a family of scalar field solutions with a parameter p characterized the strength of the scalar field. The numerical calculations showed that for black hole formation, there is a critical limit $p \to p^*$ and the mass of the resulting black holes satisfy a power law $M_{bh} \propto (p - p^*)^\gamma$, where the critical exponent γ has value of about 0.37.

The pattern that appears to be emerging from the current work on gravitational collapse is that both naked singularities and the black holes occur in several collapsing configurations from regular initial data, with reasonable equations of state such as describing radiation, dust or a perfect fluid with a non-zero pressure, or for general forms of matter. An insight that is gained from such an investigation is the final state of a collapsing star, in terms of either a black hole or a naked singularity, may not really depend on the form or equation of state of collapsing matter, but is actually determined by the physical initial data in terms of the initial density profiles and pressures. The important question then is the genericity and stability of such naked singularities arising from regular initial data. An investigation on this would enable one to reformulate more suitably the censorship hypothesis, based on a criterion that naked singularities could form in collapse but may not be generic.

References

[1] R. Penrose, 1969. *Riv. del. Nuovo Cim.* **1**, 252.

[2] R. C. Tolman, 1934. *Proc. Natl. Acad. Sci. USA,* **20** 410; H. Bondi, 1947. *Mon. Not. Astron. Soc.* **107**, 343.

[3] J. Oppenheimer and H. Snyder, 1939. *Phys. Rev.* **56**, 455.

[4] P. S. Joshi, 1993. *Global aspects in gravitation and cosmology*, Clarendon Press, OUP, Oxford.

[5] P. S. Joshi & I. H. Dwivedi, 1993. *Phys. Rev.* D **47**, 5357.

[6] A. Ori & T. Piran, 1990. *Phys. Rev.* D **42**, 1068; P. S. Joshi and I. H. Dwivedi, 1992. *Commun. Math. Phys.* **146**, 333; *Lett. Math. Phys.* **27**, 235.

[7] I. H. Dwivedi & P. S. Joshi, 1994. *Commun. Math. Physics* **166**, 117; K. Lake, *Phys. Rev. Lett.* **68**, 3129.; P. Szekeres & V. Iyer, *Phys. Rev.* D **4**, 436; P. S. Joshi & I. H. Dwivedi, 1999. *Class. Quantum Grav.*.

[8] D. Christodoulou, 1986. *Commun. Math. Phys.* **93**, 587; 1987, **109**, 591; 1987, **109**, 613; 1991, *Commun. Pure and Applied Math.* XLIV, 339; 1994, *Ann. Math.* **140**, 607; M. W. Choptuik, 1993. *Phys. Rev. Lett.* **70** 9; A. M. Abrahams & C. R. Evans, 1993. *Phys. Rev. Lett.* **70**, 2980; C. R. Evans & J. S. Coleman, 1994. *Phys. Rev. Lett.* **72**,

1782. M. D. Roberts, 1989. *Gen. Relat. Grav.* **21**, 907.; J. Traschen, 1994. *Phys. Rev.* D **50**, 7144; P. R. Brady, 1995. *Class. Quant. Grav.* **11**, 1255; 1995, *Phys. Rev.* D **51**, 4168 (1995); C. Gundlach, 1995. *Phys. Rev. Lett.* **75**, 3214.

Chapter 16

THOUGHTS ON GALACTIC MAGNETISM

Kandaswamy Subramanian
National Centre for Radio Astrophysics,
Tata Institute of Fundamental Research,
Ganeshkind, Pune 411007, India

It is a pleasure to dedicate this article to Prof. Jayant Narlikar, my first research Guru, though he may want to have little to do with toroidal and poloidal fields and the infamous 'pomega' !

—

Abstract

Magnetic fields correlated on several kiloparsec scales are seen in spiral galaxies. Their origin could be due to amplification of a small seed field by a turbulent galactic dynamo. We critically review the current status of the galactic dynamo, especially some of its problems and possible solutions. We also comment on the nature of seed magnetic fields, needed to prime the dynamo.

1. INTRODUCTION

Magnetic fields in spiral galaxies have strengths of order few $10^{-6}G$, and are coherent on scales of several kpc [1]. In several disk galaxies, like M51 and NGC 6946, they are also highly correlated (or anti-correlated) with the optical spiral arms. How do such ordered, large-scale fields arise? One possibility is dynamo amplification of a weak but nonzero

N. Dadhich and A. Kembhavi (eds.), The Universe, 169–179.

seed field $\sim 10^{-19} - 10^{-23}G$, provided the galactic dynamo can operate efficiently to exponentiate the field by a factor $\sim 30 - 40$. We critically review here some of the issues relevant to the operation of the galactic dynamo, the problems which arise and possible solutions. We also touch upon the origin of the seed magnetic field, needed for dynamo amplification.

The evolution of the magnetic field, in the MHD approximation, is described by the induction equation

$$\frac{\partial \mathbf{B}}{\partial t} = \nabla \times (\mathbf{v} \times \mathbf{B} - \eta \nabla \times \mathbf{B}), \tag{1}$$

provided one assumes the usual form of Ohms law and neglects the displacement current term in Maxwells equation. Here $\mathbf{B}$ is the magnetic field, $\mathbf{v}$ the velocity of the fluid and η the resistivity. If $\eta \to 0$ the magnetic flux through any area in the fluid is conserved during the motion of the fluid. The presence of a finite resistivity allows for a violation of such 'flux freezing' and the magnetic Reynolds number (MRN) $R_m = vL/\eta$ measures the relative importance of flux freezing versus resistive diffusion. Here v and L are typical velocity and length scales of the fluid motions. In most astrophysical contexts flux freezing greatly dominates over diffusion with $R_m >> 1$.

$\mathbf{B} = 0$ is a perfectly valid solution of the induction equation. So there would be no magnetic field generated if one were to start with a zero magnetic field. The universe probably did not start with an initial magnetic field. One therefore needs some way of violating the induction equation and produce a cosmic battery effect, to drive curents from a state with initially no current. There are a number of such battery mechanisms which have been suggested [2]. All of them lead to only small fields, much smaller than the galactic fields. Therefore dynamo action, due to a velocity field acting to exponentiate small seed fields efficiently, is needed to explain observed fields. We first briefly comment on a cosmic battery before discussing dynamos in detail.

2. COSMIC BATTERIES AND SEED FIELDS FOR THE DYNAMO

The basic problem any battery has to address is how to produce finite currents from zero currents? Most astrophysical mechanisms use the fact that positively and negatively charged particles in a charge-neutral universe, do not have identical properties. For example if one considered a gas of ionised hydrogen, then the electrons have a much smaller mass compared to protons. This means that for a given pressure gradient of the gas the electrons tend to be accelerated much more than the ions.

This leads in general to an electric field, which couples back positive and negative charges, of the form $\mathbf{E}_T = -\nabla p_e / e n_e$, where p_e and n_e are the electron pressure and number density, respectively. If such a thermally generated electric field has a curl, then by Faradays law of induction a magnetic field can grow. Taking $p_e = n_e kT$ with T the electron temperature we have $\nabla \times \mathbf{E}_T = (ck/e)(\nabla n_e / n_e) \times \nabla T$. So $\mathbf{E}_T$ has a curl only if the density and temperature gradients, are not parallel to each other. The resulting battery effect, known as the Biermann battery, was first proposed as a mechanism for thermal generation of stellar magnetic fields [3].

The Biermann battery can also lead to the thermal generation of seed fields in cosmic ionisation fronts [4]. These ionisation fronts are produced when the first ultra violet photon sources, like quasars, turn on to ionise the intergalactic medium (IGM). The temperature gradient in a cosmic ionisation front is normal to the front. However, a component to the density gradient can arise in a different direction, if the ionisation front is sweeping across arbitrarily laid down density fluctuations, associated with protogalaxies/clusters since these in general have no correlation to the source of the ionising photons. The resulting thermally generated electric field has a curl, and magnetic fields on galactic scales can grow. They turn out to have a strength $B \sim 3 \times 10^{-20} G$. This field by itself is far short of the observed microgauss strength fields in galaxies, but it can provide a seed field, coherent on galactic scales, for a dynamo. Indeed the whole of the IGM is seeded with small magnetic fields. This seed field may infact have the right symmetry properties for the galactic dynamo modes [5]. The Biermann battery has also been invoked to generate seed magnetic fields in galactic or proto-galactic environments [6].

Larger seed magnetic fields than above can arise if we combine some form of dynamo action with the battery effect. For example, if stellar dynamos work efficiently, and some stars blow out as supernovae, then they can seed the interstellar medium, with significant magnetic fields. Alternatively galactic turbulence can itself lead to a small-scale dynamo (see below) and provide a larger seed for the large-scale galactic dynamo. There have also been attempts to invoke exotic physics in the early universe to produce primordial magnetic fields [2]. (Infact a primordial field in the IGM, which redshifts to a present day value of $\sim 10^{-9} G$, and is correlated on Mpc scales, can significantly perturb structure formation, and cause detectable anisotropies in the cosmic microwave background radiation [7].) It is fair to say at present that most seed field generation mechanisms fall far short of producing large-scale correlated fields at the micro-gauss level. One does need some form of large-scale dynamo action.

3. THE LARGE-SCALE GALACTIC DYNAMO

Disk galaxies are differentially rotating systems. Also the magnetic flux is to a large extent frozen into the fluid. So any radial component of the magnetic field will be efficiently wound up and amplified to produce a toroidal component. But this results in only a linear amplification of the field. To obtain the observed galactic fields starting from small seed fields one should find a way to generate the radial component from the toroidal one. If this can be done, the field can grow exponentially and one has a dynamo.

A mechanism to produce the radial field from the toroidal field was invented by Parker [8]. The essential feature is to invoke the effects of cyclonic turbulence in the galactic gas. The interstellar medium (ISM) is assumed to be turbulent, due to for example the effect of supernovae randomly going off in different regions. In a rotating, stratified (in density and pressure) medium like a disk galaxy, such turbulence becomes cyclonic and acquires a net helicity. Helical motions of the galactic gas perpendicular to the disk can draw out the toroidal field into a loop which looks like a *twisted* Ω. Such a loop is connected to a current and because of the twist this current has a component parallel to the original field. If the motions have a non-zero net helicity, then the random current components parallel to the field, add up coherently. A toroidal current can then result from the toroidal field. Hence, poloidal fields can be generated from toroidal ones.

In quantitative terms, suppose the velocity field is the sum of a mean, large-scale velocity $\mathbf{v}_0$ and a turbulent, stochastic velocity $\mathbf{v}_T$. The induction equation becomes a stochastic partial differential equation. The equation for various moments of $\mathbf{B}$, can be derived in two ideal limits. First when $R_m << 1$, and the distortions to the mean magnetic field are small, and second when $R_m >> 1$, but the turbulence is assumed to have a delta function (or very small) correlation in time. For galaxies the latter idealisation may be more relevant.

Let us split the magnetic field $\mathbf{B} = \mathbf{B}_0 + \delta\mathbf{B}$, into a mean field $\mathbf{B}_0$ and a fluctuating component $\delta\mathbf{B}$.. Here the mean is defined either as a spatial average over scales larger than the turbulent eddy scales or more correctly as an ensemble average. Assume the turbulence to be isotropic, homogeneous and helical. The action of the turbulent velocity field $\mathbf{v}_T$, on the magnetic field, the $(\mathbf{v} \times \mathbf{B})$ term, then leads to an extra contribution to the *mean* electric field of the form $-c\mathbf{E}_0 = \alpha\mathbf{B}_0 - \eta_t\nabla \times \mathbf{B}_0$. Here $\alpha = -(1/3)\int < \mathbf{v}_T(t).(\nabla\times\mathbf{v}_T(s)) > ds$, depends on the helical part of the turbulent velocity correlation function, and $\eta_t = (1/3)\int < \mathbf{v}_T(t).\mathbf{v}_T(s) > ds$, called the turbulent diffusion depends on the non-helical part of the turbulence. Here the angular brackets $<>$ denote an ensemble average, over the stochastic velocity field. The induction equation for the mean field, with the extra turbulent component of the mean electric field, is then given by

$$\frac{\partial \mathbf{B}_0}{\partial t} = \nabla \times (\mathbf{v}_0 \times \mathbf{B}_0 + \alpha\mathbf{B}_0 - (\eta + \eta_T)\nabla \times \mathbf{B}_0) \, . \tag{2}$$

This kinematic mean-field dynamo equation, can have exponentially growing solutions, which have been studied extensively in the literature [1]. While the α-effect is crucial for regeneration of poloidal from toroidal fields, the turbulent diffusion turns out to be also essential for allowing changes in the mean field flux. Also in galaxies, differential rotation (the Ω effect) is dominant in producing toroidal from radial fields. The growth rates of the galactic 'α-Ω dynamo', are typically a few times the rotation time scales, of order 10^9 yr. Modulations of α, and η_T, due to enhanced turbulence along spiral arms, can also lead to bi-symmetric large-scale fields, correlated with the optical spirals [9].

Note that the mean field has a scale limited only by the size of the system, which can be much bigger than the scales associated with the turbulence. In this sense one has created order from chaos. One may be tempted to refer to this as an inverse cascade, a term which would suggest transfer of power from smaller to larger and larger scales. But reality is more subtle. All scales larger than the turbulent eddy scale can grow simultaneously due to the α-effect; but with larger scales growing slower. So the effect can be thought better as a long range interaction between the turbulent scales, and all larger scales.

A physics comment is in order at this stage. When one considers the effect of turbulent fluid motions on a scalar field, like say smoke, one only gets a mean diffusion of smoke, associated with the random walking nature of turbulent fluid motions. But for 'frozen' magnetic fields the induction equation has terms which not only imply a body transport due to the random motions ($\mathbf{v}.\nabla\mathbf{B}$), but also a term, $\mathbf{B}.\nabla\mathbf{v}$, which describes the generation of magnetic fields due to velocity shear. It is this qualitative difference between magnetic fields and smoke that leads to an alpha effect (provided also that the motions have a non-zero average helicity), over and above turbulent diffusion. As we see below, it also leads to the small-scale dynamo. Note that both α and η_T, depend crucially on the diffusive (random-walk) property of fluid motion. So if due to some reason (see below) the fluid motion becomes wavelike, then the above integrals average to zero, and the alpha effect and turbulent diffusion will be suppressed.

In deriving the mean-field equation, the turbulent velocities have been assumed to be given, and unaffected by the Lorentz forces due to the magnetic field; at least not until the mean large-scale field builds up sufficiently. However this does not turn out to be valid due to the more rapid build up of magnetic noise compared to the mean field, a problem to which we now turn.

4. THE SMALL-SCALE DYNAMO AND MAGNETIC NOISE

The problem with the kinematic mean field dynamo is that it is a myth. This is because, the same turbulence which contributes to the α-effect, the turbulent diffusion, and associated growth of mean fields, also leads to a more rapid growth of small-scale fields.

In incompressible turbulence, fluid particles random-walk away from each other. This leads to stretching of the field lines attached to these particles, and an exponential increase of field strength. The stretching will also be accompanied by the field being squeezed into smaller and smaller volumes. Suppose one considers a flux tube, of length l_t, cross section A, density ρ and magnetic field B. Then flux freezing implies $BA =$ constant, mass conservation $\rho A l_t =$ constant, and incompressibility $\rho =$ constant. So as l_t increases due to random stretching, the the magnetic field $B \propto l_t$ increases and the cross-section $A \propto l_t^{-1}$ decreases.

If for the moment one ignores Lorentz forces, then the sqeezing into small volumes, stops only when diffusive scales are reached. Typically the field can be thought of as being in flux ropes, curved on the eddy scale, say L, and a thickness of order the diffusive scale, say r_d (assuming only a single scale eddy is present). At this stage the energy input into the magnetic field due to random stretching would be comparable to the energy loss in diffusion. This gives $vB/L \sim \eta B/r_d^2$, implying $r_d \sim L/R_m^{1/2}$. What happens further (whether growth or diffusion wins out), can only be decided by a more quantitative treatment of the problem.

A rigorous analysis of small-scale field dynamics, was first worked out by Kazantsev [10], and elaborated extensively by many authors [11], for the simple case, where the turbulence was assumed to have a delta function correlation in time. We shall mainly draw upon our own work in Ref. [12] . It turns out that, the small-scale dynamo (SSD) can operate under fairly weak conditions; that the MRN associated with the turbulent motions be greater than a critical value $R_c \sim 100$. In particular the fluctuating field tangled on a scale l, can grow on the turn over time scale of a turbulent eddy of scale l, with a growth rate $\Gamma_l \sim v_l/l$, provided the MRN on that scale $R_m(l) = v_l l/\eta > R_c$. Here v_l is the velocity associated with eddies of scale l. For Kolmogorov turbulence, since $v_l \propto l^{1/3}$, the growth rate $\Gamma_l \propto l^{-2/3}$, and so increases with decreasing l. For galactic gas, with a significant neutral component, typically, even the eddies at the cut-off scale of the turbulence, say l_c, have $R_m(l_c) >> R_c$.

The spatial structure of the small-scale dynamo generated field is also of great interest. For this it is useful look at the behaviour of the magnetic correlation function, say $w(r,t) =< \delta\mathbf{B}(\mathbf{x},t).\delta\mathbf{B}(\mathbf{y},t) >$, where $r = |\mathbf{x} - \mathbf{y}|$. Here the averaging indicated by $<>$, is a double ensemble average over both the stochastic velocity and fluctuating magnetic fields. ¿From $\nabla.\delta\mathbf{B} = 0$, one can show that the curve $r^2 w(r)$ should enclose zero area. Since $w(0)$ is neccessarily positive, as one goes to larger r, there must be some $r \sim d$ say, when $w(r)$ becomes negative. For such r, the fluctuating field at the origin, and at a separation d are pointing in 'opposite' directions on average. This can be intepreted as saying that the field lines on average are curved on scale d.

For Kolmogorov type turbulence, and if $R_m(l_c) >> R_c$, the fastest growing mode, has $w(r,t)$ strongly peaked within the diffusive scale of the cut-off scale eddies, $r = r_d(l_c) = l_c/R_m^{1/2}(l_c)$, changing sign at $r \sim l_c$, and rapidly decaying for larger r/l_c. One can interpret such a correlation function as implying that field is concentrated into ropes of thickness $r_d(l_c)$ and curved on scales of order l_c. For slower growing modes, with growth rate $\Gamma_l \sim v_l/l$, $w(r)$ extends upto $r \sim l$, after which it decays exponentially. For these modes, the small-scale field can be thought of as being concentrated in rope-like structures with thickness of order the diffusive scale $r_d(l_c)$ and curved on a scale upto $\sim l$. In general, at the kinematic stage, the growth rate of irreducible higher order correlations, increase with order, indicating that the field becomes highly intermittent in space.

Note that the small-scale dynamo due to even the eddy at the outer scale of galactic turbulence, will lead to the exponetial growth of small-scale fields on a time $\tau = L/v \sim 10^7$ yr. (Here we have taken $L \sim 100$ pc, and a velocity scale $v \sim 10$ km s^{-1}.) This time is much shorter than the time scale $\sim 10^9$ yr for mean field growth. The magnetic field is then rapidly dominated by the fluctuating component, before the mean field has grown appreciably. If the energy in the small-scale component grows to equipartition with the turbulent energy density, the turbulence could become more wavelike 'Alfvén' turbulence, than an eddy like fluid turbulence. So diffusive effects like the α and η_T, would get suppressed, and mean field growth stopped. How does the galaxy escape this predicament?

5. SATURATION OF THE SMALL-SCALE DYNAMO

To answer this question, it is crucial to find out how the small-scale dynamo saturates. We have concentrated on the possibility that the small scale field continues to be intermittent in space, when it saturates, as it was in the kinematic regime. That it saturates as a 'can of worms'; with peak fields being limited by non-linear effects to values of order or slightly larger than equipartition fields, but with most of the space having much smaller fields. Then the average energy density of the saturated small-scale dynamo generated field, may still be sub-equipartition, since it does not fill the volume. And the turbulence will remain eddy like, and preserve diffusive effects like α and η_t. We have given one explicit realisation of the above idea, in Ref. [12], in the case of a galaxy dominated by neutral particles.

In partially ionised plasma, the Lorentz force acts on ions, which are only coupled to neutrals through collisions. This leads to a 'ambipolar' drift of ions (and hence the field) with respect to neutrals. With $\mathbf{v}$ in Eq.(1) replaced by the neutral fluid velocity, the effective diffusivity changes to $\eta_{eff} = \eta + < \delta\mathbf{B}^2 > /(6\pi\rho_i\nu_{in})$. Here ρ_i is the ion density and ν_{in} is the neutral-ion collision frequency. So, as the energy density in the fluctuating field increases, the effective magnetic Reynolds number, for fluid motion on any scale of the turbulence say $R_{ambi}(l) = v_l l/\eta_{eff}$, decreases. If R_{ambi} could decrease to R_c, this itself will lead to a SSD saturation, but for conditions appropriate to galactic gas, R_{ambi} remains much larger than R_c. So as the the small-scale field grows in strength, it continues to be concentrated into thin ropy structures, as in the kinematic regime. These flux ropes are curved on the turbulent eddy scales, while their thickness is now set by the diffusive scale determined by the effective ambipolar diffusion.

Other restraining effects have to then limit the SSD. The first of these is due to the growing magnetic tension associated with the curved flux ropes. It acts to straighten out a flux rope, at a rate determined by equating the tension force and frictional drag. Frictional drag also damps the magnetic energy associated with the wrinkle in the rope. Further, small-scale flux loops can collapse and disappear, causing an *irreversible* sink of magnetic energy into heat. These non-local effects operate on the eddy turnover time scale, when the peak field in a flux rope, say B_p, has grown to a few times the equipartion value. Their net effect is to make the random stretching needed for the SSD inefficient and hence saturate the SSD. As the field is in flux ropes which do not fill the volume. the average energy density in the saturated small-scale field is still sub-equipartition, and α and η_T are preserved.

Note that B_p has to grow to a larger and larger value, thinner the flux rope, for inefficient random stretching to operate. This is because tension is a volume force ($\propto r_d^2 l_t$) while drag acts on the surface of the rope and is $\propto r_d l_t$. But B_p cannot grow larger than $(8\pi P_{ext})^{1/2}$, where P_{ext} is the total pressure in the ISM. At the same time the thickness of flux ropes is larger, greater the ambipolar diffusion, or smaller the ion-density. So the SSD saturates as above, for a given P_{ext}, provided n_i is less than a crictical value n_i^c. In the ISM, if P_{ext} is a factor F greater than the gas pressure (the gas assumed to have $T \sim 10^4$K and a density n_n), one gets $n_i^c \sim 10^{-2} cm^{-3} (v/10kms^{-1})^{-3} (n_n/cm^{-3})^{2/3} (F/2)^{7/3}$. So for a range of 'galactic' like parameters, this picture of small-scale dynamo saturation works.

For larger ion densities $n_i > n_i^c$, the way the SSD saturates is not very clear. The peak field is still limited by the external pressure. But how the flux ropes behave in the post-kinematic stage is yet to be rigorously worked out. It is possible that, when one starts with weak fields, the field is first squeezed into small volumes until limited by magnetic pressure. Subsequently, constructive folding of the field, may lead to fusing and thickening of the flux rope, while destructive folding may lead to reconnection, and dissipation. A phenomenological model [13] which incorporates this thickening of flux ropes as the field builds up, drives the SSD into saturation, when the rope thickness becomes of order $L/R_c^{1/2}$, the peak field reaches equipartition levels, but with the average energy density of order R_c^{-1} of equipartition.

Numerical simulations of dynamo action due to mirror-symmetric turbulence [14] or convection [15] have also hinted at a saturated state of SSD as described above; a magnetic field concentrated into flux ropes, occupying a small fraction of the fluid volume, having peak fields comparable or in excess of equipartition value but average magnetic energy density only about 10% of the kinetic energy density. Such simulations are however limited by the MRN they can achieve. There have also been MHD simulations of the SSD in fourier space, adopting some form of closure approximation, like the EDQNM approximation [16], or the DIA [17]. They have also indicated that the small-scale field could saturate at sub-equipartition levels. However, why this happens and the relation of the fourier space, to the real space calculations which we have emphasised, is not at present clear.

There ofcourse remains the possibility that the SSD in a fully ionised gas saturates only when the energy density in the noise grows comparable to that of the turbulence. However, the spatial structure of the small-scale field, is still likely to be highly intermittent. The large-scale dynamo action will then depend on how such a field responds to turbulent motions, especially whether the field can reconnect efficiently [18]. If reconnection is efficient, then it may allow diffusive transport to still occur, through the forest of small-scale fields, rather like Tarzan, swinging from one rope to another crosses the jungle! Reconnection is an important issue, which deserves much more discussion than we have given (cf. [19]). Another important issue which is just beginning to be addressed [20], is the calculation of α (or η_T) in the presence of significant small-scale fields.

In summary, the survival of the diffusive effects needed for large-scale dynamo action could depend crucially on whether the SSD generated fields can saturate at sub-equipartition levels. We feel that this may indeed be possible, if the noise saturates as a 'can of worms'. Note the SSD generated field is indeed spatially intermittent in the kinematic regime. So when one starts from weak fields, kinematic evolution operates for some time, and the small-scale field is driven to an intermittent state. The important question is to what extent it remains so in the non-linear regime, when it saturates. This SSD generated field will also provide a strong seed for the large-scale dynamo [21]. Indeed, in a unified treatment of small and large-scale dynamos, large-scale correlations are produced, from small-scale fields in a way analogous to 'quantum mechanical tunnelling', of the stationary state of the SSD [13]. The helical, turbulent, α-Ω dynamo still seems to be the best bet for explaining large-scale galactic fields. Ultimately the galactic dynamo is a non-linear dynamo; but a discussion of how it operates in the final saturated regime needs more thought. Our thoughts on galactic magnetism have yielded interesting results but much remains to be done.

References

[1] Beck, R., Brandenburg, A., Moss, D., Shukurov, A. and Sokoloff, D., 1996, Ann. Rev. Astr. Astrophys. 34, 155. Ruzmaikin, A. A., Shukurov, A. M. & Sokoloff, D. D., 1988. *Magnetic Fields of Galaxies*, Kluwer, Dordrecht.

[2] Rees, M.J., 1994. *Cosmical Magnetism*, ed. Lynden-Bell, D., Kluwer, London, p155; Subramanian, K. 1995, Bull. Astr.Soc. India., 23, 481; Ratra, B., 1992, ApJ Lett., 391, L1; and references therein.

[3] Biermann, L., 1950. Zs. Naturforsch. A., 5, 65; Mestel, L. & Roxburgh, I.W., 1962. ApJ., 136, 615.

[4] Subramanian, K., Narasimha, D. & Chitre, S.M., 1994. MNRAS. 271, L15.

[5] Krause, F. & Beck, R., 1998, A&A., 335, 789.

[6] Lazarian, A., 1992, A&A, 264, 326; Kulsrud R. M., Cen, R., Ostriker, J. P., Ryu, D., 1997, ApJ, 480, 481.

[7] Wasserman, I., 1978, ApJ, 224, 337; Subramanian, K. & Barrow, J. D., 1998, Phys. Rev. D, 58, 083502; Phys. Rev. Lett., 81, 3575.

[8] Parker, E.N., 1955. ApJ., 122, 293; Steenbeck, M, Krause, F. & Radler, K-H., Z. Naturforsch., 21a, 369.

[9] Mestel, L. & Subramanian, K. 1991, MNRAS, 248, 677; Subramanian, K. & Mestel, L., 1993, MNRAS, 265, 649; Moss, D, 1996, A&A, 308, 381.

[10] Kazantsev, A. P., 1968, Sov. Phys. - JETP, 26, 1031.

[11] Zeldovich, Ya.B., Ruzmaikin,A.A. & Sokoloff,D.D., 1983. *Magnetic fields in Astrophysics*, Gordon and Breach, New York. Vainshtein, S. & Kichatinov, L. L., 1986, J. Fluid. Mech., 168, 73. Kulsrud, R.M. & Anderson, S.W., 1992. ApJ., 396, 606.

[12] Subramanian, K. 1998, MNRAS, 294, 718; 1997, preprint, astro-ph/9708216.

[13] Subramanian, K, 1999 (paper in preparation).

[14] Meneguzzi, M., Frisch, U. & Pouquet, A., 1981. Phys. Rev. Lett., 47, 1060.

[15] Brandenburg, A., Jennings, R. L., Nordlund, A., Rieutord, M., stein, R. F. & Tuominen, I., 1996. JFM, 306, 325.

[16] Pouquet, A., Frish, U. & Leorat, J., 1976, JFM, 77, 321.

[17] Chandran, B. D. G., 1997, ApJ, 482, 156; ApJ, 485, 148.

[18] Vishniac, E. T. 1995, ApJ, 446, 724; Blackman, E., 1996. Phys. Rev. Lett., 77, 2694.

[19] Lazarian, A. & Vishniac, E. T. preprint, astro-ph/9811037

[20] Field, G., Blackman, E. & Chou, H., 1999. ApJ (in press)

[21] Beck, R., Poezd, A.D., Shukurov, A. & Sokoloff, D., 1994.. A & A, 289, 94.

Chapter 17

THE BLACK HOLE IN MCG 6-30-15

Ajit Kembhavi and Ranjeev Misra
Inter-University Centre for Astronomy and Astrophysics
Pune, India

This article is dedicated to Jayant Vishnu Narlikar, who to one of us has been teacher, mentor and punching-bag, and to the other an esteemed senior colleague.

Abstract We review the evidence for the existence of black holes in active galaxies, with particular reference to the Seyfert galaxy MCG 6-30-15.

In the astrophysical context, black holes are believed to have been found on two altogether different mass scales. Black holes with mass $\sim 1 - 10M_{\odot}$ have been inferred to exist in X-ray binary systems, in which mass transfer takes place from a more or less evolved star to a compact object. A clean way to establish that a compact object is a black hole is to show that it has a mass exceeding the maximum mass that a neutron star can have, which is $\sim 3M_{\odot}$. Unfortunately, estimates of the compact object mass depend on the orientation of the plane of the binary relative to the line of sight, and the nature of the companion star. In spite of these uncertainties, in some cases the *lower limit* on the compact object mass exceeds the neutron star mass limit. The most notable case of this sort is the binary system GS2023+38, in which the mass of the companion

N. Dadhich and A. Kembhavi (eds.), The Universe, 181–190.

is $\geq 6.26 \pm 0.31 M_{\odot}$. It seems inescapable that the compact object is a black hole.

Black holes are also believed to have been detected at the centre of a variety of galaxies, ranging from a *normal* one like M 31 (Andromeda) to active galaxies. The black hole mass here is orders of magnitude larger, going upto a $\sim 10^9 M_{\odot}$. There is firm photometric and spectroscopic evidence for the existence of a large mass concentration with high mass-to-light ratio M/L at the centres of some galaxies, and it is possible to argue that these are black holes. There are also several theoretical and observational considerations which make it plausible that active galactic nuclei, amongst which we include quasars, are powered by accretion of matter on to a supermassive black hole. Supermassive black holes could also exist at the centres of all, or at least a significant fraction of all galaxies, including our own.

The disappointing element in the stellar as well as galactic contexts sketched above has been the absence of any *direct evidence* for the existence of black holes. This can come from the detection of relativistic effects which can be found in the strong gravitational field close to the Schwarzschild radius. Detailed models used in the description of these observations should be able to rule out explanations based on matter distributions of high, but finite, density like neutron stars or dense star clusters.

It is possible that such a signature of the existence of a black hole has been found in the shape of iron emission lines at X-ray wavelengths observed at high resolution in the Seyfert galaxy MCG 6-30-15. The line has the double peaked shape which is expected in the emission from a compact rotating disk. Other features of the line shape point to the effects of strong gravity and highly relativistic motion. The case for the detection of a black hole has therefore become stronger than ever, and it is possible that with some improvement in the observational data, the issue will be unambiguously settled in favour of a black hole. We shall discuss below the case for black hole detection in MCG 6-30-15, as well as in some other galaxies.

1. WHY A BLACK HOLE?

Quasars and AGN have luminosities in the range $\sim 10^{42} - 10^{48}$ erg s^{-1}, with the emission spread across an immense bandwidth, ranging from low frequency radio ($\lesssim$100 MHz) to high energy $\gamma-$rays ($\lesssim$30 GeV); TeV $\gamma-$ray emission has also been seen in a few objects like the BL Lac MRK 421. The high frequency radiation is emitted from very compact regions of size $\lesssim 10^{12}$ cm around the nucleus. The intense emission lines in quasar and AGN spectra, found in the optical band and wavelength regions close to it, arise on the parsec and kiloparsec scale, while the radio radiation comes from vast radio lobes extending to hundreds of kiloparsec from the host galaxy. There is now enough observational evidence and theoretical modelling available to argue convincingly that the emission from the extended regions is ultimately powered by radiation and energetic particle beams which arise in close vicinity of the centre of the galaxy. Production of vast quantities of energy in a compact region, with attendant relativistic beams and bulk motion, over tens or hundreds of millions of years is best facilitated by matter falling into the deep gravitational well provided by a supermassive black hole. We will summarize below a few of the standard arguments in favour of such a model.

1.1 ARGUMENTS FAVOURING BLACK HOLES

If the observed flux from an object varies on a timescale of τ, then the spatial size of the emitting region is required to be $\lesssim c\tau$. If the region were bigger than this limit, different parts of the source would not be causally related over the variability time. The subregions would therefore vary independently, and the net amplitude of variability would be reduced considerably.

It has been found that the AGN in Seyfert galaxies, as well as many BL Lacs and related objects, show rapid and high amplitude variability in their X-ray flux. In some low luminosity Seyferts variations are found on the time scale of hours or less, and the X-ray flux from the Seyfert galaxy NGC 6814 varies by a factor of ~ 2 in less than 100 s (see [1] and [2]). The size of the emitting region is therefore $\lesssim 10^{11}$ cm, which is comparable with the Schwarzschild radius $3\times10^{12}(M/10^7 M_\odot)$ cm of a $\sim 10^7 M_\odot$ mass black hole. When the energy release is due to accretion of matter, a characteristic quantity is the Eddington luminosity $L_{\rm Edd} \simeq 10^{45}(M/M_\odot)\,{\rm erg\,s^{-1}}$, at which the outward radiation pressure on the accreting matter balances the inward force due to gravity. While this limit strictly applies only to spherical accretion, there are indications that it is rarely exceeded in real situations. If the rapidly variable emission occurs close to the Schwarzschild surface, then the size of the emitting region as well as the luminosity would be roughly proportional to the mass, and there are indeed strong indications that variability timescales are proportional to the luminosity of the object.

When observed on milliarcsecond scales, many compact radio sources are found to have features which appear to move at speeds which can be as high as $\sim 10c$. The jet like structures found on these scales, as well as on much large angular scales in radio quasars are invariably one-sided, even though the extended radio structures have more or less symmetric two-sidedness, with both sides showing signs of being continuously energized. These observations can most satisfactorily be explained as being due to highly relativistic bulk motion in the source. Such motion leads to the illusion of superluminal velocities because of light travel time effects when the motion is at a small angle to the line of sight. Moreover, a feature moving towards the observer appears to be very much brighter than a receding feature, leading to observed one-sidedness. The relativistic motions required in this model are are naturally to be expected in the presence of a deep gravitational potential well generated by a black hole.

The large scale collimated beams in radio sources maintain their direction extremely well over $\gtrsim 10^7$ yr. The beams are thought to emerge from the place of creation through funnels in bloated parts of an accretion disk around the nucleus. If there is a massive spinning black hole at the centre, the coupling of the disk to the spin angular momentum of the black hole can naturally maintain the direction of the funnel. Moreover, the spin energy can be extracted to power the beam through the Blandford-Znajek process (see [3]).

The above arguments, and several others, like the the high efficiency of the energy production in quasars which rules out nuclear or atomic processes for the generation, are all indicative of the gravitational origin of the energy, but none of them unequivocally requires the existence of a supermassive black hole. It is in principle possible for the gravitational potential well to be provided by e. g. a supermassive star, a dense star cluster, a collection of stellar mass black holes or spinars. But such models are hard put to explain highly energetic outbursts, rapid variability and collimated small scale jets. Moreover, as pointed out by Rees [4], all these systems undergo a runaway as the potential well gets deeper due to accretion, and will almost inevitably collapse to a single black hole. These considerations have led to the general, but not complete, acceptance of a the existence of a spinning black hole with mass $\sim 10^6 - 10^9 M_\odot$ as the *prime mover* of AGN. If it is true that normal galaxies of the present epoch harbour a now defunct AGN, then supermassive black holes could be present in all, or at least a good fraction, of all galaxies. The question then is, how do we detect these monsters?

2. DYNAMICAL EVIDENCE FOR THE EXISTENCE OF BLACK HOLES

If a supermassive black hole of mass M is present at the centre of a galaxy, it is expected to perturb the dynamics of the nuclear region around it to a distance GM/σ^2, where σ is the stellar velocity dispersion. The distribution function of stars in the region will be affected, changing the intensity distribution of stellar light, and the velocity dispersion of the stars from what is expected in the absence of the black hole. Attempts to detect such effects were first made by Young et al. [5] and Sargent et al. [6] from observations of the powerful radio galaxy M 87 situated in the Virgo cluster. The relatively nearby location (distance $\sim 15\,\text{Mpc}$) allows regions close to the centre of the galaxy to be probed even with the limited resolution available from ground based telescopes.

Young et al. found that the radial luminosity profile of M 87 in the V band contained a barely resolved spike in the centre, with a luminosity cusp extending to $\sim 10\,\text{arcsec}$. The luminosity profile inside this region could not be fit by the isothermal King model profiles which are used in normal elliptical galaxies. Combining the photometric data with spectra obtained by Sargent et al. , which showed that the stellar velocity dispersion in the core of the galaxy was $278 \pm 11\,\text{km}\,\text{s}^{-1}$, Young et al. obtained a value of $3\times10^9 M_\odot$ for the black hole. In fact a model independent analysis of the photometric and spectroscopic data showed that the nucleus contains a compact mass, located inside a radius $< 100\,\text{pc}$, and having a mass-to-light ratio $M/\mathcal{L} > 60$. A supermassive black hole is a very plausible candidate for the compact object.

The black hole models of Young et al. and Sargent et al. have been questioned over the years. On the observational side, it has been found that M 87 like profiles with cusps are found in a number of galaxies some of which, like M 33 have low central dispersion velocities which show that a supermassive black hole cannot be present. On the theoretical side, it has been shown that the assumption of an isotropic distribution of velocities is critical to black hole interpretation. Models in which the radial component of the stellar dispersion velocity is greater than the tangential component can reproduce the observed properties of M 87 without requiring the presence of a compact dark object. Such anisotropic velocity dispersions are common in giant ellipticals (see [7] for a detailed discussion of these issues and references).

Using observations from the Hubble Space Telescope (HST), Ford et al. [8] have been able to discover what appears to be a rotating disc of ionized gas surrounding the nucleus of M 87. The disc is approximately normal to the optical jet observed in the galaxy. Harms et al. [9] have obtained spectra of the disk like structure, centred on the nucleus and at $\sim 18\,\mathrm{pc}$ on either side of it, along the major axis of the disc. The radial velocities of the gas in the disc at the two positions have been found to be $\sim \pm 500\,\mathrm{km\,s^{-1}}$ relative to the systemic velocity of the galaxy. When combined with other spectra obtained close to the nucleus, the data show that the gas disc is in Keplerian rotation about a mass of $2.4 \pm 0.7 \times 10^9 M_\odot$ within 18 pc of the nucleus with a mass-to-light ratio of $M/\mathcal{L} \simeq 170$. This is a shot in the arm for the black hole model, but observations closer still to the nucleus are required to make the case stronger. Miyoshi et al. [10] have observed line emission at radio wavelengths from water masers very close to the nucleus of the galaxy NGC 4258 with the Very Long Baseline Array. The emission appears to arise in a nearly edge on rotating disc, with the rotation velocity decreasing as $r^{-1/2}$ with distance r from the centre, as is expected in Keplerian motion. The mass of $3.6 \times 10^7 M_\odot$ around which the Keplerian motion takes place is located within a radius 0.13 pc, which corresponds to a mass density $> 4 \times 10^9 M_\odot\,\mathrm{pc}^{-3}$, which is > 40 times in excess of the density of other black hole candidates.

There are several cases of possible black holes in active as well as in normal galaxies, and a good summary about these may be found in [7]. Even in the best of these optical and radio observations, the centre is not approached close enough to measure essentially relativistic signatures which will help eliminate alternatives to the black hole model. We will see below how observations at X-ray wavelengths may provide the clinching evidence.

3. THE IRON LINE PROFILE IN MCG-6-30-15

The X-ray spectra of a majority of AGN have been represented as a power-law emission extending at least up to 200 keV. A region producing such X-rays has to be necessarily hot, i. e., with temperature around 10^8 K. It is now also known that there is colder (i.e. with a temperature less than 10^7 K) matter in the vicinity of the X-ray source. Such a medium would reflect a significant fraction of the X-rays impinging on it. In particular, the high energy photons (i.e. $E > 100$ keV) would be Compton down-scattered while the low energy photons (i.e. $E < 20$ keV) would be absorbed by the partially ionized atoms in the medium. The net result should be a broad feature centered around 50 keV. It is now confirmed that many AGN have this additional broad feature around 50 keV, thus confirming that cold matter is indeed present in the vicinity of the X-ray source (e. g. [11]).

Another important signature of the cold medium is the Iron Kα X-ray fluorescent line. The line energy here is a slowly increasing function of the ionization state of Iron, rising from 6.4 keV for Fe I to 6.45 in Fe XVII, but then increasing steeply to 6.7 keV in Fe XXV and 6.9 in Fe XXVI ([12] and references therein.). This Iron line (at 6.4 keV) has also been observed in most AGN thereby reconfirming the model.

The reflecting material is probably in the form of a cold accretion disk around a massive black hole. The Iron Line emission from the inner disk region would then be red shifted or blue shifted due to gravitational and Doppler effects. Since different regions of the disk give rise to different line energies, the net emission will be seen as broad line. The detailed calculation of the line shape for different geometries and viewing angles have been computed by Fabian et al. [13] and Laor [14]. In most cases, the shape is double peaked. Since the reflecting material is cold, the thermal broadening of the Iron line is expected to be not detectable (i. e., $\sigma << 0.1$ keV).

It was not until the launch of the Japanese-US X-ray satellite ASCA (Advanced satellite for Cosmology and Astrophysics) in February 1993 that this prediction could be tested. ASCA carries low background CCD X-ray detectors that have an unprecedented spectral resolution of 160 eV around 6.4 keV [15]. A long ($\approx$ 4.5 days) observation by ASCA of the Seyfert 1 MCG 6-30-15 revealed that the Iron line profile in this source is broad with velocity width of $> 10^{10}$ cm sec^{-1} [16]. The results also showed that the line profile has a skewed red wing which was in agreement from the inner disk line profiles calculated. The analysis showed that the spectral shape matched the predicted emission from a disk with inner and outer radii at $6r_g$ and $20r_g$ respectively, where $r_g = GM/c^2$. The inclination of the disk was also constrained to be around $30^o \pm 3$. The data and the corresponding fit to the disk line model of Laor [14] is shown in Figure 17.1.

A more detailed analysis of the data by Iwasawa et al. [17] showed that the profile is variable. They also showed that when the source intensity is minimum, the disk producing this line has to have an inner radius $\sim 1.2r_g$, which implies that the black hole is spinning close to its maximal value. If this interpretation is correct, then this is the first direct observation of the strong gravitational effects expected in the vicinity of a black hole. This would strongly constrain theoretical models since the inner disk region has to be cold in order to produce the Iron line while at the same time the hard X-ray producing region would also have to be located close to the inner edge of the disk.

Sulentic et al. [18] showed there was a disagreement between the inclination angle derived by fitting the Iron line profile and the angle

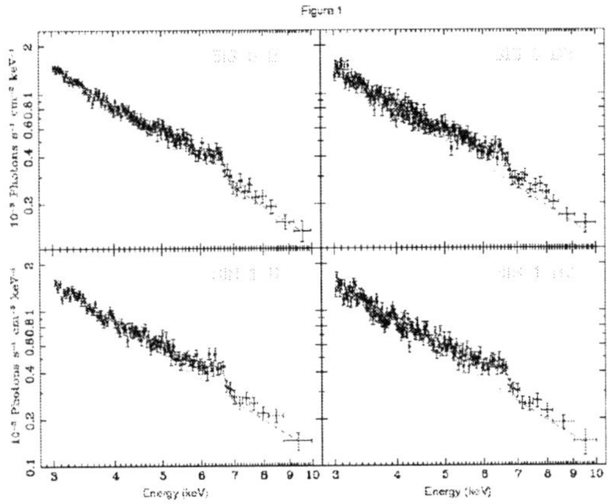

Figure 17.1 The unfolded spectra and best fit disk line model of Laor [14]. The medium intensity data from both chips (SIS0 and SIS1) for Bright (B) and Bright2 (B2) modes are shown separately. The dotted line is the power-law while the solid line is the sum of the power-law and Iron line profile.

obtained from HI and Hα measurements. Several AGN for which a broad Iron line was detected have the peak centroid close to 6.4 keV [19]. This is not expected [20] if the disks are oriented in random directions. They propose that the Iron line profile is a sum of two independent Gaussian lines in order to explain the different correlation relationships of the red and blue side with the source intensity. They also claimed that there is a blue wing in the Iron line emission which cannot be explained by the disk line model. These discrepancies and the importance of the implication of the disk line model warrant a study of alternate models to explain the phenomenon. Due to the profound implications of being able to 'observe' the immediate vicinity of a super massive black hole, it becomes important to examine models for the broad Iron line that do not require strong gravity.

As an alternative to the disk line model, Czerny, Zbyszewska and Raine [21] proposed that the line is intrinsically narrow and it gets broadened due to Compton down scattering of the photons as they pass through an optically thick cloud. This model is referred to here as the Comptonization model. Fabian et al. [22] rejected the Comptonization model by arguing that the surrounding Comptonizing cloud has to have a radius $R < 10^{14}$ cm in order that the cloud be highly ionized and does not produce strong absorption lines. For a $10^7 M_\odot$ black hole this would imply that the Iron line producing region is smaller than $50 r_g$ and gravitational effects would be important. The lack of a blue wing in the best disk-line fit to the line profile, implies that the temperature of the cloud should be $kT < 0.2\,\text{keV}$, which is in apparent conflict with the expected Compton temperature of the cloud. Further, a break in the continuum around 20 keV is expected for the Comptonization model which has not been observed. On the other hand, Misra & Kembhavi [23] pointed out that for a smaller sized black hole, the intrinsic Iron line produced may not be significantly broadened by gravitational effects. They calculated the equilibrium temperature of the cloud to be around 0.2 keV provided an intense UV source is assumed to be present in this source. They showed that the present broad band data for this source is consistent with the Comptonization model. Misra & Sutaria [24] showed that the Comptonization model indeed fits ASCA data for MCG-6-30-15 well. It was pointed out by Misra & Kembhavi (1998) that broad band data for this source (1-300 keV) will be able to distinguish between the two models. Data from the Italian BeppoSAX satellite indicates that the Comptonization model is not compatible with the broad band spectrum of the source [25]. In particular, the Comptonization model predicts a rolling off in the spectrum around 100 keV which is not observed. Thus the disk model remains as the most natural model to explain the observed broad Iron line feature in AGN.

Despite the success of the disk model, there are still some inconsistencies as noted by Sulentic et al. [20]. Firstly, the disagreement of the inclination angles derived by fitting the Iron line profile and that from HI and Hα measurements indicates that line profile is perhaps more complex than has been modeled. There are also theoretical arguments against the X-ray source being close to the marginal stable orbit rather than at around $10r_g$ where most of the gravitational energy is dissipated. Perhaps, a double component model (i.e. the emission arises from two distinct regions in the disk) can resolve the discrepancies. These questions will be answered by the next generation of X-ray satellites with still better resolution than ASCA.

References

[1] Mushotzky, R.F., Done, C. & Pounds, K.A., 1993, *Ann. Rev. Astron. Astrophys.* **31**, 717.

[2] Kembhavi, A.K. & Narlikar J.V., 1999, *Quasars and Active Galactic Nuclei*, Cambridge University Press.

[3] Blandford, R., 1990, in *Active Galactic Nuclei*, ed. Courvoisier, T.J.-L. & Mayor, M. Springer Verlag.

[4] Rees, M. J., 1984, *Ann. Rev. Astron. Astrophys.* **22**, 471.

[5] Young, P.J. et al. , 1978, *Astrophys. J.* **221**, 721.

[6] Sargent, W.L.W. et al. , 1978, *Astrophys. J.* **221**, 731.

[7] Kormendy J. and Richstone D., 1995, *Ann. Rev. Astron. Astrophys.* **33**, 581.

[8] Ford, H.C. et al. , 1994, *Astrophys. J.* **435**, L27.

[9] Harm R.J. et al. , 1994, *Astrophys. J.* **435**, L35.

[10] Miyoshi M. et al. , 1995, *Nature* **373**, 127.

[11] Nandra, K. & Pounds, K.A. 1994, *Mon. Not. Roy. astr. Soc.* **268**, 405.

[12] George, I.M. & Fabian, A.C. 1991, *Mon. Not. Roy. astr. Soc.* **249**, 352.

[13] Fabian, A.C., Rees, M.J., Stella, L. & White, N.E. 1989, *Mon. Not. Roy. astr. Soc.* **238**, 729.

[14] Laor, A., 1991, *Astrophys. J.* **376**, 90.

[15] Tanaka, Y., Inoue, H. & Holt, S.S., 1994, *Proc. Astron. Soc. Japan* **46**, L137

[16] Tanaka, Y. et al. , 1995, *Nature* **375**, 659.

[17] Iwasawa, K. et al. 1996, *Mon. Not. Roy. astr. Soc.* **282**, 1038.

[18] Sulentic, J. et al., 1998, *Astrophys. J.* **501** 54.

[19] Nandra, K., George, I.M., Mushotzky, R.F., Turner, T.J. & Yaqoob, T. 1997, *Astrophys. J.* **477**, 602.

[20] Sulentic, J., Marziani, P. & Calvani, M., 1998, *Astrophys. J.* **497**, L65

[21] Czerny B., Zbyszewska M. & Raine, D.J. 1991, in *Iron line Diagnostics in X-ray Sources*, ed. Treves A., Springer-Verlag, Berlin, p 226.

[22] Fabian, A.C. et al. 1995, *Mon. Not. Roy. astr. Soc.* **277**, L11.

[23] Misra, R. & Kembhavi, A.K., 1998, *Astrophys. J.* **499**, 205.

[24] Misra, R. & Sutaria, F.K., 1999, *Astrophys. J.* **517**, 661.

[25] Misra, R. & Sutaria, F.K., 1999, private communication.

Chapter 18

INHOMOGENEOUS COSMOLOGICAL MODELS AND SYMMETRY

S. D. Maharaj

School of Mathematical and Statistical Sciences

University of Natal

Durban 4041

South Africa

It is a pleasure to dedicate this work to Jayant Narlikar on his sixteeth birthday; Jayant's substantial contributions to to cosmology have left a lasting impact on the subject.

—

Abstract Inhomogeneous cosmological models are studied extensively in the literature, in particular when the shear vanishes. The integrability properties of the field equation $L_{xx} = F(x)L^2$ of a spherically symmetric shear–free fluid are reviewed. A first integral, subject to an integrability condition on $F(x)$, is found which generates a class of solutions which contains the solutions of Stephani (1983) and Srivastava (1987) as special cases. The integrability condition on $F(x)$ is reduced to a quadrature. The Lie procedure for this equation is considered and we list various forms of $F(x)$ and their Lie symmetry generators. A conformal Killing vector in the t–r plane is assumed to exist and for this particular case the solution to the field equation is expressible in terms of Weierstrass elliptic functions.

N. Dadhich and A. Kembhavi (eds.), The Universe, 191–199.

1. INTRODUCTION

Spherical symmetry is often assumed when seeking exact solutions to the Einstein field equations. Spherically symmetric solutions may be used as inhomogeneous cosmological models, or to model the interior of expanding or contracting spherical stars (Santos 1985, de Oliveira *et al* 1985). The shear–free condition substantially simplifies the field equations and most of the exact solutions known are not shearing (Kramer *et al* 1980, Krasinski 1997). Various approaches have been followed in seeking solutions to these equations. McVittie (1933, 1967, 1984) assumed a functional form of the metric coefficients. The solutions found by Kustaanheimo and Qvist (1948), Chakravarty *et al* (1976), Wyman (1976) and Stephani (1983) depend on a suitable choice of a function of integration. A more recent approach by Herrera and Ponce de Leon (1985), Dyer *et al* (1987), Sussman (1989), Maartens and Maharaj (1990) and Maharaj *et al* (1991) is to suppose that the spacetime is invariant under a conformal Killing vector.

One of the field equations for a spherically symmetric shear–free fluid can be reduced to the partial differential equation $L_{xx} = F(x)L^2$. The purpose of this paper is to review integrability properties of this equation. In section 2 we give the field equations for a shear–free matter distribution. A condition on the function of integration $F(x)$ is found in section 3 that yields a first integral of $L_{xx} = F(x)L^2$ and we express this condition as a third order ordinary differential equation. In section 4 we indicate that it is possible to perform a Lie analysis on this equation. The Lie point symmetries for particular forms of the function $F(x)$ are listed. Finally is section 5 we indicate, with the help of an example, how imposing a conformal symmetry on spacetime leads to new solutions.

2. FIELD EQUATIONS

For a spherically symmetric, shear–free, perfect fluid we can introduce a comoving and isotropic coordinate system $x^i = (t, r, \theta, \phi)$ such that the metric is

$$ds^2 = -e^{2\nu(t,r)}dt^2 + e^{2\lambda(t,r)}\left[dr^2 + r^2(d\theta^2 + \sin^2\theta d\phi^2)\right] \qquad (1)$$

Under these conditions the Einstein field equations become

$$e^{\nu} = \lambda_t e^{-f(t)} \qquad (2)$$

$$e^{\lambda}(\lambda_{rr} - \lambda_r^2 - \lambda_r/r) = -\tilde{F}(r) \qquad (3)$$

where $f(t)$ and $\tilde{F}(r)$ are arbitrary functions of integration for (1). The energy density μ and pressure p then assume the form

$$\mu = 3e^{2f} - e^{-2\lambda}(2\lambda_{rr} + \lambda_r^2 + 4\lambda/r) \qquad (4)$$

$$p = \lambda_t^{-1} e^{-3\lambda} \partial_t [e^{\lambda}(\lambda_r^2 + 2\lambda_r/r) - e^{3\lambda+2f})] \tag{5}$$

Note the solution of equations (2)–(5) may be simplified. The transformation $x = r^2$, $L(t,x) = e^{-\lambda}$, $\tilde{F}(x) = F/4r^2$ reduces equation (3) to

$$L_{xx} = F(x)L^2 \tag{6}$$

which is the governing equation.

In spite of the great interest generated by shear–free perfect fluids the solution of the field equations in complete generality has been studied extensively only for (Kustaanheimo and Qvist 1948)

$$F(x) = (ax^2 + bx + c)^{-5/2}$$

and (Stephani 1983)

$$F(x) = x^{-15/7} \quad \text{or} \quad x^{-20/7}$$

Many of the solutions published previously, using a variety of techniques of integration, are contained in these classes. Incidentally subclasses of the McVittie (1984) class of metrics also correspond to these forms of $F(x)$. We extend this to a new class of solutions in section 3.

3. A FIRST INTEGRAL

The technique of Srivastava (1987) may be extended to obtain a first integral of (6) without choosing an explicit functional form for $F(x)$. The first integral found is subject to an integral equation in $F(x)$ which can also be expressed as a nonlinear third order ordinary differential equation. Maharaj *et al* (1996) show that the first integral is given by

$$\psi_0(t) = -L_x + F_I L^2 - 2F_{II} L L_x + 2F_{III} L_x^2 + 2\left[(FF_{II})_I - \tfrac{1}{3}K_0\right] L^3 \tag{7}$$

when we impose the condition

$$2FF_{III} + 3(FF_{II})_I = K_0 \tag{8}$$

where $\psi_0(t)$ is an arbitrary function of integration and K_0 is a constant. It is possible to transform the integral equation (8) into an ordinary differential equation. If we make the the transformation $\mathcal{F} \equiv F_{III}$ then (8) can be written as

$$\mathcal{F}_{xxx} = K_1 \mathcal{F}^{-5/2} \tag{9}$$

where K_1 is a constant of integration. Repeated integration of (9) yields the result

$$\mathcal{F}^{-1} = K_4 + K_3\left(\int \mathcal{F}^{-3/2}dx\right) + K_2\left(\int \mathcal{F}^{-3/2}dx\right)^2 - \frac{1}{6}K_1\left(\int \mathcal{F}^{-3/2}dx\right)^3 \tag{10}$$

where K_2, K_3, K_4 are constants of integration. We now let

$$u = \int \mathcal{F}^{-3/2} dx$$

which gives the result

$$x - x_0 = \int \frac{du}{\left(K_4 + K_3 u + K_2 u^2 - \frac{1}{6} K_1 u^3\right)^{3/2}}, \tag{11}$$

where x_0 is constant. Thus we have reduced the differential equation (9) to the quadrature (11) which can be evaluated in terms of elliptic integrals in general (Gradshteyn and Ryzhik 1994).

We have established that for our class of first integrals the form of F may be given in parametric form in general. It is interesting to observe that the solution of Kustaanheimo and Qvist (1948) is not contained in the first integral (7). Note that the first integral (7) was obtained without specifying the function $F(x)$. Any $F(x)$ satisfying (8) will yield a first integral of the form (11). With the choice $F(x) = (ax + b)^n$ where $n = -15/7$ we regain the first integral reported by Srivastava (1987). With $a = 1$ and $b = 0$ equation we regain the first integral found by Stephani (1983).

4. LIE ANALYSIS

In section 3 we obtained the first integral (7) of (6) in an arbitrary fashion. We now consider a geometric technique to reduce (6) to a first order differential equation. A systematic method to determine whether a second order ordinary differential equation can be solved by quadratures is that of Lie (1912). On performing the Lie analysis (Leach 1981, Leach *et al* 1992) we find that (6) is invariant under the infinitesimal transformation

$$\mathbf{G} = A(x)\frac{\partial}{\partial x} + [B(x)L + C(x)]\frac{\partial}{\partial L}$$

provided that

$$F(x) = MA^{-5/2} \exp\left(\frac{\alpha}{2}\int \frac{dx}{A}\right) \tag{12}$$

$$A_{xxx} = 2MCA^{-5/2} \exp\left(\frac{\alpha}{2}\int \frac{dx}{A}\right) \tag{13}$$

$$B(x) = (A_x - \alpha)/2$$

$$C(x) = C_0 + C_1 x$$

where α, M, C_0 and C_1 are constants. A solution of (13) gives $F(x)$ by (12) which reduces (6) to a first order differential equation. Note that the existence of the symmetry **G** permits a reduction to first order. If two symmetries **G** are known, the second order equation (6) may be reduced to quadratures.

For a comprehensive analysis of the Lie analysis to shear–free fluids see Maharaj *et al* (1996) and Stephani and Wolf (1996). As an example we let $\alpha = 0, C = 0$. This corresponds to the simplest case and (12) yields

$$F(x) = M(ax^2 + bx + c)^{-5/2}$$

which is equivalent to the Kustaanheimo and Qvist (1948) class of solutions. As a second example we let $\alpha = 0, C_0 \neq 0 = C_1$. In this case (13) reduces to the differential equation

$$A_{xxx} = K_1 A^{-5/2}$$

where $K_1 = 2MC_0$. This is the same as equation (9) considered in section 3 and the results of that section become applicable. Thus we have established that the first integral (7) is a special class of solutions admitted by the general Lie method. Clearly the remaining cases are more complex.

There are either one or two Lie symmetries S_1 or S_2 depending on the form of $C(x)$. The various cases are listed in Leach and Maharaj (1992). The analysis to obtain the Lie symmetries S_1 or S_2 is lengthy and tedious. Here we list the functions $F(x)$ for which (6) can be solved:

$f(x) = x^m$	$S_1 = x\frac{\partial}{\partial x} - (m+2)y\frac{\partial}{\partial y}$
$f(x) = x^{-5}$	$S_1 = x\frac{\partial}{\partial x} + 3y\frac{\partial}{\partial y}$
	$S_2 = x^2\frac{\partial}{\partial x} + xy\frac{\partial}{\partial y}$
$f(x) = e^x$	$S_1 = \frac{\partial}{\partial x} - y\frac{\partial}{\partial y}$
$f(x) = (x+\alpha)^m(x+\beta)^{-(m+5)}$	$S_1 = [\alpha\beta + (\alpha+\beta)x + x^2]\frac{\partial}{\partial x}$
$(\alpha \neq \beta)$	$+[(m+3) - (m+2)\beta + x]y\frac{\partial}{\partial y}$
$f(x) = (ax^2 + 2bx + c)^{-5/2}$	$S_1 = (ax^2 + 2bx + c)\frac{\partial}{\partial x} + (ax+b)y\frac{\partial}{\partial y}$
$f(x) = x^{-15/7}$	$S_1 = -7x\frac{\partial}{\partial x} - y\frac{\partial}{\partial y}$
	$S_2 = 343x^{6/7}\frac{\partial}{\partial x} + (147x^{-1/7}y - 12)\frac{\partial}{\partial y}$
$f(x) = x^{-20/7}$	$S_1 = 7x\frac{\partial}{\partial x} + 6y\frac{\partial}{\partial y}$
	$S_2 = 343x^{8/7}\frac{\partial}{\partial x} + (196x^{1/7}y - 12x)\frac{\partial}{\partial y}$

The various functions $F(x)$ and their corresponding solutions to the Einstein field equations for shear–free spherically symmetric spacetimes are considered in a variety of publications. Systematic treatments are given in Maharaj *et al* (1996), Srivastava (1987), Stephani (1983), and Stephani and Wolf (1996). Leach and Maharaj (1992) perform a more general analysis for equations that include (6) as a special case. Their treatment can be applied to different physical applications, e.g. the study of a spherical gas cloud acting under the mutual attractions of its molecules and subject to the laws of thermodynamics.

5. CONFORMAL SYMMETRY

In trying to solve the highly nonlinear field equations of general relativity it is sometimes assumed that the spacetime admits symmetries. We analyse the example of Dyer *et al* (1987) who assume a particular form of a conformal Killing vector in the t–r plane in spherically symmetric shear–free perfect fluid spacetimes. If the spacetime is invariant under a conformal Killing vector ξ then

$$\mathcal{L}_\xi g_{ij} = 2\phi g_{ij} \tag{14}$$

where ϕ is the conformal factor. The particular ξ chosen is

$$\xi^i = t\delta^i{}_t + r\delta^i{}_r$$

The motivation for the choice of ξ is that subclasses of McVittie's solutions (1984) admit a conformal Killing vector in the t–r plane. For further details see Havas (1992) and Maharaj *et al* (1991).

The conformal Killing equations (14) become

$$\begin{aligned} r\frac{B'}{B} + t\frac{\dot{B}}{B} &= \phi - 1 \\ r\frac{A'}{A} + t\frac{\dot{A}}{A} &= \phi - 1 \end{aligned}$$

for the line element (1) where we have set $A = e^\nu$, $B = e^\lambda$. These imply

$$\frac{A}{B} = \sigma(r/t) \equiv \sigma(\mu)$$

where σ is an arbitrary function. The field equations give the condition

$$\left(\frac{B'}{B}\right)' - \left(\frac{B'}{B}\right)^2 - \left(\frac{B'}{B}\right)\frac{1}{r} = \frac{T(\mu)}{r^2}$$

We deduce that the following hold

$$\begin{aligned} T(\mu) &= (1/2\sigma)(\mu\sigma_\mu - \mu^2\sigma_{\mu\mu}) \\ \sigma &= \mu^m T_\mu \end{aligned}$$

The equation that T has to satisfy is given by

$$\mu^2 T_{\mu\mu\mu} + \mu(2m-1)T_{\mu\mu} + (m^2 - 2m + 2T)T_\mu = 0 \tag{15}$$

This third–order field equation is highy nonlinear but has a solution in terms of a Painlevé transcendent. We observe that (15) has a first integral

$$\mu^2 T_{\mu\mu} + (2m-3)\mu T_\mu + (m-1)(m-3)T + T^2 = k \tag{16}$$

where k is a constant. We make the change of variables

$$T(\mu) = \gamma y(x) + T_0, \quad x = \ln(\mu)/\beta$$

where T_0 is the constant solution of (16), given by

$$T_0^2 + (m-1)(m-3)T_0 - k = 0$$

Then (16) becomes

$$\frac{d^2y}{dx^2} + 2(m-2)\beta\frac{dy}{dx} + k\beta^2 y + \beta^2\gamma y^2 = 0 \tag{17}$$

Equation (17) is of Painlevé form (Kamke 1971, eq 6.23, p 547]). The solution for y is then given by

$$y = a^2 C_1^2 e^{-2ax}\mathcal{P}(C_1 e^{-ax} + C_2, 0, -1)$$

where C_1, C_2 are arbitrary constants. The quantity $\mathcal{P}$ is a Weierstrass elliptic integral.

This example illustrates that imposing a symmetry, in our case a conformal symmetry, enables us to integrate the field equations. This approach is an alternate mechanism, utilising a geometric structure on the manifold, to generate solutions and often proves to be fruitful.

References

[1] Chakravarty N, Dutta Choudhury S B and Banerjee A 1976 *Austral. J. Phys.* **29** 113

[2] de Oliveira A K G, Santos N O and Kolassis C A 1985 *Mon. Not. R. Astr. Soc.* **216** 1001

[3] Dyer C C, McVittie G C and Oates L M 1987 *Gen. Relat. Grav.* **19** 887

[4] Gradshteyn I S and Ryzhik I M 1994 *Table of Integrals, Series and Products* (New York: Academic Press)

[5] Havas P 1992 *Gen Relat. Grav* **24** 599

[6] Herrera L and Ponce de Leon J 1985 *J. Math. Phys.* **26** 778, 2018, 2847

[7] Kamke E 1971 *Differentialgleichungen Lösungsmethoden Und Lösungen: Band 1, Gewöhnliche Differentialgleichungen, 3. Auflage* (New York: Chelsea Publishing)

[8] Kramer D, Stephani H, MacCallum M A H and Herlt E 1980 *Exact Solutions of Einstein's Field Equations* (Cambridge: Cambridge University Press)

[9] Krasinski A 1997 *Inhomogeneous Cosmological Models* (Cambridge: Cambridge University Press)

[10] Kustaanheimo P and Qvist B 1948 *Soc. Sci. Fennica, Commentationes Physico-Mathematicae* **XIII** 16

[11] Leach P G L 1981 *J. Math. Phys.* **22** 465

[12] Leach P G L and Maharaj S D 1992 *J. Math. Phys.* **33** 465

[13] Leach P G L, Maartens R and Maharaj S D 1992 *Int. J. Nonlin. Mech.* **27** 575

[14] Lie S 1912 *Vorlesungen über Differentialgleichungen* (Leipzig und Berlin: Teubner)

[15] Maartens R and Maharaj M S 1990 *J. Math. Phys.* **31**

[16] Maharaj S D, Leach P G L and Maartens R 1991 *Gen. Relat. Grav.* **23** 261

[17] Maharaj S D, Leach P G L and Maartens R 1996 *Gen. Relat. Grav.* **28** 35

[18] McVittie G C 1933 *Mon. Not. R. Astr. Soc.* **93** 325

[19] McVittie G C 1967 *Ann. Inst. Henri Poincaré* **6** 1

[20] McVittie G C 1984 *Ann. Inst. Henri Poincaré* **40** 325

[21] Santos N O 1985 *Mon. Not. R. Astr. Soc.* **216** 403

[22] Srivastava D C 1987 *Class. Quantum Grav.* **4** 1093

[23] Stephani H 1983 *J. Phys. A: Math. Gen.* **16** 3529

[24] Stephani H and Wolf T 1996 *Class. Quantum Grav.* **13** 1261

[25] Sussman R A 1989 *Gen. Relat. Grav.* **12** 1281

[26] Wyman M 1976 *Can. Math. Bull.* **19** 343

Chapter 19

THE BLACK HOLE INFORMATION PARADOX: WHAT HAVE WE LEARNT FROM STRING THEORY?

Samir D. Mathur
Massachusetts Institute of Technology
77 Massachusetts Avenue
Cambridge MA 02139, USA

Abstract In a complete theory of quantum gravity and matter we must come to grips with the information paradox that is created when black holes form and evaporate. If the paradox is to be resolved within the framework of quantum mechanics as we know it, then we arrive at some very specific requirements from the theory of quantum gravity – the degeneracy of states must reproduce the Bekenstein entropy of black holes, and the interactions must give rise to unitarity preserving Hawking radiation. In the past few years string theory has had remarkably success in reproducings these requirements from black holes. We review some of these developments in this article.

1. INTRODUCTION

Black holes have furnished us with a very deep paradox. The path to resolving this paradox may well be the path to arriving at a consistent unified theory of matter and quantised gravity.

Let us review the basic nature of black holes, to see how the information paradox arises. We imagine a large collection of low density matter, in an asymptotically flat spacetime. For simplicity we take the starting configuration to be spherically symmetric and nonrotating - these restrictions do not affect the nature of the paradox that emerges. This ball of matter will collapse towards smaller radii under its self-gravition. At some point the matter will pass through a critical radius, the Schwarzschild radius R_s, after which its further collapse cannot be halted, whatever be the equation of state. The end result, in classi-

N. Dadhich and A. Kembhavi (eds.), The Universe, 201–212.

cal general relativity, is that the matter ends up in an infinite density singular point, while the metric settles down to the Schwarzschild form

$$ds^2 = -(1 - \frac{2M}{r})dt^2 + (1 - \frac{2M}{r})dr^2 + r^2 d\Omega^2 \qquad (1)$$

1.1 THE ENTROPY PROBLEM

Already, at this stage, one finds what may be called the 'entropy problem'. One of the most time honoured laws in physics has been the second law of thermodynamics, which states that the entropy of matter in the universe cannot decrease. But with a black hole present in the Universe, one can imagine the following process. One takes a box with some gas, say, which has a certain entropy, and then drops this box into a large black hole. The metric of the black hole then soon settles down to the Schwarzschild form above, though with a larger value for M, the black hole mass. The entropy of the gas has vanished from view, so that if we only count the entropy that we can explicitly see, then the second law of thermodynamics has been violated!

This violation of the second law can be avoided if one associates an entropy to the black hole itself. Starting with the work of Bekenstein [1] we now know that if we associate an entropy

$$S_{BH} = \frac{A_H}{4G_N} \qquad (2)$$

with the black hole (A_H is the area of the horizon, and G_N is the Newton's constant) then in any Gedanken experiment where we try to lose entropy down the hole, the increase in the black hole's attributed entropy is such that

$$\frac{d}{dt}(S_{matter} + S_{BH}) \geq 0$$

This would suggest that the proposal (2) is a nice one, but now we encounter the following problem. We would also like to believe on general grounds that the entropy of any system is the logarithm of the number of states of the system, for a given value of the macroscopic parameters of the system. For a black hole of one solar mass, this implies that there should be $10^{10^{78}}$ states!

But the metric (1) of the hole suggests a unique state for the geometry of the configuration. If one tries to consider small fluctuations around this metric, or adds in, say, a scalar field in the vicinity of the horizon, then the extra fields soon flow off to infinity or fall into the hole, and the metric again settles down to the form (1).

If the black hole has a unique state, then the entropy should be $\ln 1 = 0$, which is not what we expected from (2). The idea that the black hole configuration is uniquely determined by its mass (and any other conserved charges) arose from studies of many simple cases for the matter fields. This idea of uniqueness was encoded in the statement 'Black holes have no hair'.

As it turned out this statement was not really a theorem, and in a larger class of matter fields (which arise naturally in string theory) it is in fact not true. One may say that it is a very interesting and precise requirement on the theory of quantum gravity plus matter that there be indeed just the number of states given through (2) for a black hole with given mass.

1.2 HAWKING RADIATION

If Black holes have an entropy S_{BH}, and an energy equal to the mass M, then if thermodynamics were to be valid, we would expect them to have a temperature given by

$$TdS = dE = dM, \quad T = \frac{dS}{dM} = \frac{1}{8\pi G_N M} \tag{3}$$

Again assuming thermodynamical behavior, the above statement implies that if the hole can absorb at a given wavelength with absorption cross section σ, then it must also radiate at the same wavelength at a rate

$$\Gamma = \sigma \frac{1}{e^{\frac{\omega}{kT}} - 1} \frac{d^3k}{(2\pi)^3} \tag{4}$$

Classically, nothing can come out of the black hole horizon, so one may be tempted to say that no such radiation is possible. But quantum mechanically we find that the vacuum for the matter fields has fluctuations, so that pairs of particles and antiparticles are produced and annihilated continuously. Because of the gravitational field of the hole, one member of this pair can fall into the hole, where it has a net negative energy, while the other member of the pair can escape to infinity as real positive energy radiation [2]. The profile of this radiation is found to be thermal, with a temperature

$$T = \frac{1}{8\pi G_N M}$$

in accord with the expectation (3).

1.3 THE INFORMATION PROBLEM

But the above 'Hawking radiation' was produced from the quantum fluctuations of the matter vacuum, in the presence of the gravitational field of the hole. The gravitational field near the horizon, where the particle pairs are produced in this simple picture, is given quite accurately by the classical metric (1). The curvature invariants at the horizon are all very small compared to the planck scale, so quantum gravity seems to not be required. Further, the calculation is insensitive to the precise details of the matter which went to make up the hole. Thus if the hole completely evaporates away, the final radiation state cannot have any significant information about the initial matter state. This circumstance would contradict the assumption in usual quantum mechanics that the final state of any time evolution is related in a one-to-one and onto fashion to the initial state, through a unitary evolution operator. Worse, the final state is in fact not even a normal quantum state. The outgoing member of a pair of particles created by the quantum fluctuation is in a mixed state with the member that falls into the hole, so that the outgoing radiation is highly 'entangled' with whatever is left behind at the hole. If the hole completely evapotates away, then this final state is entangled with 'nothing', and we find that the resulting system is described not by a pure quantum state but by a mixed state.

If the above reasoning and computations are correct, one confronts a set of alternatives, none of which are very palatable. The semiclassical reasoning used in the derivation of Hawking radiation cannot really say if the hole continues to evaporate after it reaches planck size, since at this point quantum gravity would presumably have to be important. Thus the hole may not completely evaporate away, but leave a 'remnant' of planck size. The radiation sent off to infinity will remain entangled with this remnant. But this entanglement entropy is somewhat larger than the black hole entropy S_{BH}, which is a very large number as we have seen above. Thus the remnant will have to have a very large number of possible states, and this number will grow to infinity as the mass of the initial hole is taken to infinity. It is uncomfortable to have a theory in which a particle of bounded mass (planck mass) can be allowed to have an infinte number of configurations. One might worry that in any quantum process one can have loops of this remnant particle, and this contribution will diverge since the number of states of the remnant is infinite. But it has been argued that remnants from holes of increasingly large mass might couple to any given process with correspondingly smaller strength, and then such a divergence can be avoided.

Another possibility, advocated most strongly by Hawking, is that the hole does evapotate away to nothing, and the passage from an intial pure state to a final mixed state is a natural process in any theory of quantum gravity. In this view the natural description of states is in fact in terms of density matrices, and the pure states of quantum mechanics that we are used to thinking about are only a special case of this more general kind of state. Some investigations of this possibility have suggested, however, that giving up the purity of quantum states causes difficulties with maintaining energy conservation in virtual processes [3].

The possibility that would be most in line with our experience of phyiscs in general would be that the Hawking radiation does in some fashion manage to carry out the information of the collapsing matter [4]. The hole could then completely evaporate away, and yet the process would be in line with the unitarity of quantum mechanics. The Hawking radiation from the black hole would then be not really different from the radiation from a lump of burning coal - the information of the atomic structure of the coal is contained, though in a very hard to deciph er form, in the radiation and other products that emerge when the coal burns away.

1.4 DIFFICULTIES WITH OBTAINING UNITARITY

Let us review briefly the difficulties with having the radiation carry out the information. We can study the process of formation of the hole, where we see the matter collapse from a large radius down through the Schwarzschild radius.

To study the evolution, we must choose a foliation of the spacetime by spacelike hypersurfaces. We can find a foliation where the hypersurfaces themselves are smooth, which means that the curvature length scale is large enough that quantum gravity effects will not be important. Note that this requires that the spatial slices be smooth as well as that the embedding of neighbouring slices change in a way that is not too sharp.

As we evolve along this foliation, we see the matter fall in towards the center of the hole, while we see the radiation collect at spatial infinity. It is important to realise that the information in the collapsing matter can not also be copied into the radiation - in other words, there can be no quantum 'xeroxing'. The reason is that suppose the evolution process makes two copies of a state

$$|\psi_i> \quad \rightarrow |\psi_i> \times |\psi_i>$$

where the $|\psi_i>$ are a set of basis states. Then, as long as the linearity of quantum mechanics holds, we will find

$$|\psi_1> + |\psi_2> \quad \rightarrow |\psi_1> \times |\psi_1> + |\psi_2> \times |\psi_2>$$

and not

$$|\psi_1> + |\psi_2> \quad \rightarrow (|\psi_1> + |\psi_2>) \times (|\psi_1> + |\psi_2>)$$

Thus a general state cannot be 'duplicated' by any quantum process.

The infalling matter cannot lose its information suddenly as it crosses the horizon in this foliation, since it sees nothing special happen at the horizon. Thus the radiation that collects at infinity cannot contain the information of the infalling matter , and we are forced to a scenario where information is lost in the process of black hole formation and evaporation.

To bypass this conclusion we have to find that effects of quantum gravity invalidate the above semiclassical analysis of spacetime. Thus if we can show that there is no information loss in black hole evaporation, we will probably also find at the same time a nontrivial change in our understanding of spacetime physics, and not just at scales shorter than planck length.

2. STRING THEORY

String theory had its origins in quantum chromodynamics, where the phenomenon of confinement made the flux lines between quarks behave like a flux tube or string. Such a string was 'open' since it had the quarks at the ends. It was then realised that while open strings could describe gauge theories, closed strings would necessarily contain the graviton in their spectrum. The closed string can carry travelling waves both clockwise and anticlockwise along its length. The state with one quantum of the lowest harmonic in each direction is the graviton: if the the transverse directions of the vibrations are i and j then we get the graviton h_{ij}.

The extended nature of the string removes the loop divergences in the interaction of gravitons, thus bypassing a major problem with formulating a quantum theory of gravity. But the theory had several features that made it unpalatable to many physicists. For one thing, The string could be consistently quantised only in 10 dimensions, so that to obtain a 4-dimensional spacetime one had let 6 dimensions to be compact and small. Another feature was that the string could be excited to any energy level, so that there was an infinite spectrum of massive particles above the massless graviton. Further, the change from a point-like particle to a string seemed to some to be somewhat arbitrary - if strings then why not extended objects of other dimensionalities?

Over the past few years, as non-perturbative string theory has developed, it has come to be realised that these features are actually very natural, and also perhaps essential to any correct theory of quantum gravity. Before the advent of strings as a theory of quantum gravity, there was an attempt to control loop divergences in gravity by letting the theory have supersymmetry. The more the number of supersymmetries, the better was the control of divergences. But in 4 dimensions the maximal number of supersymmetries is 8, since more supersymmetries would force the theory to have fields of spin higher than 2 in the graviton supermultiplet, which leads to inconsistencies at the level of interactions. Such D=4, N=8 supersymmetric theories appear to be complicated, but can be obtained in a simple way from dimensional reduction of a D=11, N=1 theory or a D=10, N=2 theory. The gravity multiplet in the higher dimensional theory gives gravity as well as matter fields after dimensional reduction to lower dimensions, with specific interactions between all the fields.

Thus higher dimensional theories had naturally arisen in the study of quantum gravity without any direct connection to string theory. But even the supersymmetric theories with the maximum number of allowed supersymmetries had divergences at some loop order, and thus were not satisfactory quantisations of gravity.

If we take string theory and restrict ourselves to the massless fields, then we get either the dimensional reduction of the D=11 supergravity theory to 10 dimensions, or the above mentioned supergravity theory in 10 dimensions. (These two cases correspond to the type IIA and type IIB string theories respectively.) The presence of the infinite tower of massive modes also present in the string theory smooths out all the loop divergences. But a closer look at the supergravity theories leads to the observation that the existence of extended objects is actually very natural within those theories, and in fact is essential to completing them to unitary theories at the quantum level.

Consider the case of 11 dimensional supergravity. The supercharge Q_α is a spinor, with $\alpha = 1 \ldots 32$. The anticommutator of two supercharge components should lead to a translation, so we write

$$\{Q_\alpha, Q_\beta\} = (\Gamma^A C)_{\alpha\beta} P_A$$

where C is the charge conjugation matrix. Since the anticommutator is symmetric in α, β, we find that there are $(32 \times 33)/2 = 528$ objects on the LHS of this equation, but only 11 objects (the P_A) on the RHS. Suppose we write down all the possible terms on the RHS that are allowed by Lorentz symmetry, then we find [5]

$$\{Q_\alpha, Q_\beta\} = (\Gamma^A C)_{\alpha\beta} P_A + (\Gamma^A \Gamma^B C)_{\alpha\beta} Z_{AB} + (\Gamma^A \Gamma^B \Gamma^C \Gamma^D \Gamma^E C)_{\alpha\beta} Z_{ABCDE}$$

where the Z are totally antisymmetric. The number of Z_{AB} is ${}^{11}C_2 = 55$, while the number of Z_{ABCDE} is ${}^{11}C_5 = 478$, and now we have a total of 528 objects on the RHS, in agreement with the number on the LHS.

While for example $P_1 \neq 0$ implies that the configuration has momentum in direction X^1, what is the interpretation of $Z_{12} \neq 0$? It turns out that this can be interpreted as the presence of a 'sheet-like' soliton stretched along the directions X^1, X^2. It is then logical to postulate that there exists in the theory a two-dimensional fundamental object (the 2-brane). Similarily the charge Z_{ABCDE} corresponds to a 5-brane in the theory. The 2-brane has a $2+1=3$ dimensional world volume, and couples n aturally to the 3-form gauge field present in 11-dimensional supergravity, just like a particle with 1-dimensional worldline couples to a 1-form gauge field as $\int A_\mu dx^\mu$. The 5-brane is easily seen to be the magnetic dual to the 2-brane, and cou ples to the 6-form that is dual to the 3-form gauge field in 11 dimensions.

Thus we see that it is natural to include some specific expended objects in the quantisation of 11-D supergravity. But how does this relate to string theory, which lives in 10 dimensions? Let us compactify the 11-D spacetime on a small circle, thus obtain ing 10-D noncompact spacetime. Then if we let the 2-brane wrap this small circle, we will get what looks like a string in 10-D. This is exactly the type IIA string which had been quantised by the string theorists! The size of the small compact circle turns out to be the coupling constant of the string.

We can also choose to not wrap the 2-brane on the small circle, in which case there should be a 2-dimensional extended object in IIA string theory. Such an object is indeed present - it is one of the the D-branes shown to exist in string theory by Polchin ski [6]. Similarily, we may wrap the 5-brane on the small circle getting a 4-dimensional D-brane in string theory, or leave it unwrapped, getting a solitonic 5-brane which is also known to exist in the theory.

Thus one is forced to a unique set of extended objects in the theory, with specified interactions between them - in fact there is no freedom to add or remove any object from the theory, or the freedom to change any couplings. Now we would like to see what is predicted about black holes in this theory.

3. BLACK HOLES IN STRING THEORY

3.1 ENTROPY FROM STRING STATES

An idea of Susskind [7] has proved very useful in the study of black holes. Since the coupling in the theory is not a constant but a variable field, we can take a state of the theory, and study it at weak coupling, where we can use our knowled ge of string theory. Thus we may compute the 'entropy' of the state, which would be the logarithm of the number of states with the same mass and charges. Now imagine the coupling to be tuned to strong values. Then the gravitational coupling also increases, and the object must becomes a black hole with a large radius. For this hole we can compute the Bekenstein entropy from (2), and ask if the microscopic computation agrees with the Bekenstein entropy.

For such a calculation to make sense, we must have some assurance that the density of states would not shift when we change the coupling. In a supersymmetric theory we have a special class of states - the BPS states, whose mass is indeed determined solely from their charges. Such states give, at strong coupling, black holes that are extremal - they have the minimal mass for their charge (if the metric is not to have a naked singularity).

Let some of the directions of the 10-D spacetime be compactified on small circles. Take a string and wrap it n_1 times around one of these circles. From the point of view of the noncompact directions, this looks like a massive point object carrying 'win ding charge'. From the microscopic viewpoint, the state of such a string is unique (it does have a degeracy 256 due to supersymmetry, but we can ignore this - it is not a number that grows with n_1). Thus the microscopic entropy is zero. If we increase the coupling, we expect a black hole, but this hole turns out to have a horizon area zero. This happens because the tension of the string 'pinches' the circle where the string was wrapped. Thus we get an entropy zero both from the microscopic and from th e black hole viewpoints, which is consistent but not really interesting.

To prevent this 'pinching', we can put some momentum along the string, which amounts to having travelling waves move up the string. The momentum modes have an energy that goes as the reciprocal of the length of the compact circle, so now this circle attai ns a finite size. The number of microscopic states is now large, since the same total momentum can be obtained in many ways: for example two quanta of the lowest harmonic have the same momentum as one quantum of the second harmonic for vibrations of the s tring. The entropy of microstates is $\approx 2\sqrt{2}\sqrt{n_1 n_2}$ where the the momentum was $p = n_2/R$, with R the radius of the compact circle.

From the viewpoint of the noncompact directions, we have an object with two charges. At strong coupling this gives a black hole with Bekenstein entropy $\sim \sqrt{n_1 n_2}$, which agrees with the microscopic count [8]!

The black hole result here was not exact, since this hole is also singular: the coupling constant diverges on the horizon. It turns out that one needs three kinds of charges for a 5-D hole, and four kinds of charges for a 4-D hole, before a completely non singular hole is reached and an exact comparison is possible. Since the 5-D case is thus slightly easier, we discuss that here.

We compactify $M^{10} \rightarrow M^5 \times T^5$. On this T^5 we can wrap the 5-branes that we find in the theory. Let there be n_3 5-branes, and further take as before n_1 strings wrapping one of the directions in the T^5, with n_2 units of momentum along the string. The string can vibrate ins ide the 5-branes and thus carry the momentum as travelling waves along it. A count of microstates this time gives the entropy $S_{micro} = 2\pi\sqrt{n_1 n_2 n_3}$.

If we go to large coupling, we get a nonsingular black hole carrying three charges. The Bekenstein entropy computed for this geometry from (2) gives $2\pi\sqrt{n_1 n_2 n_3}$, in complete accord with the microscopic calculation [9][10]!

3.2 HAWKING RADIATION FROM THE STRING MICROSTATE

In the above calculation we had allowed all the momentum to flow in one direction along the string, thus obtaining the maximum possible momentum charge for the given energy, and getting a BPS state which became an 'extremal' black hole at strong coupling. To study Hawking radiation we must take waves travelling in both directions along the string: these waves can collide and result in emission of a graviton or other massless quantum of string theory. While at weak coupling this would be a computable stri ng theory process, at strong coupling it is expected to go over into Hawking radiation. Note however that if we can reproduce the Hawking radiation in this manner from a microscopic process, then it will be unitarity preserving, and no different from the radiation from the lump of coal referred to earlier.

First we must know what we expect from the macroscopic calculation. The relation (4) says that the radiation rate follows from the absorption cross section. We will be performing computation at low energies, so we need σ at long wavelengt hs. The method of doing this calculation is an old one, but it was found in [11] that at leading order in the energy a massless minimally coupled scalar is absorbed into *any* black hole (in any dimension) with a cross section

$$\sigma = A_H$$

where A_H is the area of the horizon,

On the microscopic side, we must find the amplitude for vibrations of the string to couple to, say, gravitons. This coupling will arise from the action

$$S = \frac{1}{16\pi G}\int R\sqrt{-G} + S_{string}$$

$$S_{string} = [Tension]\,[Area\ of\ `world\ sheet']$$

Note that the area of the world sheet sees the embedding metric, and thus couples the vibrations of the string to gravity. Suppose a vibration in the direction i transverse to the string, carrying an energy $\omega/2$ collides with a vibration in the tr ansverse direction j travelling the opposite way, also carrying energy $\omega/2$. Then we emit the graviton h_{ij} with energy ω with an amplitude per unit time

$$[\frac{4\pi G_N}{V}]^{\frac{1}{2}} \quad \omega^{\frac{1}{2}}$$

We must combine this factor with the number of vibration quanta that can collide. We finally get an emission rate

$$\Gamma \;=\; (4G)\;[2\pi\sqrt{n_1 n_2 n_3}]\;\frac{d^3k}{e^{\omega/T}-1} = A_H\;\frac{d^3k}{e^{\omega/T}-1}$$

which agrees exactly with the semiclassical Hawking radiation rate [12]!

Further, note that the string vibrates inside the 5-branes, which means that the directions i, j in the above are among the compact directions of spacetime. Thus from the viewpoint of the noncompact spacetime, the emitted quantum h_{ij} is a scalar, an d the emission of vectors and gravitons is suppressed in the microscopic calculation in this low energy limit. But this agrees exactly with what one expects from the semi-classical computation of Hawking radiation from the black hole: higher spin emission is suppressed at low energies due to an angular momentum barrier.

4. CONCLUSIONS

We thus find that the details of the microscopic state in the compact directions somehow reproduce exactly the correct degrees of freedom to yield the Bekenstein entropy for the hole that is seen from the non-compact point of view. They also reproduce th e gross properties required of the Hawking radiation, but by a process that is manifestly unitary and thus information preserving.

What is lacking this far is a picture of what exactly can go wrong in the reasoning suggested by Hawking where information loss is seen to occur. This requires us to understand the changes that occur to the picture of a string state as we tune the coupli ng from small (where we understand the state) to large (where the black hole radius becomes classical). Progress in this direction is expected to be swift. Recently it has been suggested that spacetime can be understood as a large N limit of gauge theorie s [13]; this should provide us with a direct view of the string state at strong coupling.

In any case it is fair to say that the results of string theory have illuminated many issues about black holes, and black holes in turn have given us reason to believe that string theory has the essence to be a correct theory of quantum gravity.

References

[1] Bekenstein J.D., 1973. *Phys. Rev.* **D7**, 2333.

[2] Hawking, S. 1975. *Comm. Math. Phys.* **43**, 199.

[3] Banks, T., M. Peskin & L. Susskin, L., 1984. *Nucl. Phys.* **B244**, 135.

[4] 't Hooft, G., 1985. *Nucl. Phys.* **B256**, 727.

[5] Townsend, P., 1995, Proceedings of March 95 PASCOS/John Hopkins Conference.

[6] Polchinski, J., 1995. *Phys. Rev. Lett.* **75**, 4724.

[7] Susskind, L¿, 1993. hep-th 9309145.

[8] Sen, A., 1995. *Nucl. Phys.* **B440**, 421.

[9] Strominger, A. & Vafa, C., 1996. *Phys. Lett.* **B379**, 99.

[10] Callan, C. & Maldacena, J. 1996. *Nucl. Phys.* **B472**, 591.

[11] Das, S.R., Gibbons, G. Mathur, S.D., 1997. *Phys. Rev. Lett.* **78**, 417.

[12] Das, S.R. & Mathur, S.D., 1996. *Nucl. Phys.* **B478**, 561.

[13] Maldacena, J., 1998. *Adv. Theor. Math. Phys.* **2**, 231.

Chapter 20

THE COUNTING OF RADIO SOURCES: A PERSONAL PERSPECTIVE

Jayant V. Narlikar
Inter-University Centre for Astronomy and Astrophysics
Pune 411 007, India

Abstract

This article gives the author's personal perspective on the continuing efforts by radio astronomers to determine the nature of the cosmological model by counting radio sources in the universe out to different levels of faintness. Although initially the source counts were expected to reveal the underlying geometry of space and time, subsequent experience showed that the issue is mixed up with the physical properties of the sources and their evolution with epoch. It is shown, how the earlier claims of disproof of the steady state model through source sounts, turned out to rest on very uncertain evidence.

1. INTRODUCTION

When Naresh Dadhich asked me to write an article for this volume, I was hesitant, as I was not aware that festchrifts as a rule permit self-action. However, he then produced a few examples, where this had happened, and I therefore agreed to contribute an article. It describes an area of astronomy to which I was drawn willy-nilly from my early research student days, and to which I have returned from time to time. I refer to the counting of radio sources in order to test the validity of a cosmological model. From a perspective four decades later, the issues involved look different from what they seemed in the late fifties and the early sixties.

This is therefore a historical account, and a highly personal one, and consequently open to the charge of being biased. But in this somewhat

N. Dadhich and A. Kembhavi (eds.), The Universe, 213–226.

contoversial field, it will be very difficult to find an account that is completely neutral!

So let me begin with the basic cosmological test itself, as ideally conceived, and then come to the trials and tribulations of translating it into reality.

Suppose we live in a uniform Euclidean universe which has a uniformly distributed class of sources of radiation, each with a luminosity L. Let n denote the number density of such sources in the universe. We therefore expect the number of sources within a distance R from us, to be

$$N = \frac{4\pi}{3} n R^3, \tag{1}$$

and the faintest of these will be those on the periphery of the sphere of radius R centred on us. The flux received from each of these sources will be

$$F = \frac{L}{4\pi R^2}. \tag{2}$$

In general if we count the number N of sources brighter than F for a range of values of F, we will get a plot of points in the log N-log F plane, lying on a straight line given by

$$\log N = -1.5 \log F + \text{constant}. \tag{3}$$

In other words, we expect the slope of the log N-log F line to be -1.5. In our later discussion of radio source count we shall have frequent occasions to describe the slope of the source count curve in the above sense. We will refer to the 'slope' by magnitude: thus in the above example, the slope is 1.5. Likewise, a slope of 1.8 is steeper than a slope of 1.5, although in source count equations like (3) above, the slopes are negative.

In optical astronomy, we may wish to count galaxies, in which case the appropriate quantity for F will be the apparent magnitude m. Since

$$m = -2.5 \log F + \text{constant}, \tag{4}$$

the above relation becomes modified to

$$\log N = 0.6\, m + \text{constant}. \tag{5}$$

In other words, the curve of log N plotted against m should be a straight line with a slope of 0.6. Likewise, if we are counting radio sources, then the relevant measure of flux received is the flux density S, which measures the flux received in a relatively narrow bandwidth, usually 1 Hz. In this article we are mainly concerned with the counts of radio sources and hence will be discussing the log N-log S relation.

If we wish to extend this result to *non*-Euclidean cosmological models, such as those used for describing the expanding universe of modern cosmology, naturally the prediction will be different. We keep all other assumptions the same but change the Euclidean spacetime to the Robertson-Walker spacetime given by the line element

$$ds^2 = c^2dt^2 - a^2(t)\left[\frac{dr^2}{1-kr^2} + r^2(d\theta^2 + \sin^2\theta d\phi^2)\right], \qquad (6)$$

where we have used the standard notation for the coordinates, and $a(t)$ denotes the expansion factor. The space-curvature is denoted by the parameter k which can take values, $0, 1, -1$. The coordinates r, θ, ϕ are the constant comoving coordinates of a typical galaxy and t denotes the cosmic time.

The relations derived above for the Euclidean geometry can be obtained for the above models also. In the standard Friedmann cosmology, the simplest generalization is that the number density of sources in the comoving coordinates is constant with respect to t. For details see some standard text in cosmology (e.g. Weinberg 1972, Narlikar 1993). The question is, can the actual number count tell us whether any of the wide ranging cosmological models described by the RW-line element above comes closest to reality?

This is the basic issue to be discussed here.

2. HISTORICAL BACKGROUND

One of the first attempts to try this test with galaxies as the sources to be counted, was made in the 1930s by Edwin Hubble, who hoped to measure the curvature of space (the parameter k) through this observation. However, for the data to be decisive enough, one needs to go to high redshifts, and hence counts of a large number of galaxies is involved. Moreover, the other assumption in the test is that all sources are equally powerful, which also is not the case in reality. The sheer enormity of the operation rendered the test impractical and Hubble eventually abandoned it. However, one salutary effect this abortive operation had on observational astronomy was that the proposed test provided motivation for building a large telescope, and that is how the 5-metre telscope got built on the Palomar Mountain!

The radio astronomers got into this game in the mid-fifties, when they realized that a substantial part of the radio source population is extragalactic, and that their number density is considerably lower than that of galaxies. Thus by counting relatively fewer number of sources, there was the chance of obtaining the answer Hubble was looking for.

Hubble's procedure was to compare the predicted galaxy count with the observed one. Instead of the number-magnitude relation, the radio astronomers had a number-flux density relation for radio galaxies. What does a typical relation look like?

The qualitative signature of a typical expanding universe model is to flatten the log N-log S curve as one goes from high to low flux levels, that is to progressively reduce the slope from 1.5 to lower values, 1.4,1.3,1.2,..., because of the redshift effect on flux densities and volumes. Thus, it was claimed that the test was a powerful tool for distinguishing between cosmological models.

Martin Ryle (1955) from the Mullard Radio Astronomy Observatory of the Cavandish Laboratory in Cambridge announced the first result based on his early catalogue of radio sources in the Halley Lecture delivered on May 6, 1955, where he got a slope of the log N-log S curve to be 3.0. Thus, with a magnitude exceeding 1.5, the curve was *steeper*, instead of flatter than the Euclidean value.

Prima facie, the result seemed to disprove all expanding universe (big bang) models. However, there is a loophole in such models. Because the universe is evolving, one could argue that the number density of radio sources in the past was greater than at present. By suitably adjusting the number density n as a function of the cosmic time t, *any* slope can be accommodated. The alternative also exists that the luminosity L is a function of t, and between the two variations the fit to the observed data could be achieved.

There is one model, however, which does not have this freedom of choice. This is the steady state model, wherein the universe has the same physical properties at all epochs. Thus the n and L values for any source population cannot be epoch dependent. For such a model, prima facie, a slope as steep as 3 spelt doom. Indeed, Ryle, who never liked the steady state theory, took pains to underscore this conclusion.

There was, however, the alternative possibility that the counts could have been in error. Indeed, in 1958, Ryle and his colleagues (Archer, et al 1959) revised the index down from 3 to 2.2. Although the credibility of the claim suffered somewhat by this drastic come-down, the revised finding was again projected with great certainty as clearly disproving the steady state theory.

In the meantime, the Australian radio astronomers had been conducting their own surveys largely of the southern sky, although having an overlap region with Ryle's northern sky. Mills and Slee (1957) announced that their results showed a slope not significantly different from the Euclidean slope of 1.5. Based on their study of the overlap region, they pointed out possible sources of errors, which might have affected the data reduction by the Cambridge group leading to their claimed steeper slope.

These results related to the 3C (Third Cambridge) Survey. In 1960, Ryle announced a new result related to the 4C Survey which had fainter (and hence assumed to be more distant) sources. The slope this time was claimed to be 1.8, and because the new survey had more sources, it was claimed to be more accurate. Surely, argued Ryle, this finding disproved the steady state theory conclusively.

This is where I was drawn into the controversy.

3. A DEFENCE OF THE STEADY STATE THEORY

In June 1960, when I approached Fred Hoyle with the request to be my Ph.D. guide, he readily agreed and suggested a number of interesting lines of investigation in astronomy and astrophysics. However, when discussing cosmology, he did not mention the steady state theory, which I had found an attractive approach to the study of the universe. He said that although there were many challenging problems in that theory, he wished to keep a research student away from controversy. Consequently, I set to work on an idea proposed by Heckmann and Schucking on spinning universes. The problem was to see if spin allows non-singular cosmological models, models which oscillate with finite upper and lower scales of size. I was then to look at the problem of primordial nucleosynthesis in such models.

Within six months, however, I had reached a dead end, i.e., I could see that the Heckmann-Schucking models would not lead to non-singular oscillating models as their authors had claimed. So the next part of my investigation did not arise. I thus found myself somwhat at a loose end in January, 1961. This was when the Hoyle-Ryle controversy broke out.

For, of the three originators of the steady state theory, Hermann Bondi and Tommy Gold did not take Ryle all that seriously, dismissing the claim as just one along the line of earlier claims in which the announced slope had steadily come down from 3 to 2.2 to 1.8. Perhaps further errors may be discovered in the future which would lower the slope further to the Euclidean 1.5, which the Australians were happy with, anyway. It was only Hoyle who took the results seriously enough to realize that a threat to the steady state theory indeed existed.

The Hoyle-Ryle confrontation took place at a press conference given by the latter. As recalled by the former in a recent book (Hoyle, et al 1999), it did not help in making their personal interaction any easier. Ryle, however, realized that for a peer-understanding of his findings, he had to present his work not before the media but to a body of scientists. Accordingly, he arranged to describe his results during the February 10 meeting of the Royal Astronomical Society. It was expected, that Hoyle would reply to his claimed disproof of the steady state theory. Indeed there was great expectation of a lively scientific confrontation in a society which had previously witnessed controversies between Milne and Eddington, Eddington and Chandrasekhar, etc.

Hoyle, as mentioned earlier, had not been dismissive of Ryle's data; rather his attitude had been that given the observational uncertainties, it was still possible to fit a realistic steady state model to the observed source counts. This was when he asked me to work out on an idea that might possibly serve the purpose. To begin with we needed to get the data from Professor Ryle and his colleagues. This was, however, not so easy, as the Cavendish group was very secretive about its findings. Rather than get a catalogue of sources with flux densities from which one could prepare tables and plot curves, all we got from Ryle after a tea-time discussion was a hand-drawn log N-log S curve, with a few points marked for numbers and flux densities. The curve had a slope of 1.8 which Ryle challenged us to reproduce within the cosmological framework of the steady state model.

However, it was already mid-January and in order to have a viable model ready for reporting we had to work 'overtime' during those three weeks or so. I shall come to the model and its aftermath in the following section. Here I recall using the EDSAC computer with punched paper tape and machine language programming as well as hand operated Facit calculators to churn out the numbers. The numbers came out fine: the theory could indeed reproduce Ryle's steep slope within the framework of the steady state theory. We thus had a counter-example to Ryle's claim. We then had to persuade the Engineering Labs to make at short notice a few 'lantern slides' based on our calculations. We managed to get everything organized with a couple of days to spare.

However, in the meantime, Fred Hoyle had set off a bomb-shell so far as I was concerned. He had discovered that he had a prior engagement to speak at a college in London on February 10, and so he would not be able to attend the RAS meeting at all. Instead, *he asked me to reply to Ryle!*

So here was I, a raw research student with barely six months of research experience now launched into the limelight of a major controversy. However, Hoyle assured me that with the mathematical backing we had for our model, I should be able to handle any counterattack. He drilled me, nevertheless, in speaking concisely so that I could convey the salient features of our model within ten minutes. He made sure that the RAS would allot me that much time for presenting our counter-example.

In the end, my presentation went well. Ryle raised a minor protest that this seemed a new version of the steady state theory. However, Bondi who was quick to see the point behind our approach rose to its defence. In any case I came back from the meeting considerably elated and with a newly acquired confidence that I had now been groomed into participating in scientific debates. This experience has stood me in good stead in facing other astronomical controversies.

4. RADIO SOURCE COUNT IN THE STEADY STATE COSMOLOGY

Let me recall the model we had proposed early in 1961. It rested on two premises:

1. The universe is inhomogeneous on the scale of 50-100 Mpc, being made of superclusters and voids.

2. The probability of a galaxy becoming a radio source increases with its age τ, being proportional to $\exp[4H\tau]$.

The rationale behind these two assumptions as perceived then was as follows. The 'hot universe' model of the steady state universe proposed by Gold and Hoyle (1959) envisaged that galaxies would form typically in large groups with characteristic dimensions of 50-100 Mpc. Thus the theory envisaged an inhomogeneity on this scale. In the late 1950s and the early 1960s only one observational astronomer G. deVaucouleurs was talking of the 'Local Supercluster', and he was generally not taken seriously by the majority which believed in universal homogeneity beyond the cluster scale. There was, however, evidence already for 'second order clustering' from the work of George Abell (1958), who had done an extensive analysis of distribution of clusters on the sky. Thus there was both theoretical and observational support for the first assumption.

The second assumption was based on the finding then emerging, that the property of radio emission seemed confined largely to elliptical galaxies, which are generally considered old. Thus the correlation of radio property with age was conjectured through the second assumption. At the time, the radio astronomers tended to believe that radio emission arose from collisions of galaxies. Which is why, there was a ready acceptance of the Cambridge belief that the number density of radio sources was significantly higher in the past: in a big bang universe, the density of galaxies was higher in the past and hence the chances of collisions were greater. The collision hypothesis had already been demonstrated as theoretically untenable by the work of Geoffrey Burbidge, whose estimates of the energy of a typical radio source emitting synchrotron radiation, were far higher than the energy of collision of two galaxies (Burbidge 1959). Subsequently, in a few years, the collision idea received a decent burial; however, belief in the notion that the process of radio emission had to be more frequent in the past persisted.

In the steady state theory, however, no appeal could be made to an epoch-dependent process, as the word 'steady' forced one to regard the average state of the universe to be the same at all epochs. The second assumption, however, coupled with the first one, led to an apparently evolutionary effect *in a local statistical sense*, as follows.

In the steady state theory, the age-distribution of galaxies follows the formula:

$$Q(\tau)d\tau = \exp(-3H\tau)d\tau, \tag{7}$$

where $Q(\tau)d\tau$ denotes the number density of galaxies in the age range $[\tau, \tau + d\tau]$. Thus to observe very old galaxies, a typical observer would have to sample a larger volume, and hence look out to farther distances. Hence, one expected that a generic observer would begin to see an increasing density of radio sources (which by assumption 2 were more likely to be found in older galaxies) at larger distances. This effect was in a sense a statistical fluctuation from the 'average' situation which was represented by the completely homogeneous Robertson Walker line element for the steady state theory, with $a(t) = \exp Ht$. In other words, after a somewhat local steepness, the log N-log S curve would revert to the standard progressively flattening form described earlier.

In 1961 we published a detailed version of this model (Hoyle and Narlikar 1961), followed by another carrying out computer simulations of the real universe, in the following year (Hoyle and Narlikar 1962). In retrospect, I think these papers were pathbreaking on the following counts, although because of the general feeling of hostility against the steady state theory these aspects went unnoticed at the time.

1. They introduced the idea of a universe inhomogeneous on the scale of superclusters and voids with typical length scales 50-100 Mpc, an idea that became accepted as reality two decades later, although at the time it was seen as introducing unnecessary complications into cosmology.

2. The idea of counting a source population which was evolving with age became standard practice in the big bang cosmology, from mid-sixties onwards, although it was first introduced here in the framework of the steady state theory.

3. The second of these papers used Monte Carlo techniques to simulate source distributions on a computer, and these were counted by random observers to demonstrate fluctuating source counts at high flux levels. I believe, this was the first simulation of its kind in cosmology, and was made possible because Fred Hoyle had rented time on the IBM 7090 machine in London. In today's desktop workstation environment it is difficult to imagine the mode in which we were operating, viz. going to London once a week with our punched cards which were to be handed over to the computer staff in the morning and the results collected in the evening. If the programme had a serious bug, one had to wait for a week to sort it out!

4. Although a super-Euclidean slope was seen as a clear indication of support for big bang, it could clearly not be sustained at low flux densities. Our model naturally led to a flatter curve at low flux densities, which was borne out by later surveys.

5. We had estimated the majority of sources to be of medium power at modest redshifts, whereas Ryle and his colleagues believed them to be typically very powerful and very distant. The issue could not be settled till the sources could be optically identified and their redshifts measured. This is a slow process, and to date only the 3C Revised catalogue has all sources optically identified and their redshifts measured. As we shall see later on this account, the data have turned out to be closer to our interpretation rather than to that of the Cambridge radio astronomers.

5. QUASARS VS RADIO GALAXIES

While we were working on this paper, Fred Hoyle arranged to visit Hanbury Brown at the other premier radio observatory in Britain, the Nuffield Observatory at Jodrell Bank. Hanbury Brown had just carried out a study of radio sources, especially their angular sizes. He had noticed that there were quite a few which were too compact for their angular size to be measured by interferometric techniques. What were these 'chaps', as he called them? In any case their presence indicated that the population of radio sources was by no means homogeneous, and hence basing cosmological deductions on them might be misleading. [As a sound precaution, it is best to understand the class of objects you are counting, before drawing profound conclusions from them.]

To study these compact sources, it was essential to optically identify them, measure their redshifts and other physical features. The positions given by radio astronomers needed to be made more precise for optical identification to be attempted.

During 1962, the special technique of lunar occultation used in Australia enabled the position of the compact source 3C 273 to be measured accurately. In 1963, the source was optically identified and its spectrum examined. The resulting finding of a redshift of ~ 0.16, despite the extraordinary optical brightness (13^m) of the source, suggested that here we are looking at a new class of radio sources. Later months brought to light several of these objects which, because of their starlike appearance, eventually came to be called *Quasi-Stellar Objects*, (QSOs) or *quasars*. Clearly the radio astronomers were looking at mixed populations of radio galaxies and quasars. It made more sense to separate the two before counting them in order to draw cosmological conclusions.

Using the maximum likelihood method of determining the slope of the log N-log S curve, Jauncey (1967) pointed out that the 3CR catalogue had three kinds of sources, (i) radio galaxies, (ii) quasars and (iii) unidentified sources. Of these, the radio galaxies had a slope not significantly different from the Euclidean 1.5, while the quasars showed a steeper slope ~ 2. So far as quasars are concerned, there are reasons, not widely accepted but still not completely disproved either, castinging doubts on the cosmological interpretation of their redshifts.

In his lecture at the Royal Society, Fred Hoyle (1968) discussed the source counts as he perceived them at the time. The steepness of the source count curve for the three types of sources turned out to be not significantly different from 1.5 for (i) radio galaxies and (ii) quasars, but was 2.5 for (iii) the unidentified sources. For radio galaxies, which do not have very large redshifts, the slope 1.5 is not cosmologically significant; nor is it so for quasars if their redshifts are not cosmological. *The slope is, however, cosmologically significant and inconsistent with the steady state value for the quasars if their redshifts do follow Hubble's law.* What about the sources of class (iii)?

The number of unidentified sources was however, small and the implication of their steep slope was like this, as pointed out by Hoyle. There are 10 unidentified sources at $S = 12.5$ Jy and 93 at $S = 5$Jy, which gives the 2.5 slope. However, suppose that we don't know the value of N at the high flux level and wish to determine it on the assumption that the slope is 1.5. Thus the number at the high flux level $S = 12.5$

Jy is 23. Had there been 23 instead of 10 sources at the high flux end, the observed slope would have been 1.5. The observed deficit is thus of 13 sources over 3 steradians, i.e., about 4-5 sources per steradians. The Hoyle-Narlikar model of radio source evolution described above, allowed for this deficit by the probability law proportional to $\exp[4H\tau]$. In short, we are in a local hole which has a deficit of radio source activity. Since our model did allow for local inhomogeneity (–which is now being observed), such a local hole was not inconceivable.

As pointed out by Hoyle, Ryle's interpretation of the above data would be different, however. Suppose we have a 1.5 law with the high flux value given, i.e., $N = 10$ at $S = 12.5$Jy. What is the expected number at $S = 5$? The answer is 40. Thus the observed number 93 represents an excess of 53 sources, i.e., about 18 sources per steradian. A strong evolution is required to explain this rise, as per the big bang cosmology, an evolution that cannot be accommodated within the steady state theory.

The issue thus became one of local fluctuations (limited to say 50-100 Mpc) if Hoyle's interpretation is accepted versus a cosmologically significant evolution, if Ryle's view were adopted.

6. IS EVOLUTION NECESSARY?

The case of the 3CR survey was finally resolved when almost all of its sources were optically identified and had measured redshifts, thanks to the efforts of Spinrad, et al (1985). In 1985, of the 298 sources of the Spinrad compilation of redshifts in the 3CR catalogue, 195 were radio galaxies, 53 were QSOs and 38 were unidentified. If one avoided the low galactic latitude sources with $\|b\| > 7\,\mathrm{deg}$, $S \geq 10$Jy, there were 163 radio galaxies. DasGupta, et al (1988) studied this sample to see if there were any need to postulate evolution over and above the standard Friedmann evolution of spacetime geometry. The procedure adopted was simple and straightforward. First, we take a generic Friedmann model which is characterized by the deceleration parameter q_0. Writing the bolometric luminosity distance in this model as $D = (c/H_0)x$, (H_0=Hubble's constant) we have the well known relation

$$x = \frac{1}{q_0^2}[q_0 z + (q_0 - 1)(\sqrt{1 + 2q_0 z} - 1] \tag{8}$$

between the luminosity distance and redshift. This relation can be inverted to give

$$z = q_0 x - (q_0 - 1)[\sqrt{1 + 2x} - 1]. \tag{9}$$

Note that, the above relation is valid also for radio sources with spectral index 1.

Next we see how the radio luminosity function $g(L)$ is completely determined if we assume no physical evolution. We define the number of sources per unit proper volume brighter than a given luminosity L as

$$F(L) = \int_L^\infty g(l)dl, \tag{10}$$

with the expectation that $F(L) \to 0$ as $L \to \infty$. If the survey is limited to flux densities $S \geq S_0$, say, then the sources of luminosity L will appear in the survey, provided

$$L > 4\pi(c/H_0)^2 S_0 x^2. \tag{11}$$

A little manipulation with the Friedmann geometry then gives the number of sources with redshifts in the range $z, z + dz$ as

$$dN = \left(\frac{c}{H_0}\right)^3 \Omega \frac{x^2}{(1+z)^3\sqrt{1+2q_0 z}} \times F\left[4\pi\left(\frac{c}{H_0}\right)^2 S_0 x^2\right] dz. \tag{12}$$

Writing $G(z) = dN/dz$, which can be determined by observations, we can invert the above relation to write:

$$F(L) = \left(\frac{H_0}{c}\right)^3 \frac{G(z)(1+z)^3\sqrt{1+2q_0 z}}{\Omega x^2}, \tag{13}$$

where x is determined from:

$$x^2 = \frac{L}{4\pi(c/H_0)^2 S_0}, \tag{14}$$

and z is given by the equation (9) in terms of x.

Thus, the important conclusion is that *the radio luminosity function is completely determined by the observed* $G(z)$. The RLF can therefore be used to compute the number of sources expected in a typical small cell $z_1 \leq z \leq z_2, S_1 \leq S \leq S_2$ in the (z, S) plane. These predicted numbers can be compared with the observed ones through the χ^2–test. Dasgupta et al (1988) carried out the test and found that the fit is statistically good. At a more sophisticated level, they also applied the Kolmogorov-Smirnov test adapted to two-dimensional distributions along the lines discussed by Peacock (1985). Again, the fit is good, without having to introduce any evolution of luminosity and/or number density of sources.

However, DasGupta (1988), also found that a similar analysis applied to the steady state theory produces not only a good fit, but a *better fit* than for the Friedmann cosmologies. Both the χ^2– and the Kolmogorov-Smirnov tests turn up better figures for the steady state theory than for the non-evolving standard Friedmann models.

7. CONCLUSION

Thus I believe, so far as the 3CR catalogue of radio sources is concerned, the Hoyle-Ryle controversy has been laid to rest. The technique followed in the previous section can be used only for the complete flux-limited samples which have all redshifts known. So far no other such samples are available. The log N - log S curve does not contain information on redshifts and is thus likely to be less definitive.

In any case, the standard big bang approach involving fitting the observed curve to the theoretical one which folds in evolutionary parameters, defeats the original purpose of Hubble, that of determining the geometry of the universe by counting sources.

Acknowledgments

I thank Naresh Dadhich and Ajit Kembhavi for giving me this opportunity of recollecting my past light cone.

References

[1] Abell,G.O. 1958, *Astrophys. J. Suppl.* **3**, 211.

[2] Archer, S., Baldwin, J., Edge, D., Elsemore, B., Scheuer, P. and Shakeshaft, J., 1959, *Paris Symposium on Radio Astronomy*, Eds. R.N. Bracewell & Stanford ,487.

[3] Burbidge, G. 1959, *Paris Symposium on Radio Astronomy*, Eds. R.N. Bracewell & Stanford, 541.

[4] DasGupta, P. 1988, *Ph.D. Thesis, Bombay University.*

[5] DasGupta, P., Narlikar, J.V. & Burbidge, G. 1988, *Astron. J.* **95**, 5.

[6] Gold, T. and Hoyle, F. 1959, *Paris Symposium on Radio Astronomy*, Eds. R.N. Bracewell & Stanford, 583.

[7] Hoyle, F., 1968, *Proc. Roy. Soc. Lond. A* **308**, 1.

[8] Hoyle, F. & Narlikar, J.V., 1961, *Mon. Not. Roy. astr. Soc.* **123**, 133.

[9] Hoyle, F. & Narlikar, J.V., 1962, *Mon. Not. Roy. astr. Soc.* **125**, 13.

[10] Hoyle, F., Burbidge, G. & Narlikar, J.V., 1999, *A different approach to cosmology*, Cambridge University Press.

[11] Jauncey, D.L. 1967, *Nature* **216**, 1967.

[12] Mills, B.Y. & Slee, O.B., 1957, *Austr. J. Phys.* **10**, 162.

[13] Narlikar, J.V., 1993, *Introduction to Cosmology, 2nd Ed.*, Cambridge University Press.

[14] Peacock, J.V., 1985, *Mon. Not. Roy. astr. Soc.* **217**, 601.

[15] Ryle, M., 1955, *The Observatory***75**, 137.

[16] Spinrad, H., Djorgovski, S., Marr, J. & Aguilar, L., 1985 *Proc. Astr. Soc. Pacific* **97**, 932.

[17] Weinberg, S., 1972, *Gravitation and Cosmology*, John Wiley.

Chapter 21

A VARIATIONAL PRINCIPLE FOR TIME OF ARRIVAL OF NULL GEODESICS

Ezra T. Newman
Department of Physics and Astronomy, University of Pittsburgh Pittsburgh, PA 15260, USA

Simonetta Frittelli
Physics Department, Duquesne University, Pittsburgh, PA 15282, USA

Dedicated to Jayant Narlikar.

Abstract We show how to construct a generating family for the singularities of the null surface that is obtained by following null geodesics normal to a spacelike closed two-surface. The construction is based on the priciple of least time of arrival of light signals from a source to a localized observer.

1. INTRODUCTION

In the study of the wavefronts and their related characteristic (or null) surfaces in Lorentzian spacetimes one is often confronted with difficulties in the analytic description of the development of caustics and crossover

N. Dadhich and A. Kembhavi (eds.), The Universe, 227–237.

regions. Probably the most powerful technique for their study is V.I. Arnold's theory of Lagrangian and Legendre submanifolds and the associated Lagrange and Legendre maps [1]. One of the main ingredients in this theory is the construction of what has been refered to as *generating families*. They are, in general, two point functions, $F(x^i, s^J)$ (chosen from, perhaps, different spaces with different dimensions), that are constructed from physical arguments and which are stationary with respect to variations in one of the two different spaces. In what follows we give a particularly important example of this construction where we consider the time-of-arrival function of light rays which begin from points on a two-surface, embedded in a four dimensional spacetime, (thought of as a source of radiation), and which end at points on a curve (thought of as the worldline of an observer of that radiation) also embedded in the same four-space. This example appears to be, in principle if not in practice, of generic use in the theory of gravitational lensing [2] in any Lorentzian spacetime.

In Section 2. we state the time of arrival theorem and prove it via a modification of Schrödinger derivation of gravitational frequency shifts in the cosmological context. In Section 3. we reobtain our result via Arnol'd's generating families.

2. THE TIME OF ARRIVAL FUNCTION

We are concerned with the travel time of light signals from an extended source to a localized observer. For our purposes, the source lights up instantaneously, in its own rest frame, emitting photons in all directions from every point on its (closed) surface. There is one photon that arrives first at the observer's location, in the observer's proper time. If the metric is stationary, then this photon is, intuitively, the one that takes the shortest spatial path, perpendicularly to the surface of the source. In the following, we make these notions more precise, extending them to the case of arbitrary metrics.

Consider, in an arbitrary Lorentzian four-dimensional manifold, a given closed spacelike two-surface, $\mathcal{S}$, described by

$$x^a = x_0^a(s^1, s^2) \tag{1}$$

where x^a are spacetime coordinates in the neighborhood of the source, and $s^J = (s^1, s^2)$ parametrize the surface $\mathcal{S}$. In addition, consider a timelike worldline, $\mathcal{L}$. In the neighborhood of the worldline, with no loss of generality, let the local coordinates be such that $\mathcal{L}$ is given by (τ, X^i) where the X^i are three constants, the spatial location of the observer, and τ is the proper time along the worldline.

¿From each point s^J of $\mathcal{S}$, construct its future lightcone, $\mathcal{C}_{s^J}$. In general, in the absence of horizons, the line $\mathcal{L}$ intersects each $\mathcal{C}_{s^J}$ at least once. The intersection takes place at a particular value of the proper time τ for each point s^j on the surface. This means that there is a two-point function

$$\tau = T(X^i, s^J). \tag{2}$$

that represents the proper time of arrival at $\mathcal{L}$ of light signals from $\mathcal{S}$. One explicit way of constructing such a function is as follows. The lightcone $\mathcal{C}_{s^J}$ is foliated by lightrays from the point $x_0^a(s^J)$, which are solutions

$$x^a = \gamma^a(r; s^J, \theta, \phi) \tag{3}$$

of the geodesic equation with initial data labeled by the initial point s^J and the initial direction (θ, ϕ) of the ray. Here r can be thought of as an affine parameter along the null geodesics. The intersection of the lightcone with the worldline $\mathcal{L}$ takes place at points where

$$\gamma^i(r; s^J, \theta, \phi) = X^i \tag{4}$$

and the time $(x^0 = \tau)$ at which the lightray reaches the observer is

$$\tau = \gamma^0(r; s^J, \theta, \phi) \tag{5}$$

where the values of $(r; s^J, \theta, \phi)$ in the right-hand side are restricted by (4). In cases where (4) is invertible for every value of s^J, it provides (r, θ, ϕ) as functions of (X^i, s^J), which can be inserted into (5) to yield (2).

For large distances between the source and observer, a worldline intersects any generic future lightcone several times, due to the folds in the individual lightcones produced by spacetime curvature (see Fig. 21.1). This is the case where (4) is not invertible, since for every fixed value of s^J there would be several values of the set (r, θ, ϕ) corresponding to the same spatial location X^i. This means that there are several photons, shot in different directions, that reach the observer's location at different times. Therefore, for large distances the function $T(X^i, s^J)$ is multivalued. We restrict attention to such cases where $T(X^i, s^J)$ is single valued. In this case, there exists the following theorem, mentioned by Arnol'd (see [1], p. 251, and [3], p. 298):

Theorem. *The proper time of arrival, T, at $\mathcal{L}$, is extremized by those rays that leave $\mathcal{S}$ perpendicularly to it. In other words, the points s^J such that*

$$\frac{\partial T}{\partial s^J}(X^i, s^J) = 0 \tag{6}$$

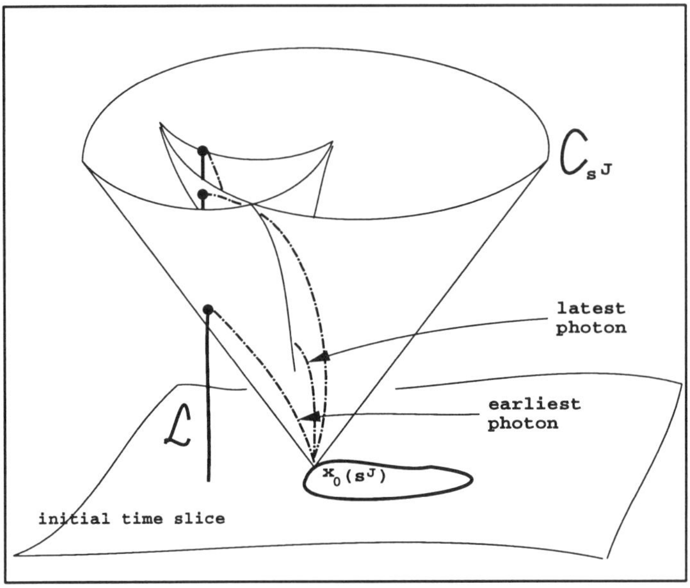

Figure 21.1 Origin of the multiplicity of the time of arrival function. Due to the folding of the lightcone in the presence of curvature, there are several null geodesics that reach the same spatial location, at different times. In our picture, three null geodesics, leaving x_0 in different directions at the same time, reach the same worldline.

are connected to $\mathcal{L}$ by lightrays that are normal to $\mathcal{S}$.

Proof: The proof is based on the standard variational principle for null geodesics, and is an extension of a similar result in [4]. Consider the action

$$I(p,q) = \int_0^1 g_{ab}\dot{x}^a\dot{x}^b dr, \tag{7}$$

where $\dot{x}^a \equiv dx^a/dr$ is the tangent vector to an affinely parametrized null geodesic between the points $p = x^a(0)$ and $q = x^a(1)$, and s is the affine parameter. Consider the variation of I constructed by taking the difference between two neighboring null geodesics with different initial points, p_1 and p_2, and different final points, q_1 and q_2 Since I evaluates identically to zero in both instances, its variation is zero as well,

$$0 = \Delta I = I(p_2, q_2) - I(p_1, q_1) \tag{8}$$

The variation is

$$\begin{aligned} \Delta I &= \int_0^1 g_{ab,c}\dot{x}^a\dot{x}^b\delta x^c + 2g_{ab}\dot{x}^a\delta\dot{x}^b \, dr \\ &= \int_0^1 \left((g_{ab,c} - 2g_{cb,a})\dot{x}^a\dot{x}^b - 2g_{cb}\ddot{x}^b\right)\delta x^c + 2\frac{d}{dr}\left(g_{ab}\dot{x}^a\delta x^b\right) \, dr \end{aligned} \tag{9}$$

(10)

Since the curves are null geodesics, the term proportional to δx^c in the integrand vanishes, and we are left with

$$\Delta I = 2\left(g_{ab}\dot{x}^a \delta x^b\right)\Big|_0^1 . \tag{11}$$

By (8) and (11), we have

$$g_{ab}\dot{x}^a \delta x^b\Big|_{r=1} = g_{ab}\dot{x}^a \delta x^b\Big|_{r=0} \tag{12}$$

In (12), δx^b represents an arbitrary (up to the condition that p and q can be connected by a null geodesic) displacement at $r = 0$ and $r = 1$ between the null geodesic with tangent $\dot{x}^b$ and a neighboring one. We now particularize (12) to our case of interest, in which, initially, neighboring null geodesics are connected by displacements on the surface $\mathcal{S}$; *i.e.*;

$$\delta x^b|_{r=0} = \left(\frac{\partial x_0^b}{\partial s^J}\right) ds^J \tag{13}$$

where $\frac{\partial x_0^b}{\partial s^J}$ are the two coordinate tangent vectors to $\mathcal{S}$ and ds^J is arbitrary. The final displacement must be tangent to $\mathcal{L}$, *i.e.*,

$$\delta x^b|_{r=1} = v^b d\tau \tag{14}$$

with v^b the tangent vector to the curve $\mathcal{L}$. However, because the two null geodesics arriving at $r = 1$ and separated by $\delta x^b|_{s=1}$ must be the same pair of null geodesics leaving $r = 0$ separated by $\delta x^b|_{r=0}$ then $d\tau$ is not arbitrary, but

$$d\tau = dT|_{X^i} = \frac{\partial T}{\partial s^J} ds^J . \tag{15}$$

With (13), (14) and (15), Eq. (12) reads

$$\left(g_{ab}\dot{x}^a \frac{\partial x_0^b}{\partial s^J} - g_{ab}\dot{x}^a v^b \frac{\partial T}{\partial s^J}\right) ds^J = 0. \tag{16}$$

Since ds^J is arbitrary, and since $g_{ab}\dot{x}^a v^b$ can not vanish as long as $\dot{x}^a$ and v^b are tangent to a null and a timelike curve, respectively, then (16) is equivalent to

$$\frac{\partial T}{\partial s^J} = \left(\frac{1}{g_{cd}\dot{x}^c v^d}\right) g_{ab}\dot{x}^a \frac{\partial x_0^b}{\partial s^J} . \tag{17}$$

This implies that, at each point of $\mathcal{S}$, $\partial T/\partial s^J$ vanishes if and only if the null ray $\dot{x}^a$ is normal to the surface at that point. This proves the theorem.$\triangle$

Note that since no property of the line $\mathcal{L}$ was used, the theorem can be restated as follows. Given a time foliation of a Lorentzian manifold with local coordinates chosen as (τ, X^i), and a source, a closed two-surface that "lights up", the time $\tau = T(x^i, s^J)$ of arrival at any spatial point X^i, of light signals from a surface point s^J, is extremized by the lightrays leaving the surface perpendicularly.

3. EIKONALS AND THE TIME OF ARRIVAL

In the following, we provide an alternative method for obtaining the time of arrival function which is based entirely on the use of the eikonal equation - with Arnold's generating families - and specifically on knowledge of a two-parameter family of solutions of the eikonal equation,

$$g^{ab}(x^c)\partial_a Z \partial_b Z = 0; \tag{18}$$

i.e., it is assumed that a solution, with the two parameters $\alpha^A = (\alpha^1, \alpha^2)$

$$u = Z(x^a, \alpha^A) \tag{19}$$

to Eq.(18) is known. Then for each value of α^A the level surfaces of Z are null (i.e., $\partial_a Z$ is a null covector). Furthermore it is assumed that at each point $x^c, \partial_a Z$ sweeps out the entire null cone at x^a as α^A goes through its range.

Remark. We point out and emphasize that the level surfaces of the solutions to Eq.(18) though referred to as "null or characteristic surfaces" are not strictly speaking surfaces; they can have self-intersections and in general are only piece-wise smooth. Though Arnold refers to them as "big-wave-fronts" we will continue to call them null surfaces. The intersection of a big wave front with a generic three surface yields a two-dimensional (small) wave front.

The first thing that we want to show is that the light-cone, $\mathfrak{C}_{x_0}$, from an arbitrary space-time point x_0^a can be constructed from knowledge of the function Z of Eq.(19).

One sees immediately, from Eqs.(18) and (19), that the function

$$S^*(x^a, x_0^a, \alpha^A) = Z(x^a, \alpha^A) - Z(x_0^a, \alpha^A) = 0 \tag{20}$$

defines a two-parameter set of surfaces which all pass thru the point x_0^a and which, furthermore, are all null surfaces. The envelope of this family is constructed by demanding that

$$\partial_A S^*(x^a, x_0^a, \alpha^A) = 0 \tag{21}$$

where ∂_A denote the derivatives with respect to the α^A. Assuming for the moment that (21) could be solved for the $\alpha^A = \alpha^A(x^a)$, then when they are substituted into (20) one obtains the function

$$S(x^a, x_0^a) = Z(x^a, \alpha^A(x^a)) - Z(x_0^a, \alpha^A(x^a)) = 0. \tag{22}$$

Using (21) it is easy to see that $\partial_a S = \partial_a S^*$ so that again $S(x^a, x_0^a)$ is a null surface thru the point x_0^a; its gradient at x_0^a, namely $\partial_a S = Z(x_0^a, \alpha^A)$,spans the light-cone at x_0^a [at x_0^a, Eq.(21) can not be solved for the $\alpha^A = \alpha^A(x^a)$; all values of α^A are allowed.] We thus see that Eq.(22) represents the light-cone $\mathfrak{C}_{x_0}$. The assumption that Eq.(21) could be solved for $\alpha^A = \alpha^A(x^a)$ depended on the non-vanishing of the determinant J of the matrix $J_{ij} \equiv \partial_i \partial_j S^*(x^a, x_0^a, \alpha^i)$. J does vanish at the singularities of the "surface" $S(x^a, x_0^a)$, e.g., at the apex $x^a = x_0^a$. In general, however even when $J = 0$, Eqs.(21) and (20) can be solved for other variables, namely *some* set of three (say x^α;which might be different in different regions) of the four x^a, in terms of the fourth one (say x^*) and the α^A, i.e.,

$$x^\alpha = x^\alpha(x_0^a, x^*, \alpha^A). \tag{23}$$

Note the important point that if the coordinates x^a are such that three of them are space-like and one of them is a time coordinate, x^0, then Eq.(23) has a stronger version, namely

$$x^j = x^j(x_0^a, x^*, \alpha^A), \tag{24}$$
$$x^0 = x^0(x_0^a, x^*, \alpha^A) \tag{25}$$

where the two x^j and the x^* are the three space-like coordinates. That one can solve for the $x^0 = x^0(x_0^a, x^*, \alpha^A)$ follows from the fact that Eq.(22) can always be solved, from the implicit function theorem, for x^0 since S^* satisfies the eikonal equation and hence $\partial S^*/\partial x^0 \neq 0$

Eqs.(24) and (25) are a parametric representation of $\mathfrak{C}_{x_0}$ via the null geodesics that rule it. For the different given values of the α^A, they are the null geodesics thru x_0^a.

We thus have the result that the $\mathfrak{C}_{x_0}$can be given either via the surface (22) or by its geodesics (24) and (25). We will return to Eq.(25) later.

If we now allow the x_0^a to lie on a space-like two-surface S described by $x_0^a = x_0^a(s^J)$, parametrized by the two parameters s^J , then the previous construction of light-cones yields the family of light-cones of

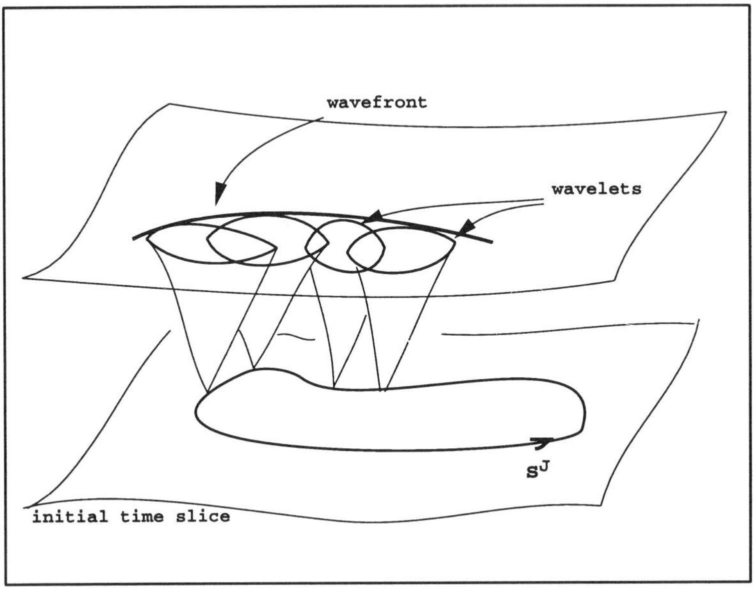

Figure 21.2 The construction of the wavefront as the envelope of the individual wavelets.

all the points of S via $x_0^a \Rightarrow x_0^a(s^J)$. The intersection of the set of all the light-cones with a constant-time slice $x^0 = constant$, is a family of individual (small, two-dimensional) wavefronts emanating from each point on the surface; they are denoted as "Huygen's wavelets". By Huygen's principle, the envelope of all the wavelets, at $x^0 = constant$, is the two dimensional wavefront from the source S. (see Fig. 21.2) The evolution, as x^0 changes, of these wavefronts yields a new characteristic surface (big wave-front). It is equivalent to the envelope of the family of light-cones of all the points of S; the envelope corresponding to the stationary variation of the family of light-cones with respect to variations in the s^J.

More precisely, the envelope is the three-surface defined, first by Eqs.(20) and (21), [the conditions for the light-cones from $x_0^a(s^J)$], i.e.,

$$S^*(x^a, x_0^a(s^J), \alpha^A) = Z(x^a, \alpha^A) - Z(x_0^a(s^J), \alpha^A) = 0, \qquad (26)$$

$$\partial_A S^*(x^a, x_0^a(s^J), \alpha^A) = 0 \qquad (27)$$

augmented by the s^J variations, i.e., by

$$\partial_J S^*(x^a, x_0^a(s^J), \alpha^A) = 0. \qquad (28)$$

These are five conditions on the eight variables (x^a, s^J, α^A) thus forming a three-surface in the eight dimensional space; this when projected down to the space-time results in the aforementioned envelope. It is easily seen from Eqs.(27) and (28) that this surface, which we will denote by

$$N(x^a) = 0 \tag{29}$$

is a characteristic surface and hence satisfies the eikonal equation, (18). Though almost everywhere it can be given in the form of the vanishing of a function of x^a, i.e., by Eq.(29), there will be lower dimensional regions where it must be given parametrically. See e.g., Eq.(23) or Eqs.(24) and (25).

Before looking at the time of arrival function, we first look at Eq.(28) more closely. Substituting Eq.(26) into Eq.(28) and taking the required derivatives we have

$$\partial_a Z(x_0^a(s^J), \alpha^A) \frac{\partial x_0^a}{\partial s^J} = 0 \tag{30}$$

which is the statement that for the null ray leaving S at the point $x_0^a(s^J)$, $\partial_a Z(x_0^a(s^J), \alpha^A)$ must be normal to the tangent vectors $\frac{\partial x_0^a}{\partial s^J}$ at S and thus normal to S. Eq.(30), hence, chooses among all the rays forming the light-cone at $x_0^a(s^J)$, i.e., the rays parametrized by α^A, just the appropriate α^A so that the ray is the (unique) normal to S. We thus have the result that (30) can be solved by

$$\alpha^A = \alpha^A(s^J). \tag{31}$$

Using Eq.(31), we have that Eqs.(26) and (27) become

$$S^*(x^a, x_0^a(s^J), \alpha^A(s^J)) = Z(x^a, \alpha^A(s^J)) - Z(x_0^a(s^J), \alpha^A(s^J)) = 0, \tag{32}$$

$$\partial_A S^*(x^a, x_0^a(s^J), \alpha^A)|_{\alpha^A = \alpha^A(s^J)} = 0 \tag{33}$$

Using the same argument that led to Eq.(25), namely the implicit function theorem and $\partial S^*/\partial x^0 \neq 0$, we see that Eq.(32) is equivalent to

$$x^0 = T(x^\alpha, x_0^a(s^J), \alpha^A(s^J)) \tag{34}$$

If we take the three $x^\alpha = X^\alpha$ as the "constant spatial position" of the world-line of Sec. II, we have the time of arrival function. Since, interpreting Eq.(32) as defining Eq.(34) implicitly, we have that

$$\frac{\partial S^*}{\partial x^0}\partial_J T + \partial_J S^* = 0, \tag{35}$$

which, since $\partial S^*/\partial x^0 \neq 0$, implies that

$$\partial_J S^* = 0 \Rightarrow \partial_J T = 0. \tag{36}$$

Thus the extremization of S^* implies the extremization of $T(x^\alpha, x_0^a(s^J), \alpha^A(s^J))$ as was to be proved. This proof is not affected by the difficulties in Sec. II of the possible multivaluedness of the earlier T.

In the terminology of Arnold, these results follow from his theory of Legendre submanifolds and maps, where

$$Z(x^a, \alpha^A) - Z(x_0^a(s^J), \alpha^A) = 0 \tag{37}$$

from Eq.(26), defines a generating family $F(x^a, \alpha^A, s^J)$ and Eqs.(27) and (28) define the Legendre map.

4. DISCUSSION

We have given two derivations of a variational principle for the time of arrival of null geodesics at an observer. Superficially, it appears as if it were a version of Fermat's principle; in actuality it is quite different. Fermat's principle leads to local evolutionary laws for the rays while here we have from the start assumed that the rays are given by null geodesics. Our variational principle gives the initial direction of the ray. Often Fermat's principle is invoked to derive the equations of gravitational lensing [5, 6]. A paper is now in preparation, using the techniques discussed here, in which a universal lensing equation valid in all situations is obtained.

Acknowledgments

This work received the support of the NSF grant No. PHY-9722049.

References

[1] V. I. Arnol'd, *Mathematical Methods of Classical Mechanics*, 2nd ed. (Springer-Verlag, New York, 1978).

[2] P. Schneider, J. Ehlers, and E. E. Falco, *Gravitational Lenses* (Springer-Verlag, New York, 1992).

[3] V. I. Arnol'd, S. M. Gusein-Zade, and V. A. N, *Singularities of Differentiable Maps* (Birkhäuser, Boston, 1985), Vol. I.

[4] E. Schrödinger, *Expanding Universes* (Cambridge University Press, Cambridge, 1956), pp. 41–45.

[5] V. Faraoni, Astrophysical Journal **398**, 425 (1992).

[6] V. Perlick, Class. Quantum Grav. **7**, 1849 (1990).

Chapter 22

CONCEPTUAL ISSUES IN COMBINING GENERAL RELATIVITY AND QUANTUM THEORY

T. Padmanabhan
Inter-University Centre for Astronomy & Astrophysics
Ganeshkhind, Pune 411 007, India

Abstract

Points of conflict between the principles of general relativity and quantum theory are highlighted. I argue that the current language of QFT is inadequete to deal with gravity and review attempts to identify some of the features which are likely to present in the correct theory of quantum gravity.

1. INTRODUCTION

The question of bringing together the principles of quantum theory and gravity deserves to be called "the problem" of theoretical physics today. The history of failures in this attempt illustrates not only the conceptual complexity of the problem but also the sociology of science in the late twentieth century. Since Jayant will be sympathetic to my — rather heretical — way of thinking about this issue, I thought a description of my views on this subject will be appropriate for this volume.

2. THE MIRACLE OF QUANTUM FIELD THEORY

In proceeding from classical mechanics [with finite number of degrees of freedom] to quantum mechanics, one attributes operator status to various dynamical variables and imposes the commutation relations among them. These relations, and an expression for the hamiltonian operator

N. Dadhich and A. Kembhavi (eds.), The Universe, 239–251.

$\hat{H}(\hat{q},\hat{p})$, allow us to write down the equations for the time evolution of the operators. If these equations can be solved, then we can determine the full structure of the theory. Often, it is conveneient to provide a repesentation for the operators in terms of normal differential operators so that the problem can be mapped to solving a partial differential equation — say, the time-dependent Schrodinger equation — with specific boundary conditions. Such problems are mathematically well defined and tractable, allowing us to construct a well defined [though, in general, not unique] quantum theory for a classical system with finite number of degrees of freedom.

The geralisation of such a procedure to a *field* with infinite number of degrees of freedom is *not* straightforward and is fraught with conceptual and mathematical problems. Given a classical field with some dynamical variables, one can attempt to quantise the system by elevating the status of dynamical variables to operators and imposing the commutation rules. But finding a well defined and meaningful representation for this commutator algebra is a nontrivial task. Further, if one tries to extend the approach of quantum mechanics [based on Schrodinger picture] to the field, one obtains a *functional* differential equation instead of a partial differential equation. The properties — let alone solutions! — of this equation are not well understood for any field with nontrivial interactions. Somewhat simpler (and better) approach will be to use the Heisenberg picture and try to solve for the operator valued distributions representing the various dynamical variables. Even in this case, one does not have a systematic mathematical machinary to solve these equations for an interacting field theory. The procedure to quantise an arbitrary [but well defined] classical field theory fails right at the outset due to inadequete mathematical apparatus. We have no right to expect quantum field theories to exist!

It is, therefore, quite surprising to me that quantum field theories indeed could be developed and used to make veriafiable predictions. To see how this miracle was achieved, let us look at the prototype of quantum field theory, viz. QED. The evolution equations for operators in QED [in 3+1 dimensions] cannot be solved exactly; however, it is possible to set up a perturbation expansion for these variables in powers of the coupling constant $(e^2/\hbar c) \approx 10^{-2}$. The lowest order of the perturbation series, in which all interactions are switched off, defines the so called *free field* theory. It is possible to map this theory to one describing infinite number of noninteracting harmonic oscillators and solve for the dynamics of any one of the oscillators completely. The perturbation expansion can be then used to obtain the "corrections" to this free field theory. Several issues crop up when such an attempt is made:

(a) To begin with, the decomposition of the field in terms of the harmonic oscillators is not unique and there exists infinite number of inequivalent representations of the basic commutator algebra for the system. This shows that "physical" quantities like ground state, particle number etc. will depend on the specific representation chosen and will not be unique.

(b) Since the system has infinite number of degrees of freedom, quantities like total energy can diverge. The actual form of the divergence depends on the representation chosen for the algebra and the differences between infinite quanties may retain a representation dependent [finite] value, unless one is careful in regularising such expressions. In some cases, one may be forced to choose particular set of harmonic oscillators because of the boundary conditions. Then, the difference between two infinite quantities could be physically relevant (and even observable as in the case of, for example, Caisimir effect).

(c) The situation becomes worse when the perturbation is switched on. In general, the perturbation series will not converge and has to be interpreted as an asymptotic expansion. Further, the individual terms in the perturbation series will not, in general, be finite creating a far more serious problem. This arises because the amplitude for propagation of a free field quanta, of mass m and euclidian momentum p varies as $(p^2+m^2)^{-1}$, which does not die down sufficiently fast at large p. This, in turn, is related to the fact that virtual quanta of *arbitrarily* high energy are allowed to exist in the theory [incorporating Lorentz invariance at arbitarirly small length scales] and still propagate as free fields.

(d) Perturbation theory completely misses all effects which are non-analytic in the coupling constant. In QED, for example, perturbation theory cannot lead the result that an external electromagnetic field can produce $e^+ - e^-$ pairs, since this effect has nonanalytic dependecy on e. [1] One cannot even estimate the seriousness of this problem since very few nonperturbative results are known.

How does one cope up with these difficulties? Issue (a) is handled by choosing one particular representation for the free field theory by fiat, and working with it — and ignoring all other representations which are not unitarily equivalent to the same. This also dodges the issue (b) provided some means of regularisation can be found to discard the infinities of the free field theory. Once a representation for the harmonic oscillators is chosen, this can be implemented by a procedure like normal ordering. Issue (d) is accepted as a failure of the method [at least by the honest researchers!] and then ignored. Most of the successful effort was concentrated on handling the problem of infinities *in the individual terms* of the perturbation series, that is, on issue (c). The pardigm for handling these infinities can be stated in terms of the concept of renomalization which, by itself, has nothing to do with any divergence. In the simplest terms, renormalization expresses the fact that the interactions will change the values of the various coupling constants in the theory; that is, the physically observed coupling constants are the "renormalized" ones and not the "bare" ones which appear in the original Lagrangian. The phenomena of renormalization exists, for example, in condensed matter theories where both the bare and renormalized coupling constants can be finite. In the context of field theory, renormalization can provide a means to eliminate divergences, if *all* the divergent terms of a perturbation expansion can be eliminated by redefining the coupling constants in the theory. For an arbitrary field theory, we have no assurance that all the divergences can be so eliminated; in fact, it is quite easy to construct well defined classical field theories for which divergences cannot be eliminated by this process.

The unexplained miracle of 20th century quantum field theory lies in the fact that several physically relevant field theories — describing quantum electrodynamics, electro-weak interactions and QCD — belong to this special class of *perturbatively renormalisable* theories. For such theories, perturbation series can be developed as an algorithmic procedure to evaluate matrix elements for transitions between asymptotic states of the free field theory, to any order in perturbation theory. The agreement of such predictions with observations led to (several nobel prizes and) a religious faith in perturbative renomalization as the paradigm of quantum field theory by late 60's - early 70's. Nobody knows why this mathematically non-rigorous, conceptually ill-defined, formalism of perturbative quantum field theory works. The miracle becomes even more curious when we notice that the bag of tricks fail miserably in the case of gravity.

3. THE EDIFICE OF GENERAL RELATIVITY

Until early seventies, most of the hardcore particle physicists used to ignore general relativity and gravitation and the first concrete attempts in putting together principles of quantum theory and gravity were led by general relativists (see e.g. ref.[2]). It was clear, right from the beginning, that this is going to be a formidable task since the two "theories of principle" differed drastically in many aspects. The key features of gravity which are of relevence in this context are the following:

(a) The Lagrangian describing classical gravity, treated as a function of $h_{ik} = g_{ik} - \eta_{ik}$, is *not* perturbatively renormalizable; in fact, there does not exist any simple redefinition of the field variables which will lead to a perturbatively renormalizable theory. So the most straight forward approach, based on the belief that nature will continue to be kind to us, is blocked. The miracle fails.

(b) The principle of equivalence implies that any resonable description of gravity will have a geometrical structure and that gravitational field will affect the spacetime intervals in a specific manner. This inescapable conclusion leads to several corollaries, all of which make gravity an odd-man-out: (i) To begin with, this makes the spacetime itself a dynamical entity and not something which can be prescribed beforehand. (ii) Secondly, the description of gravitational field in terms of the metric tensor g_{ik} translates into a *constrained* dynamical system; that is, the true degrees of freedom of gravity are only 2 per event rather than the full set of 10 functions contained in g_{ik}. Understanding the nature of constraints in general relativity — and implementing it in different descriptions of quantum theory — turn out to be a very non trivial task. (iii) Thirdly, the geometrical description leads to a fairly unique (class of) Lagrangian(s) for the gravitational field. The equivalent Hamiltonian formulation of the theory in terms of 3-geometries lead to a degree of freedom (conformal factor) which is unbounded from below. (iv) The geometrical structure also implies that there is no prefered coordinate system in the presence of gravitational field. In fact, there is no unique and meaningful separation of the various effects as those due to gravity and those due to noninertial forces, if we stick to the metric tensor as the fundamental physical variable. For a general gravitational field, there will be no way of choosing a special class of spacelike hypersurfaces or a time coordinate.

(c) Gravity affects the light signals and hence determines the causal structure of spacetime. In particular, gravity is capable of generating regions of spacetime from which no information can reach the outside world through classical propagation of signals. This feature, which may be loosely called 'the existence of trapped surfaces' has no parallel in any other interaction.

(d) Since all matter gravitates, the gravitational field becomes more and more dominant at larger and larger scales. In the limit, the asymptotic structure of spacetime is determined by global, smoothed out distribution of matter in the cosmological context. In such a case, the spacetime will not be asymptotically flat in the spatial variables at any given time. The behaviour of the spacetime for $t \to \pm\infty$ will also be highly non-trivial and could be dominated by very strong gravitational fields.

(e) All energies gravitate thereby removing the ambiguity in the zero level for the energy, which exists in non-gravitational interactions. This feature also suggests that there is no such thing as a free, non-interacting field. Any non trivial classical field configuration will possess certain amount of energy which will curve the spacetime, thereby coupling the field to itself indirectly. Gravitational field is not only nonlinear in its own coupling, but also makes *all other fields* self-interacting.

(f) The coupling constant governing gravitational interaction has a non trivial dimension in the language of quantum field theory; $E_P \equiv (G/\hbar c)^{-1/2}$ has the dimensions of energy in contrast to $(e^2/\hbar c)$ which is dimensionless. Simple power counting arguments based on this result will show that gravity will be perturbatively non renormalizable. Further, one can construct a quantity with dimensions of length, $L_P \equiv (G\hbar/c^3)^{1/2} \approx 10^{-33}$ cm, from the gravitational coupling constant. Though no formal proof exist, it is very likely that quantum gravitational effects will modify the spacetime structure at length scales comparable to $L \approx L_P$. In fact, simple thought experiments combining the principles of quantum theory and gravity show that the planck length acts as a 'zero-point- length' to any spacetime. (see e.g. ref [3]) Any correct formulation of quantum gravity must have the infrastructure to incorporate this feature just as the operator description of quantum mechanics is capable of incorporating the uncertainty principle.

(g) The truely remarkable feature of classical general relativity is that *this theory is fundamentally wrong.* This is most easily seen from the fact that one can ask questions — in the form of thought experiments — to which the theory cannot provide sensible answers. One such question could be the following: "A neutron star of mass $6M_\odot$ collapses to form a blackhole. How will the physical phenomena appear with respect to a hypothetical observer on the surface of the neutron star at arbitrarily late times as measured by the observer's clock?" Such questions cannot be answered in classical general relativity because the relevant equations lead to an infinte curvature singularity. Such a theory must clearly be wrong and has to be replaced by a better formulation at very strong curvatures.

The features (a) to (d) already suggest that there are fundamental contradictions between the formulation of quantum field theory and that of general relativity. Given the result (a), one could have taken two separate routes: (i) How can gravity be made to conform to the tenets of QFT ? or (ii) Why did QFT work in the case of other interactions and how should QFT be modified to handle gravity ? Historically, most of the effort went into route (i) and led to a blazing trail of failures. This is in spite of the fact that many of the features listed above show that contradictions of language surface even when one tries to develop a quantum field theory in an external gravitational field (without worrying about the quantization of gravity itself). Since gravity does not allow a prefered slicing of the spacetime, quantum field theory needs to be formulated without using any prefered representation for the operator algebra. Loosely speaking, this implies that there is no generally covariant definition for the vacuum state (or particle excitations) in a generic curved spacetime. Infinite number of inequalent representations exist and we have no means of choosing any one of them as 'more physical' than another. It is clear that such a description — based on a ground state and the particle-like excitations — is of very limited value and will not survive the transition to the next layer, say, the one in which we need to take the back reaction of the particle production into account.

An abstract way of stating the same conclusion is as follows: Gravity is inherently local (local coordinate charts, observers, freely falling frames ...) while the standard formulation of QFT is global (global spacelike hypersurface, global mode functions, ...). There is no such thing as 'one-particle-state- at-the-event-$\mathcal{P}$' in QFT and there are serious problems in defining any such concept.

More difficulties arise from the feature (c) listed above. When gravity makes certain regions inaccessible, the data regarding quantum fields in these regions can "get lost". This requires reformulation of the equations of quantum field theory, possibly by tracing over the information which resides in the inaccessible regions — something which is not easy to do either mathematically or conceptually. Trapped surfaces also highlight the role of boundary conditions in QFT. The structure of a free field propagating in an arbitrary spacetime can be completely specified in terms of, say, the Feynman Greens function $G_F(x, y)$ which satisfies a local, hyperbolic, inhomogeneous, partial differential equation. Each solution to this equation provides a particular realization of the theory. In other words, there exists a mapping between the realizations of the quantum field theory and the relevant boundary conditions to this equation which specify a useful solution. When trapped surfaces exists, the differential operator governing the Greens function will be singular on these surfaces (in some coordinate chart) and the issue of boundary conditions become far more complex. It is, nevertheless possible — at least in simple cases with compact trapped surfaces — to provide an one-to-one correspondence between the ground state of the theory and the boundary conditions for G_F on the compact trapped surface. In fact, the Greens function connecting events outside the trapped surface can be completely determined in terms of a suitable boundary condition on the trapped surface, indicating that trapped surfaces acquire a life of their own even in the context of QFT in CST. In a way, the procedure is reminiscent of renormalisation group approach, but now used in real space to integrate out information inside the trapped surface and possibly replace it by some boundary condition.

In this connection, it is worth noting that effects like particle production by a blackhole (or expanding universe) are *infrared* phenomena and arises due to the coupling of modes at large scales. [The conflict between local GR and global QFT is again apparent]. The ultraviolet modes are comparitively local and decoupled. This is somewhat different from standard situations in QFT where the ultraviolet modes get coupled due to interaction and the infrared ones get a free ride. Integrating out the information inside a trapped surface in real space might also translate into a renormalisation group approach in fourier space *with infrared modes integrated out.*

The importance of cosmological solutions in classical gravity [item (d)] led to the investigations in quantum cosmology and the possibilities of 'wave function of the universe'. Two features emerged from these attempts: (i) It may be possible to circumvent the classical cosmological singularity in quantum cosmological models. (ii) If the ground state of the universe is globally determined, the boundary conditions could also lead to specification of the ground state for matter fields, thereby providing a quantum version of Mach's principle [4]. Both these results are tentative and nonrigorous but go to show the richness of possibilities. The feature (d), however, creates problems in formulating quantum field theory in terms of scattering amplitudes or asymptotic "in" "out" states. Such concepts are meaningful when the global spacetime structure is externally specified but not when dynamics determines the structure of asymptotic universe.

I think the key *physical* message from some of these investigations is the following: Fields are more important than particles and could be more robust entities. In fact, this conclusion is apparent even from the existence of a phenomena like Casmir effect which cannot be explained in terms of virtual particles and is independent of the coupling constant ($e^2/\hbar c$) of the perturbative theory. *This is in sharp contradiction with the phisolophy of perturbative gauge theories in which the particle physicist uses fields just as a tool to obtain an algorithm for computation of, say, S-matrix elements.* The baggage we carry from Lorentz invariant, perturbatively renormalizable, quantum field theory — like the concepts of quanta, vacua, in-out states, Smatrix.... etc. — is probably to be abandoned.

Features (e) to (g) make the situation worse. The fact that all matter gravitates [see (e)], once again stresses the need to abandon description based on free field theory to handle virtual excitations with arbitrarily high energies. An excitation with energy E will probe length scales of the order of $(1/E)$ and when $E \to E_P$, the nonlinearity due to self gravity cannot be ignored. The same conclusion is applicable even to vacuum fluctuations of any field, including gravity. If we attempt to treat the ground state of the gravitational field as the flat spacetime, we must conclude that the spacetime structure at $L \lesssim L_P$ will be dominated by quantum fluctuations of gravity and the smooth macroscopic spacetime can only emerge when the fluctuations are averaged over larger length scales.

The difficulties mentioned above should caution one against approaching the problem of quantum gravity as one of mathematics requiring a better technical apparatus. There is very strong indication that the basic language of field theory is inadequate to grapple with the complications introduced by gravity. Perturbative language which — at best — gives an algorithm to calculate S-matarix elements, is not going to be of much use in understanding the quantum structure of gravitational field. Most of the interesting questions — possibly *all* the interesting questions — in quantum gravity are non perturbative in character; whether a theory is perturbatively renormalizable or not is totally irrelevent in this context. Conventional quantum field theory works best when a static causal structure, global Lorentz frame, asymptotic in-out states, bounded Hamiltonians and the language of vacuum state, particle excitations etc., are supplied. The gravitational field removes all these features, strongly hinting that we may be working with an inadequate language. The gradual paradigm shift in the particle physics community from perturbative finiteness of supergravity (in early 80's) to non perturbative description of superstrings (in late 90's) represents a grudging acceptance of the lessons from gravity. The history of these failures indicates that we have not been ruthless enough in attacking the problem.

4. QUANTUM GRAVITY FROM PURE THOUGHT?

Given the above results, is it possible to describe the key features which must be present in any future, successful, theory of quantum gravity? I believe this can be done to certain extent thereby providing some useful pointers.

The fact that there will exist violent spacetime fluctuations at small scales suggests that the macroscopic, continuum, description of spacetime can only be approximate and valid when quantum fluctuations are averaged over large scales. The description of continuum spacetime in terms of, classical, Einstein's equation is similar to the description of a solid by elastic constants or the description of a gaseous system by an equation of state. While the knowledge of microscopic quantum theory of atoms and molecules will allow us, in principle, to construct the description in terms of elastic constants, the reverse process is unlikely to be unique. What one could hope is to take clues from well designed thought experiments, thereby identifying some key generic features of the microscopic theory.

One might assume that the microscopic description is in terms of certain [as yet unknown] variables q_i and that the conventional spacetime metric is obtained from these variables in some suitable limit. Such a process will necessarily involve coarse-graining over a class of microscopic descriptors of geometry. I will now outline an argument which suggests that there are *infinite* number of microscopic descriptors which are "integrated out" in proceeding from the fundamental description to spacetime description, [5]. The argument proceeds in three steps: (1) Among all systems dominated by gravity, the universe possess a very peculiar feature. If the conventional cosmological models are reasonable, then it follows that *our universe proceeded from quantum mechanical behaviour to classical behaviour in the course of dynamical evolution defined by some intrinsic time variable.* It can be shown that a system with bounded Hamiltonian can never make such a transition if classicality is defined in terms of behaviour of a suitable Wigner function. It follows that the quantum cosmological description of our universe, as a Hamiltonian system, should contain atleast one unbounded degree of freedom. It can also be shown that the unbounded mode — which, in the case of FRW universe, corresponds to the expansion factor — will go classical first, as is experienced in the evolution of the universe. (2) Let us next address the task of obtaining an unbounded Hamiltonian for an effective theory when the original theory contained a larger set of dynamical variables. It can again be shown that, if one starts with a bounded Hamiltonian for a system with finite number of quantum fields and integrate out a subset of them, the resulting Hamiltonian for the low energy theory cannot be unbounded. (3) Assuming that the original theory is describable in terms of a bounded Hamiltonian for some suitable variables, it follows that an infinite number of fields have to be involved in its description and an infinite subset of them have to be integrated out in order to give the standard low energy gravity. This feature is indeed present in one form or the other in the descriptions of quantum gravity based on strings [6] or Ashtekar variables [7]. My argument suggests that this is indeed inevitable.

If the description in terms of continuum spacetime is like theory of elasticity, and we do not know the fundamental descriptors of spacetime, is there any way of bridging the gap between the two? It turns out that this is possible by using the properties of macroscopic spacetime near the trapped surfaces. I have given detailed arguments elsewhere [8] to show that the event horizon of a Schwarzschild blackhole acts as a magnifying glass, allowing us to probe Planck scale physics. Consider, for example, a physical system described by a low energy Hamiltonian, $H_{\rm low}$. By constructing a blackhole made from the system with this Hamiltonian and requiring that the blackhole should have a density of states that is immune to the details of the matter of which it is made, one can show that the Hamiltonian, $H_{\rm true}$ describing the interactions of the system at transplanckian energies must be related to $H_{\rm low}$ by $H^2_{\rm true} = \alpha E^2_P \ln[1 + (H^2_{\rm low}/\alpha E^2_p)]$ where α is a numerical factor. Of course, the description at transplanckian energies cannot be in terms of the original variables in the rigorous theory. The above formula should be interpreted as giving the mapping between an effective field theory (described by $H_{\rm true}$) and a conventional low energy theory (described by $H_{\rm low}$) such that the blackhole entropy will be reproduced correctly.

In fact, one can do better and construct a whole class of effective field theories [9] such that the one-particle excitations of these theories possess the same density of states as a Schwarzschild blackhole. All such effective field theories are non local in character and possess a universal two-point function at small scales. The nonlocality appears as a smearing of the fields over regions of the order of Planck length thereby confirming ones intuition about microscopic structures, trapped surfaces and blackhole entropy.

If the physical description above Planck energy (or equivalently below Planck length) changes drastically, how can one modify the low energy description such that the singularities in spacetimes and the perturbative divergences in quantum field theory are removed? This question cannot be answered rigorously without knowing the microscopic structure of spacetime. However two broad class of theories can be distinguished in terms of a general criterion. In the first class of theories, the low energy ($E \ll E_P$) and high energy ($E \gg E_P$) behaviour are not related by any manner and the high energy sector of the theory *does* affect the low energy behaviour significantly. If nature is built along these lines, then we cannot predict much without knowing the full theory. On the other hand, one can think of another class of theories in which the high energy and low energy descriptions are related in a specified manner and are not completely independent. The simplest form of such a relation will be a 'duality' in which the behaviour at a scale E is related to a behaviour at scale (E_P^2/E), or — equivalently — the behaviour at length scales l and (L_P^2/l) are related. Implementing this duality in the path integral representation for a propogator, say, leads to a remarkable result [10]: The effect of this duality is the same as assuming that the spacetime possesses a 'zero-point-length' and replacing the flat spacetime interval $(x-y)^2$ by $(x-y)^2 + L_P^2$. I suspect that the converse is also true: if the structure of the theory is such that planck length acts as a minimal length to the spacetime, then the theory will possess a duality between length scales l and (L_P^2/l). String theories do show related — though not the same — features. If nature is built along these lines, then transplanckian physics is dual to the low energy theory and must possess a description in terms of some effective field theory.

The key conclusions which emerge from all these are the following: (i) It is unlikely that one will make genuine progress, unless the language of quantum field theory is expanded to be capable of handling the features listed in section 3. The question to understand is *not* why gravity is difficult to quantise but *why the perturbative approach was so unreasonably successful in dealing with other interactions ?*. This must be because the conventional QFT is a wrong way of looking at physics though it accidentally incorporated several features of the right [as yet unknown] approach — as was in the case of, say, old quantum theory. Rethinking about QED in a possibly new language might offer hints on how to proceed further. (ii) Given the unlikely event of experimental confirmation of quantum gravity, it is necessary to attempt a top-down approach [classical gravity $\rightarrow$ effective field theory $\rightarrow$ microscopic spacetime descriptors] using, say, well-defined thought experiments. In this regards, spacetimes with trapped surfaces will be valuable. (iii) It is also important to worry, at a conceptual level, the effect of transplancian physics on low energies. If this effect is not to be unreasonably strong, thereby killing predictability, it is necessay that the low energy theory is protected by some kind of duality mapping relating transplancian energies to low energies. A progam for quantum gravity, along these lines, holds promise.

Acknowledgments

I thank Apoorva Patel for several illuminating discussions.

References

[1] Schwinger, J (1951) Phys. Rev **82**, 664

[2] DeWitt, B S (1967) Phys. Rev **160**, 1113; **162** 1195; **162**, 1239

[3] Padmanabhan, T (1987) Class. Quan. Grav., **4** L 107

[4] Padmanabhan, T (1990) Pramana, **35**, 317

[5] Padmanabhan, T (1998) paper in preparation

[6] Polchinski, J H (1996) Rev. Mod. Phys **68**, 1245

[7] Rovelli, C (1997) gr-qc/ 9710008; A. Ashtekar and K. Kransov, gr-qc/9804039

[8] Padmanabhan, T (1998) hep-th 9801138; IUCAA preprint 4/98

[9] Padmanabhan, T (1998) Phys. Rev. Letts, **81** 4297

[10] Padmanabhan, T (1997) Phys. Rev. letts, **78**, 1854; Phys. rev. D (1998) **57**, 6206

Chapter 23

OPEN INFLATION IN HIGHER DERIVATIVE THEORY

B. C. Paul and S. Mukherjee
Physics Department, North Bengal University
Siliguri, Dist. : Darjeeling, 734 430, India

Abstract We look for Hawking-Turok (HT)instanton solutions in a higher derivative theory of the type $R + \alpha R^2 - 2\Lambda$, which describes the creation of an open inflationary universe. Converting the R^2 -theory into a theory of a scalar field minimally coupled to Einstein gravity by a conformal transformation, we obtain a singular HT instanton solution for a class of R^2 -theory, with parameters $\alpha < 0, \Lambda = f(\alpha) > 0, 8|\alpha|f(\alpha) < 1$. Nonsingular de Sitter type instantons are also present in this case.. The HT instanton solutions are not very generic in nature in R^2 -theory.

1. INTRODUCTION

In two recent papers, Hawking and Turok (HT) [1,2] have suggested that Hartle-Hawking [3] no boundary proposal provides for the creation of an open inflationary universe in a generic sense. Till recently, it was believed that all inflationary models lead to $\Omega_o \sim 1$ to a great accuracy. This view was modified after it was discovered that there is a special class of inflaton effective potentials which may lead to a nearly homogeneous open universe with $\Omega_o \leq 1$ at the present epoch. These potentials should have a metastable minimum followed by a small slope region which permits a slow roll inflation. The inflaton is supposed to be initially trapped in the false vacuum leading to a period of inflation, which gives an almost de Sitter space with small quantum fluctuations. The inflaton field eventually undergoes a quantum tunneling, nucleating bubbles within which the inflaton field slowly rolls down to the true vacuum. It was pointed out by Coleman and De Luccia [4] that the interior of such a bubble is actually an open universe. However, this

N. Dadhich and A. Kembhavi (eds.), The Universe, 253–259.

scenario can be realised only at the cost of making very fine tunning. To keep the quantum fluctuations small, a very flat potential is required while one also needs a metastable minimum. The mechanism suggested by HT does away with the requirement of a false vacuum and it also leads to a universe created with minimal quantum fluctuations.

Vilenkin [5], however, questioned the validity of the HT mechanism by raising the following points :

- The instanton of HT is singular in the sense that both the curvature and the scalar field become singular at one point.
- The field equations are not satisfied at the singularity and it is, therefore, not a stationary point of the Euclidean action.
- It is not clear that HT instanton will give the dominant contribution to the Euclidean path integral.
- There is also a counter example : Consider the case of a massless scalar field interacting minimally with gravity, which also gives a singular instanton, which is asymptotically flat. The nucleation probability of such an instanton is not suppressed. But, if accepted as legitimate it leads to the unacceptable results that the singular bubbles expand rapidly to engulf the universe.

In response, HT [2] argue that the instanton considered by them is legitimate since it is integrable and the Euclidean action is finite. Moreover, it avoids the catastrophe which Vilenkin's example predicts. However, the question whether it is proper to use the singular instanton in evaluating the path integral remains to be answered. Considerable work has already been done [6-11] on various aspects of this problem.

Consideration of models which give singular as well as non-singular instantons may be quite useful in this connection. We present here the higher derivative theory as a model for such a calculation. The presentation of the paper is as follows : in section 2, we discuss the non-singular instanton solutions in R^2 - theory. The scalar field theory, obtained by a conformal transformation of the R^2 - theory is discussed in section 3. Instanton solutions, both singular and non-singular, have been determined. We discuss our results in section 4.

2. HIGHER DERIVATIVE THEORY

We consider here a generalized theory of gravity to explore the possibility of an instanton solution for creation of an open-inflationary universe. It is well-known that the R^2 -theory has a number of good features. It is known that with suitable counter terms viz., $C^{\mu\nu\rho\delta}C_{\mu\nu\rho\delta}$, R^2

, Λ added to the Einstein action, one gets a perturbation theory which is well behaved, formally renormalizable and asymptotically free [12].

Let us consider the following Euclidean action

$$I_E = -\frac{1}{16\pi}\int d^4x\sqrt{g}[R+\alpha R^2 - 2\Lambda] - \frac{1}{8\pi}\int_{\partial M} d^3x\sqrt{h}k(1+2\alpha R). \quad (1)$$

where g is the determinant of the 4-dimensional metric, R is the scalar curvature, Λ is cosmological constant, h_{ij} is the metric induced on ∂M and $K = h^{ij}K_{ij}$ is the trace of the second fundamental form.

We consider here O(4) symmetric Euclidean metric which is

$$ds^2 = d\sigma^2 + b^2(\sigma)\left(d\psi^2 + sin^2\,\psi\; d\Omega_2^2\right) \quad (2)$$

where $d\Omega_2^2$ is the metric of 2-sphere. The scalar curvature is

$$R = -6\left(\frac{b"}{b} + \frac{b'^2}{b^2} - \frac{1}{b^2}\right). \quad (3)$$

where prime denotes derivative with respect to σ. We treat b and R as independent variables and rewrite the action, including the constraint (3) through a Lagrange multiplier β,

$$I_E = \frac{\pi}{4}\int d\sigma\left[(R+\alpha R^2 - 2\Lambda)b^3 - \beta\left(R + 6\frac{b"}{b} + 6\frac{b'^2}{b^2} - \frac{6}{b^2}\right)\right]$$
$$-\frac{1}{8\pi}\int_{\partial M} d^3x\sqrt{h}\;K\;(1+2\alpha R). \quad (4)$$

Varying with respect to R, we determine

$$\beta = b^3(1+2\alpha R).$$

Substituting this in eq.(4) we now obtain

$$I_E = \frac{\pi}{4}\int_{\tau=0}^{\tau_{\partial M}}\left[(-2\Lambda - \alpha R^2)b^3 + 6b(1+b'^2)(1+2\alpha R) + 12\alpha b^2 b' R'\right] d\sigma$$
$$+3\pi[b'b^2(1+2\alpha R)]_{\tau=0}. \quad (5)$$

The field equation for $b(\sigma)$ is given by

$$(1+2\alpha R)\left[\frac{b"}{b} + \frac{1}{2}\frac{b'^2-1}{b^2}\right] + \alpha R" + 2\alpha\frac{b'}{b}R' + \frac{1}{4}\alpha R^2 + \frac{\Lambda}{2} = 0. \quad (6)$$

The eq.(6) admits an instanton solution

$$b = H_o^{-1}\sin H_o\sigma \quad (7)$$

with $R = 12H_o^2$ and $\Lambda = 3H_o^2$. The solution is non-singular and independent of the parameter α. The Euclidean action (4) in this case can be calculated by integrating over half of the S^3 :

$$I_{S^3} = - \left[\frac{3\pi}{2\Lambda} + 12\pi\alpha \right] . \tag{8}$$

The dependence of the probability of creation of the de Sitter type universe on the two parameters Λ and α can be read out from equation (8). Note that the higher derivative term with $\alpha > 0$ enhances the probability. A smaller Λ also does the same.

An open inflationary universe may be obtained by analytic continuation of the metric (2) to the Lorentzian region, $\psi = \frac{\pi}{2} + i\,\tau$,

$$ds^2 = d\sigma^2 + b^2(\sigma) \left(- d\tau^2 + cosh^2 \tau d\Omega_2^2 \right) \tag{9}$$

which is a spatially inhomogeneous de Sitter like metric. However, if one sets $\sigma = i\,t$ and $\tau = i\,\frac{\pi}{2} + \chi$, the metric (8) becomes

$$ds^2 = - dt^2 + a^2(t) \left(d\chi^2 + \sinh^2 \chi \, d\Omega_2^2 \right) \tag{10}$$

where $a(t) = - i\ b(i\ t)$. If the model admits Hawking-Turok singular instantons, the creation of the open universe may be a generic phenomenon. It is, therefore, interesting to see if such instantons are permitted in this model. To see this we need to consider a conformal transformation which converts the R^2 -theory into a scalar field theory. This will be taken up in the next section.

3. INSTANTON WITH A SCALAR FIELD

We consider a generalised action given by

$$I = \int f(R)\sqrt{g}\, d^4x \tag{11}$$

where f(R) is a function of scalar curvature R. To write the action in terms of Einstein gravity coupled to a scalar field we use a conformal transformation of the form

$$\tilde{g}_{\mu\nu} = \Omega^2 g_{\mu\nu} \tag{12}$$

where $\ln\Omega = \frac{\phi}{\sqrt{6}} = \frac{1}{2}\ln|f'(R)|$, prime denoting a derivative with respect to R. The action given by (11) gets transformed into

$$I = - \int \sqrt{\tilde{g}} \left[R(\tilde{g}) - \frac{1}{2}(\tilde{\partial}\phi)^2 - V(\phi) \right] , \tag{13}$$

with $V(\phi) = \frac{1}{2} e^{-2\sqrt{\frac{2}{3}}\phi} \left[R\frac{\partial f}{\partial R} - f(R) \right]$. Comparing with the action given by (1), considered in the previous section, we have $f(R) = R + \alpha R^2 - 2\Lambda$. The scalar field potential in this case takes the form

$$V(\phi) = \frac{1}{8\alpha} e^{-2\sqrt{\frac{2}{3}}\phi} \left[e^{\sqrt{\frac{2}{3}}\phi} - 1 \right]^2 + \Lambda e^{-2\sqrt{\frac{2}{3}}\phi}. \tag{14}$$

The potential depends on two parameters α and Λ. The Einstein field equations corresponding to the action (13) are derived using the equations $\frac{\partial I}{\partial \tilde{g}^{\mu\nu}} = 0$ and $\frac{\partial I}{\partial \phi} = 0$. For $\tilde{g}$ described by the metric

$$ds^2 = d\tilde{\sigma}^2 + a^2(\tilde{\sigma}) \left(d\psi^2 + \sin^2 \psi \, d\Omega_2^2 \right) \tag{15}$$

the field equations are

$$3 \left[\frac{a'^2 - 1}{a^2} \right] = \left[\frac{1}{2} \phi'^2 - V(\phi) \right], \tag{16}$$

$$\phi'' + 3\frac{a'}{a}\phi' = \frac{\partial V}{\partial \phi}, \tag{17}$$

with prime denoting derivative with respect to $\tilde{\sigma}$. We now look for non-singular as well as singular instanton solutions of these equations :

(1) Non-singular Instanton solutions exist for $\alpha > 0, \Lambda > 0$ as well as for $\alpha < 0, 0 < \Lambda < \frac{1}{8|\alpha|}$. The solution is

$$\phi = \phi_o = \sqrt{\frac{3}{2}} \ln(1 + 8\alpha\Lambda),$$

$$a(\tilde{\sigma}) = \sqrt{\frac{3(1 + 8\alpha\Lambda)}{\Lambda}} \sin \sqrt{\frac{\Lambda}{3(1 + 8\alpha\Lambda)}} \tilde{\sigma}. \tag{18}$$

The corresponding Euclidean action evaluated in this case becomes

$$I_E = - \left(\frac{3\pi}{2\Lambda} + 12\pi\alpha \right) \tag{19}$$

which is the same as in the original R^2 -theory.

(2) HT instantons : The HT instantons cannot be obtained for all values of α and Λ. The only class that permits HT instanton is given by $\alpha < 0$, $\Lambda = f(\alpha), 8|\alpha|\Lambda < 1$. We note the general features of the potential $V(\phi)$ in this case (with $\alpha' = -\alpha$) :

(i) $V(\phi)$ has two zeros, ϕ_+ and ϕ_- , given by

$$\phi_\pm = \sqrt{\frac{3}{2}} \ln(1 \pm \sqrt{8\alpha'\Lambda}) \tag{20}$$

(ii) It has a maximum at $\phi = \phi_m = \sqrt{\frac{3}{2}} \ln(1 - 8\alpha'\Lambda))$ with $V(\phi_m) = \frac{\Lambda}{1-8\alpha'\Lambda}$.

(iii) $V(\phi) \to -\frac{1}{8\alpha'}$ as $\phi \to \infty$.

The choice of initial values play a dominant role in the evolution of these solutions. Let us assume that at $\tilde{\sigma} = 0$, $\phi(0) = \phi_+$ and $V(\phi_+) = 0$. It then follows that $\phi'(0) = 0$ and $\frac{dV}{d\phi}|_{\phi_+} = -\frac{\sqrt{\Lambda}}{\sqrt{3\alpha'}(1+\sqrt{8\alpha'\Lambda})}$. Since the point $\tilde{\sigma} = 0$, is a non-singular point, the manifold looks locally like R^4 in spherical polar coordinates and we may assume $b(\tilde{\sigma}) = v_o\tilde{\sigma} + 0(\tilde{\sigma}^2)$ where v_o is the initial velocity, i.e., $b'(o) = v_o$. The potential has a negative gradient at $\tilde{\sigma} = 0$.. The initial conditions along with the field eqs.(16) and (17) determine the evolution of b and ϕ.

To see if a singular solution of HT type is present we assume that close to the singularity $\tilde{\sigma}_f$, ($\tilde{\sigma}_f - \tilde{\sigma} < 1$)

$$\phi = q \ \ln(\tilde{\sigma}_f - \tilde{\sigma}) \tag{21}$$

$$b \sim (\tilde{\sigma}_f - \tilde{\sigma})^n \tag{22}$$

The eq.(17) then determines $q = \sqrt{\frac{3}{2}}$ for $n < 1$. The eq.(16) now determines

$$n = \frac{3}{4} = \frac{1}{3}\left[1 + \frac{1 - 8\alpha'\Lambda}{6\alpha'}\right], \tag{23}$$

which also determines Λ in terms of α',

$$\Lambda = f(\alpha') = \frac{2 - 15\alpha'}{16\alpha'}$$

and hence $\alpha' < \frac{2}{15}$. Thus HT instanton can be obtained only for a special domain of the parameter space ($\alpha < 0, \Lambda = f(\alpha)$ and $8|\alpha|\Lambda < 1$). Note that in this model, $V(\phi)$ cannot be neglected even close to the singularity, in contradiction to the speculation of HT. Although the existence of a HT type instanton is indicated by the above calculation, two possibilities may emerge :

(a) The scalar field may move uphill and get stabilised at ϕ_m, giving the non-singular instanton solution (18).

(b) The universe may end up in a singularity at $\tilde{\sigma} = \tilde{\sigma}_f$, giving the HT instanton. Further studies are required to decide which of the two possobilities is open to the scalar field for the given potential $V(\phi)$. An approximate calculation of the Euclidean action for the HT instanton can be done by following the suggestions of HT[2]. This will be taken up elsewhere.

4. DISCUSSION

We have seen that both singular and non-singular instantons are permitted by field equations in R^2-theory which satisfy some constraints. The corresponding scalar field theory clarifies the special features of the two types of solutions. In the non-singular type, the scalar field sits on top of the maximum of the potential. This gives an indication that the Lorentzian continuation of this solution (with $\alpha < 0$) may be unstable, as is the case with the de Sitter solution in the R^2 - theory [13]. The fact that HT instantons cannot be obtained in R^2 - theory for $\alpha > 0$ shows that these objects are indeed not very generic. There is also no HT instanton for $\Lambda < 0$.

Acknowledgments

One of the authors (SM) would like to thank the Inter-University Centre for Astronomy and Astrophysics (IUCAA), Pune for a visit under the Associateship programme, during which a part of the work was done. The work was also supported partially by a research grant of DST, New Delhi.

References

[1] S. W. Hawking & N. G. Turok, 1998. hep-th/9802030.

[2] N. G. Turok & S. W. Hawking, 1998. hep-th/9803156.

[3] J. B. Hartle & S. W. Hawking, 1983. *Phys. Rev.* D **28**, 2960.

[4] S. Coleman & F. De Luccia, 1980. *Phys. Rev.* D **21**, 3305.

[5] A. Vilenkin, 1998. *Phys. Rev.* D **57**, R7069.

[6] M. S. Bremer, M. J. Duff, H. Lu, C. N. Pope & K. S. Stelle, 1998. hep-th/9807051.

[7] J. Garriga, 1998. hep-th/98030210.

[8] P. F. Gonzalez-Diaz, 1998. hep-th/9805012.

[9] R. Bousso and A. D. Linde, 1998. gr-qc/9803068.

[10] W. Unruh, gr-qc/9803050.

[11] R. Bousso and A. Linde, 1998. gr-qc/9803068.

[12] K. S. Stelle, *Phys. Rev.* D **16**, 953.

[13] J. D. Barrow and A. C. Ottewill, *J. Phys.* A **16**, 2757.

Chapter 24

THE NON-HOMOGENEOUS AND HIERARCHICAL UNIVERSE

Jean-Claude Pecker
Collège de France,Paris

Abstract The cosmological principle of homogeneity and isotropy, widely used by all mathematical cosmologies, is discussed in several aspects, in which it leads to misunderstandings. One first describes some implications of the observed fractal distribution of matter, continuous or not. One thus covers the discussion of Olbers and Seeliger's effects. Some suggestions are made in order to use the observed distribution of matter as a starting point for an extrapolation of the present to the past, towards... who knows? Second, one examines the consequences of a non-continuous distribution of matter, and in particular Narlikar's reflections upon a quantized phase of the Universe. Finally, one gives a short account of the recent *scale-invariant relativity* of Nottale, which may be another way of looking at these problems.

1. THE COSMOLOGICAL PRINCIPLE IN QUESTION

One of the basic principles of both the standard big bang cosmologies (for example, read Tolman, 1934), and of the Quasi Steady State Cosmology of Burbidge, Hoyle & Narlikar (1998), is the Einstein's *cosmological principle* (implied in Einstein,1 917; see also Weinberg, 1972), of homogeneity and isotropy of the Universe, hence considered as a continuum, of fractal dimension 3. It assumes that all parts of the Universe are essentially equivalent. Of course it is much more constraining than that, when one expresses it in mathematical terms.

The reason for adopting this principle is not altogether so obvious. It is practical for solving the equations of General Relativity. One could almost say it is necessary for it. But it has no obvious physical basis.

N. Dadhich and A. Kembhavi (eds.), The Universe, 261–276.

However, it can be justified, as said Weinberg (1971, p. 407-408) just in the same way as one justifies the use of gas equation in physics, although their molecular structure is known: Let us quote Weinberg: "Of course, the homogeneity of the Universe has to be understood in the same sense as the homogeneity of a gas. It does not apply to the Universe in detail, but only to a *smeared-out* Universe averaged over cells of diameter 10^8 to 10^9 light years, which are large enough to include many clusters of galaxies"...... "The real reason, though, for our adherence to the Cosmological Principle is not that it is surely correct, but rather, that it allows us to make use of the extremely limited data provided to cosmology by observational astronomy". These statements invite some comments.

First of all, one notes that amongst the "significant" data (significant from the point of view of cosmology) are mentioned "extremely limited data". They are not listed at this place in Weinberg's book; but we know well which they are: (a) Hubble's linear law,(b) the background black body radiation and (c) the abundance of light elements. This is the case for the two groups of theories we have mentioned, and for some others as well.

But is not the tremendous inhomogeneity of the Universe (10^{13} gm cm^{-3} in neutron stars, not to speak of black holes, to 10^{-30} gm cm^{-3} in intergalactic space), not an obvious essential astronomical datum, quite significant from the point of view of cosmology, and perhaps even more significant than some of the three others, if we note that several different types of theories can account for these, but not for the inhomogeneity?

The distribution of mass is not only inhomogeneous but also hierarchical as shown by Charlier (1896, 1908, 1922), by de Vaucouleurs(1971), by others (for example, Nottale 1993, Pecker 1998). Is not this fractal distribution of mass (between some scale limits of course, not necessarily extending to very large scales or to very small ones) also a significant observed datum for cosmology?

Let us start this discussion by the Weinberg's remark and its biases. At a scale smaller than 10^8 or 10^9 light-years, the assimilation of the matter distributed in the Universe to a gas is misleading, and wrong. But what consequence does it have on the physical situation near the so-called big bang (say, at $z = 5$, as observed by the Hubble telescope) ? Galaxies are then much closer to each other than at $z = 0$, the equation of state of the *universal gas* is not valid anymore; we are in a situation comparable to that of an ordinary gas, when one must take into account the molecular structure. The ideal gases approximations are not any more valid; we have got to find some equivalence to the Van der Waals theory for the equation of state, and we are far from the ideal gas situation. So, are the classical solutions of the General Relativity equations still valid? I doubt it very much.

Actually, from a certain degree of condensation, the assumption of the continuous nature of a gas, easy to treat and to include in the equations, is not valid. The equation themselves are not valid anymore. We could call the needed physics a *quantum physics of the condensed Universe.* But we do not know this quantum physics, except for some earlier discussions due to Narlikar (1992)

I would like to claim that one should start from this observed distribution of matter, hierarchical, fractal, inhomogeneous, and even discontinuous, as we see it now, in our light cone, and go back to what the situation may have been earlier. A first step is perhaps to discuss the implications of the fractal distribution of matter, as it has been done earlier by Charlier.

1.1 OLBERS AND SEELIGER'S PARADOXES

The habit, in "standard cosmology" is to treat very lightly Olbers' paradox, and to ignore Seeliger's paradox. Let us express these two paradoxes simply.

Olbers (and his predecessors, including Kepler, Halley and Loÿs de Chézeaux) assumed an Euclidean space in which sources of light are distributed, perhaps discontinuously, as stars, but evenly in the whole space. Then, a simple integration shows that the intensity of the light coming from the shining matter to the observer located at the origin of coordinates (or at any other point in space), is extremely large. In other words, the sky should be uniformly bright, would this description be correct. The intensity of light is a *scalar*, decreasing like r^{-2}, r being the distance of the shining surface to the observer.

Olbers paradox was discussed by Charlier, along similar lines as Lambert, in the XVIII-th Century, assuming that the observer is located at the edge of a mass distribution of successive structures, imbedded in each other, like a series of Russian dolls. So he considered only the local effect of only one half of the space. It explains why he had not seen that Seeliger's paradox is of an entirely different nature (see Figure 1).

Seeliger's paradox concerns the gravitational field of these evenly distributed stars. It was first implicit in the work of Carl Neumann, in 1879, as noted by Solomon, but was formulated by Seeliger (1894). The gravitational field due to any one star de creases as r^{-2}, like the light intensity. But it is a vectorial quantity, not a scalar quantity. Hence one half of the sky exerts an infinite gravitational force, but this is compen-

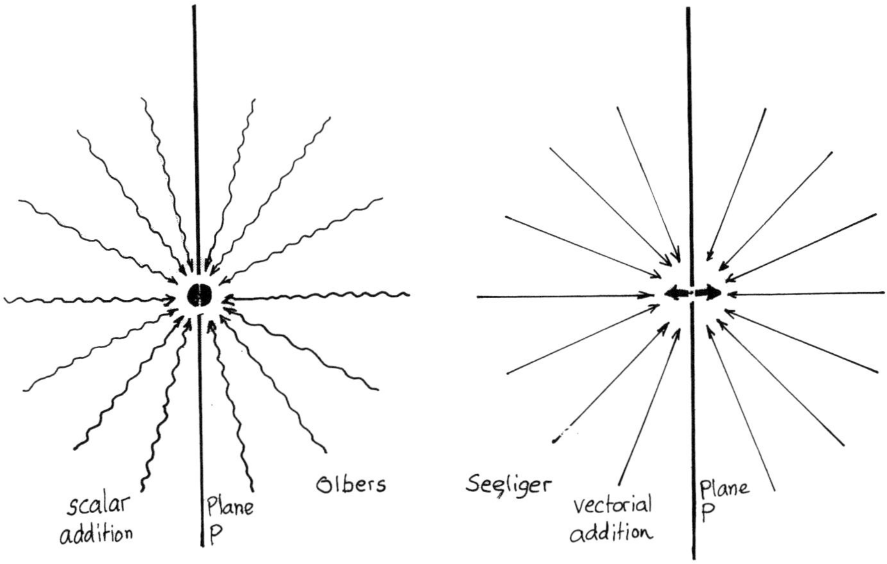

Figure 24.1 Olbers and Seeliger's paradoxes. On the left, is an illustration of Olbers paradox: the light comes from all directions. At the origin the observer receives a positive quantity of energy, the same from each side of the (artificial, introduced by Olbers, for whom the observer is located a the edge of a distribution) dividing plane P of the space. In an Euclidean, homogeneous, non-evolving Universe, this quantity of energy is "infinite". On the right, is an illustration of the Seeliger's paradox: At the origin, the mass m is attracted towards every direction. It results, on each side of the plane P, in two forces; each of the two has an "infinite" value in an Euclidean, homogeneous, non evolving Universe. But each of them strictly annihilates the other. The paradox would exist only in the Olbers construction, where the observer is located at the edge of the mass distribution, only on one side of the plane P.

sated by the force exerted on the observer by the other half of the sky. This is the reason why the paradox had been ignored. But even from one half of the sky, an infinite force is a hard thing to admit, even compensated. Why should the compensation be exact? So Seeliger introduced in the Newton law a complementary term, decreasing like exp $-Kr$. This indeed was an alternate way to solve some of the paradoxes which led to special relativity.

Olbers paradox gave place (Charlier, de Vaucouleurs) to what Weinberg calls *naive models* (Weinberg, page 611-613). Of course, as said by Weinberg, in a big bang cosmology, there is obviously no paradox since the integral of the light intensity from $r = 0$ to $r \to \infty$ is effectively cut off at a lower limit, $t = 0$. Weinberg completes the discussion by stating that the steady state cosmology requires a luminosity tending to zero with time for each star, so that the integration extends only over the life-time of the star, proper account being of course taken of the stellar evolution. The "only difficult case" seems tor Weinberg to be that of oscillatory models (such perhaps as that of the QSSC). In this case, "absorption occurs during the highly contracted era, and the red shift during the subsequent expansion saves us from an intolerable bright sky. From this point of view the 2.7 K microwave background appears as pale image of the fiery furnace with which we were threatened by de Cézeaux and Olbers". Other authors have invoked the displacement to the red of the spectrum to show that the apparent luminosity is decreasing more rapidly than r^{-2}, until its complete disappearance from the visible domain. These arguments of course are fallacious when one looks at Seeliger's paradox. There is no screen against the gravitation (a principle of cosmology which in my view is much more important than the cosmological principle)! So only the big bang models (because of the limit of integration at $t = 0$, i. e., at the origin of the expansion).

There is however a way to look at these paradoxes which differs basically from the ones above, which is that of Charlier(1908, 1922, not even mentioned by Weinberg !). In a hierarchical Universe (as well noted much later by de Vaucouleurs, 1971, 1972) with a fractal distribution of matter (see Figure 2) with a dimension D different from 3, one can show easily that the Charlier's integral converges for $D < 2$ (strict inequality!). Hence fractal models such as those of Fournier d'Albe or Hoyle ($D = 1$) satisfy this condition. The value $D = 1.3$ found in the observed world by de Vaucouleurs, from neutron stars to superclusters of galaxies, is equally satisfactory. In Charlier's hierarchical Universe, there is no Olbers paradox any more.

What about the Seeliger's paradox? Quite clearly, it requires a very different treatment than Olbers' paradox, contrary to what was believed

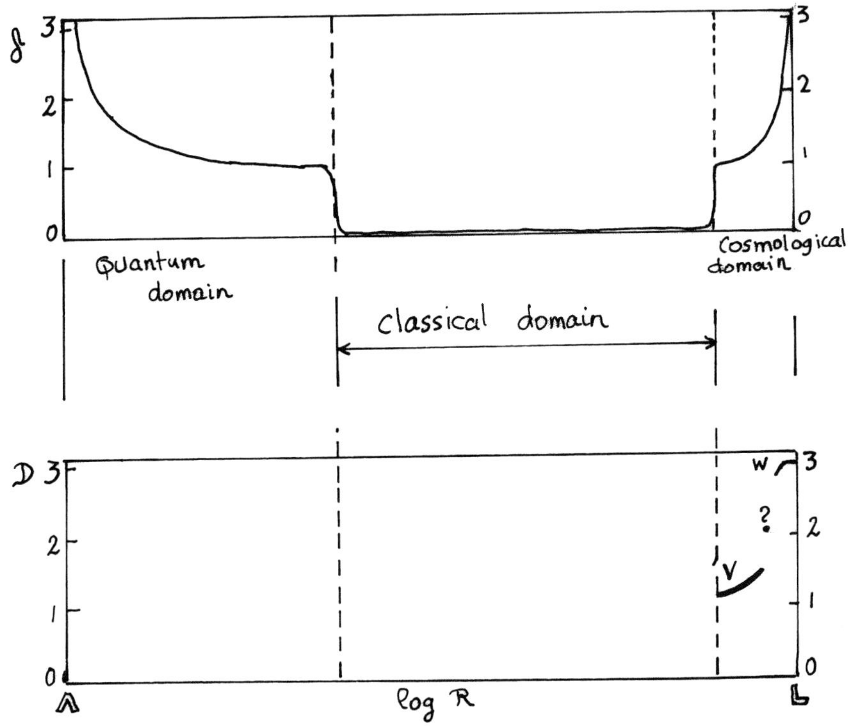

Figure 24.2 Definition of the dimension D of a fractal but continuous distribution of matter. On the abscissae is indicated the radius (logarithmic scale) of the various structures, or steps, of an hierarchical distribution of mass, numbered : 1, 2, 3...6. On ordinate is indicated the average density (in logarithmic scale). If the density is independent of the radius of the volume in consideration, the distribution is 3-dimensional; the *fractal dimension* is $D = 3$ (chain-dotted line). For a point-mass surrounded by vacuum (at happens in first approximation in all steps of the hierarchical Universe), the average density decreases as r^{-3}. The fractal index is the quantity: $x = D - 3$. It varies from step to step, and between the steps, of course; but the successive steps may follow a fractal law. One has $D = 0$ for the point-mass surrounded by empty space (here represented ideally in step 2, by a straight line of slope -3), until the next step, towards which the average density tends to a constant, as in a distribution of dimension $D = 3$. D is generally < 3. See Figure 24.3 for actual values.

by Charlier, who treated the problem linked with matter located only on one side of a plane containing the observer. We must now introduce the fluctuations in the distribution of light sources responsible for the gravitational attraction. Whatever the distribution, or the spectrum of sizes of this distribution, the local gravitational field at the location of the observer, is infinite under the classical Olbers conditions. But if we assume the hierarchical model, with a fractal distribution of matter, the decrease with the distance to the observer of the gravitational force resulting from the fluctuations is faster than if we do assume an homogeneous distribution ($D = 3$). To solve the problem, and to get a finite force, we need (Pecker, in preparation) to have $D < Df$, where Df is the limit dimension in a fluctuating distribution of matter, and depends upon the typical size scale f of the distribution, an obvious parameter of the problem. Df differs definitely from 2 and the restriction appears as more restrictive than in the case of Olbers paradox (I hope to be able to publish soon more definite results); therefore, in the Charlier Universe, there is perhaps still an *infinite* Seeliger's paradox. But of course, we are reasoning in a Euclidean space, and this discussion is merely academic.

But the gravitational field locally still exists, due to the actual fluctuations. Hence, locally, the matter must be attracted by some attractor. Actually, there is probably a residual force, acting to create the dispersion of velocities in clusters of galaxies. The scale f of the fluctuations is then involved. This scale is commanded by the size of clusters, superclusters, etc. Such force could explain, quite in agreement with the Machian point of view, motions other than expansion; i.e. proper motions of the clusters of galaxies with respect to each other, of galaxies with respect to each other, in other terms the proper motions of the galaxies. We know such motions to exist at velocities of the order of $1000\,\mathrm{km\,s^{-1}}$. The evaluation of this velocity field for a cluster of, say, 1000 galaxies, allows to determine a kinetic *quasi-thermal* energy within the cluster. It would be quite interesting to analyse the data with this idea in mind. The very existence of this energy might strongly affect any model implying a big bang.

We should also note, as a logical consequence of the hierarchical structure of the Universe, on the scales from stars to that of clusters of galaxies, of a variation of the apparent expansion rate with scale, as the expansion rate should be a function of t he average density of the expanding matter. This actually was suggested by Pecker & Vigier (1974), and it should be tested through more complete observations of clusters.

2. A STEP-TO-STEP TRIP TOWARDS THE PAST.

If the dispersion of velocities did not exist, as observed, but also as more or less predicted by the Seeliger's paradox, it would be very difficult to understand the extrapolation backwards of the presently observed expansion, an extrapolation which is basic to all models such as the standard one or the QSSC. These extrapolations backwards lead, in the standard cosmology, to a point-Universe, at the time $t = 0$. If we follow this backwards extrapolation, climbing up the flow, in what could be called a *de-pansion*, all galaxies get closer and closer to each other, and finally dissolve in a cloud of amorphous matter; all matter coalesces, before the reversal of the usually assumed inflation (*deflation*), and the the usually admitted reunification of forces (if we follow backwards the standard model). The conservation of energy and momentum imposes the conservation of this one-way motion, completed by an accelerated inflow, according the equations of GR.

In the QSSC, the same type of extrapolation backwards leads to galaxies getting closer and closer to each other; but we have to introduce a rate of *de-creation* of matter when the Universe has a small enough size. In both cases the flow of energy is centrally directed at the time zero of the simulation, i.e. at the present epoch ($t = t_0$).

But if we admit, in clusters, a dispersion of velocities, as suggested by the above discussion of the Seeliger's paradox, as soon as their mutual distances are such that transverse motions are comparable to de-pansion motions, galaxies can go away each from each other.

The previous qualitative attempt by the author (Pecker 1997) should perhaps be looked at again; it implies a sort of continuity condition, in the piling-up of structures, during the de-pansion extrapolation backwards. In other terms, a quickly collapsing cluster should not become smaller than the possibly not so quickly collapsing galaxies it contains. A more quantitative simulation is certainly very much in need.

There are models for which the problem has to be put in entirely different terms: such is the Gödel's (1949) model, which implies a general anisotropy of the whole Universe; but this model is not acceptable for several reasons. But some *quasi-turbulent* motions in the Universe are perfectly acceptable. Locally, they may appear as rotations.

3. FROM CONTINUUM TO QUANTIZATION: THE QUANTUM UNIVERSE OF NARLIKAR[1]

It is clear that, whether one does or does not accept the expansion, whether or not one does or does not accept the inflationary phase, and the grand unification (GUT), there stays a basic question: what was the Universe before? Before what? A certain time at which all these theories seems to fail, for very basic physical reasons. If people often quote the time $t_{\mathrm{Pl}} = (Gh/c^5)^{1/2} = 5.4\times10^{-44}$ second (the so-called *Planck time*) as the time after which all is "clear" (at least for the fans of the standard cosmology), and before which one just does not know (we do not even know what to know, or whether there is anything to know), we have still to ask the question- Could we know? Still, the time, which is a physical quantity, as it enters the equation of General Relativity, has a clear meaning, - but which one? Has this past (before t_{Pl}) left any observable relics, as one sometimes claims have been left by the following period?

One obvious question, which we have already alluded to many times, is that a simple formula on which both the macroscopic gravity (through G), and the quantities such as h, typical of the quantization at small scales of the Universe, are intervening, brings in itself the problem of quantization of gravitational energy. That the gravitational energy be transmitted through gravitons is the expression of a partial reply to this fundamental interrogation.

Another question, which arises at this point is : what is then the meaning of time? We know well what is the time of GR -an absolute time, in an absolute reference frame, defined in the Mach's way, by the ensemble of all masses present at once in the Universe. Obviously, the GR equations imply a continuous time, which enters the equations as t, and fits at the present the time as defined by the usual clocks, be they astronomical, or more precisely, atomic. There are no unexplained gravitational or astronomical phenomena which require a change in this point of view.

The quantization of gravitation, the introduction of *gravitons* in physics, is linked primarily to the need for insuring the grand unification (GUT) of the four fundamental forces, - the theories of the three other interactions (electromagnetic, weak, and strong) being already well-established quantum theories. As clearly stated by Narlikar(1992), the Universe, around the hypothetical big bang epoch, is so small as to make the clas-

[1] This section and the following one are very similar to the corresponding paragraphs of the book *Understanding the Universe*, by the author (Springer c. 1999).

sical description impossible, in term of *action*, the classical notion used for example in the (macroscopic) *principle of least action*, or of *stationary action*. One can actually demonstrate the GR from the macroscopic principle of stationary action, as shown by Hilbert as early as in 1915. But it results also from Hilberts's demonstration that classical laws of physics cannot be applied for $t < t_{\mathrm{Pl}}$.

What to say then? We feel immediately embarrassed by the fact that in GR the force of gravitation is merely a geometrical effect. In quantum gravity, the very process of quantization must therefore affect the structure of space-time, and this may distort the causality linking two events A & B, as well described by Narlikar.

A solution might lie in the *conformal quantization*, which keeps the angles of the light cones unchanged. Without entering into many details in a difficult theory, we may give, as a hint to more general problems (always according to Narlikar), the description of the hydrogen atom. In the classical (Newtonian!) description, the electron, massive, accelerated, loses energy (all accelerated particles radiate) and therefore spirals inwards and falls onto the proton, in a time of the order of 10^{-23} s. Obviously, this does not occur. By quantizing only r, the distance from proton to electron, the electron can exist for an extremely long time in a *stationary* orbit of radius $r = h^2/mc^2$. This could be done as well at the scale of the hypercondensed Universe of the Friedmann's models, before the Planck epoch. We could introduce conformal transformations that would not satisfy the classical GR equations of Einstein.. Narlikar concluded from this discussion that, generally, these new models do not have any singularity, the big bang models being actually extremely unlikely. So the question is: can stationary states thus exist for the Universe? It has been demonstrated that it is indeed the case, and that their characteristic scale is (not surprisingly) the one associated with the Planck time, i.e. the Planck length $L_{\mathrm{Pl}} = (Gh/c^3)^{1/2} = 10^{1.6}-33\,\mathrm{cm}$. It is interesting that such models eliminate the need for inflation (in the sense that the horizon problem does not occur anymore, in particular). And one may conclude that the pre-Planck era may be a very important phase in the history of the Universe - although unknown.

4. NOTTALE'S MODEL OF SCALE-INVARIANT COSMOLOGY.

We have mentioned the important suggestions made by Narlikar, as to the possibility of a quantum phase in the life of the Universe. But the fact that the Universe has a fractal structure does not imply in itself any quantum phase, any quantization. However, we should be very careful about this concept. The standard cosmology and the QSS cosmology assume a continuous distribution of matter; the fractal distribution may be also continuous, and the QSS Cosmology satisfies the Charlier's condition. But it is one thing to speak about hierarchical and fractal distribution, or fractal but not hierarchical (structures being not imbedded in successive steps), and quite another one to speak about quantized Universe, - not only at the very small scale addressed by the preceding paragraph. In addition, one may state that the Universe may be quantized and hierarchical (fractally) in a certain interval of size scale, but that it may be homogeneous at a larger scale (this is what Weinberg says explicitly), and completely quantized at a much smaller scale (according the type of description given by Narlikar).

It is obviously not easy to reconcile these points of view, and to treat all the scales in a unique way, like existing cosmologies generally do. This is why it seems interesting here to give an account of the *scale-invariant* cosmology of L. Nottale. Actually, the physical conditions and their mathematical expressions are different from one scale to another (say: the scales of the quantum domain and of the classical domain ,of the cosmological domain). What is actually the scale determining the physics? The basic physics, and mathematical formulation of Nottale's theory assume that it is not the case. There is for him a basic unity in the physical laws, whatever the scale, and the inhomogeneous structure is for him an essential ingredient of the observable Universe.

Nottale's theory is difficult; we shall describe it only briefly. It is founded on an extension of Einstein's principle of relativity (that was up till now applied only to motion transformations) to scale transformations. It proceeds as follows: One first gives up the arbitrary hypothesis of the differentiability of space-time coordinates, while keeping their continuity (an hypothesis which was implicitly basic in all previous cosmologies). Such non-differentiable space-time must be fractal, and it must be *resolution-dependent.* Therefore the space-time resolution becomes inherent to the physical description, and it is defined as an essential variable which characterises the *state of scale* of the reference system. One can then set a principle of scale relativity, according to which the laws of nature apply whatever the state of scale of the reference system. Its mathematical translation consists in writing the equations of physics in a scale-covariant way.

The *motion relativity* evolved, as well known, from the Galilean relativity to Einsteinian special relativity. In an analogous way, the *scale relativity* evolves from the simplest scaling laws to much more evolved ones.

The simpler scale-relativity laws have the structure of the Galilean group; they correspond to the *standard* fractal power laws, with constant and uniform fractal dimensions. From such scale laws, one can recover the main axioms of quantum mechanics. In other words, one can demonstrate that the quantum behaviour implies a non-differentiable and fractal geometry of the micro-space-time, - quite in the same way that, at large scale, gravitation is a manifestation of the Riemannian geometry of the large-scale space-time.

But the standard scale laws are only a very particular case. More generally, one can show that the laws of transformation from a scale to another have the structure of the Lorentz group of transformations. In this framework (which can be called *special- scale-relativity*), the fractal dimension is no longer uniform, but varies with scale. The effect of two successive dilations is not any more their direct product. Instead of the *zero* and the *infinite*, there appear two minimal and maximal scales, invariant under dilations while keeping the properties of the previous zero and infinite. This is in a way similar to what happens in special motion relativity: there the velocity of light is finite, and a limit that cannot be passed beyond; but it keeps the physical properties of the Galilean infinite velocity. If one wants also to give up some other aspects of the usual treatment, such as its linearity, even more general scale laws can be considered.

Nottale's new theory accounts through unique formalism, and in terms of an unique fundamental constant, the structures observed in our own solar system (Titius-Bode-like laws). It also accounts for the Tift effect of redshift quantization, studied by Arp, and Napier & Guthrie, as well as for gravitational hierarchical structures observed on a range of scales covering 10^{15} orders of magnitude.

Another cosmological consequence of the scale-invariant relativity theory applies to the *primeval* universe. The minimal, impassable scale is naturally defined by the Planck length and Planck time scale. In the special scale relativistic framework, this scale is invariant under dilations, so that the whole of the universe is connected at the Planck epoch, which solves the horizon-causality problem, and therefore makes inflation quite unnecessary. The expansion would start asymptotically from the Planck length scale. Scale invariant gravitational structures are predicted by the theory, even in the absence of initial fluctuations.

A third class of consequences arises from the suggestion of the existence of a maximal scale, also invariant under dilations. Such a scale is naturally identified with the cosmological constant scale $\mathcal{L} = \Lambda^{-1/2}$, where Λ is the usual Einstein cosmological constant. This suggestion provides a meaning for the cosmological constant; it implies it is non-

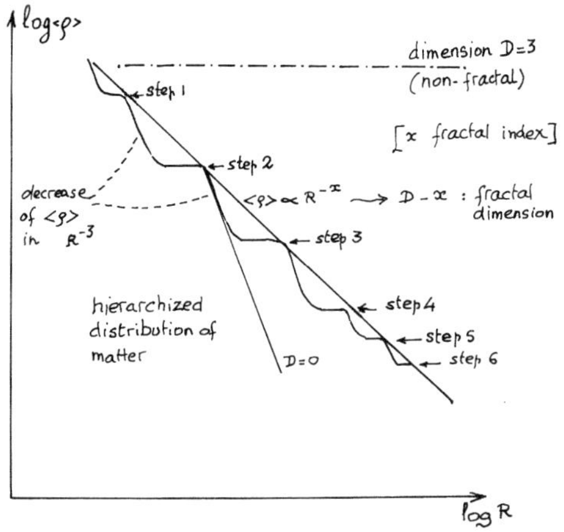

Figure 24.3 The Hierarchical Universe, according G. de Vaucouleurs. We reproduce here, somewhat modified, Figure 3 from the original de Vaucouleurs's paper. The symbols 1, 2, 3..., of the successive steps of the hierarchy represent, respectively: 1: neutron stars and white dwarfs; 2: stars 3: protostars; 4: compact galaxies and (cross) globular clusters; 5: spiral galaxies, and (cross) compact groups; 6: groups of galaxies (cross) and clusters; 7: local supercluster, 8: *largest explored universe* in the 50s and 60s. The chain-double-dotted line (C) corresponds to the Charlier limit, $D = 2$, the chain-dotted line (FA-H) to the abstract construction of the Universe of Fournier d'Albe and to the fragmentation process of Hoyle ($D = 1$). The density of the growing volume, when passing from one step to the next is represented as in Figure 24.2, but only to represent the interplanetary (IPM), interstellar (ISM), intergalactic (IGM), and intercluster (ICM) medium (dotted lines). The overall distribution can reasonably well be represented by a fractal distribution of $D = 1.2 - 1.3$ (index : $x = 1.8 - 1.7$), according to the computations of de Vaucouleurs.

zero, and it provides several new ways to measure it. The resulting values are consistent and solve the *age problem*. The ratio of the two minimal and maximal fundamental scales is then found to be of the order of $5{\times}10^{60}$, from several different tests. It yields perhaps a basis for the physical understanding of the Dirac's large number coincidences. Moreover, as in the microphysical domain, the fractal dimension is expected to vary with scale (see Figures 3 and 4). This allows both domains to recover the observed value of the fractal dimension of the distribution of matter r at scales 10 kpc to 100 Mpc, and to predict a transition to uniformity at a scale of about 1 Gpc.

5. THE QSS COSMOLOGY AND THE HIERARCHICAL UNIVERSE.

In the standard cosmology, the structures grow and they cluster through gravitational interactions, during the expansion. To grow, they need the

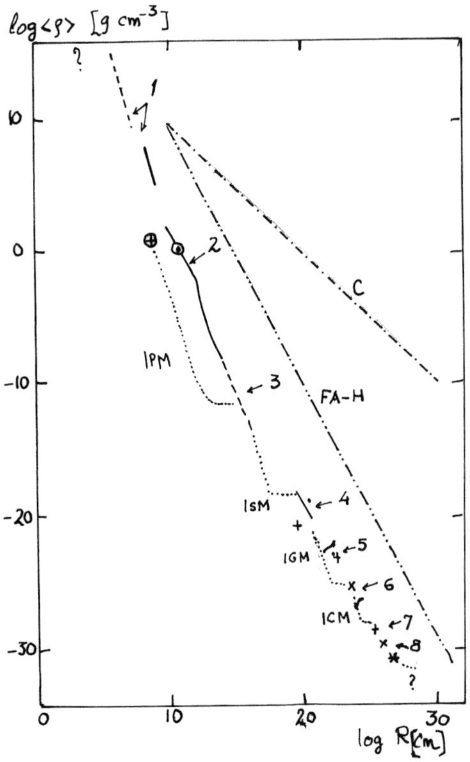

Figure 24.4 The dimensional aspect of the scale invariant relativity theory. On the upper part of this figure, according Nottale (from his Figure 7.2, somewhat simplified), the trend of the *anomalous* dimension *delta*, from the smallest possible scale (L_{Pl}) to the largest possible scale ($\mathcal{L}$), through the quantum domain to the cosmological domain. The classical domain corresponds to a physics of scale independence, whenever both quantum and cosmological domains corresponds to scale dependence. The scale on the abscissa is logarithmic. In the cosmological case, the anomalous dimension δ is equal to the fractal dimension D (in may be generally not so, as ? is linked with a two-point correlation coefficient). On the lower part of the figure, one has reproduced the trend of the fractal dimension D according Figure 3, and according Weinberg's ideal gas logic, quoted in the text. Nottale gives the following estimates: $\mathcal{L} = \Lambda{-}1/2$, where $\Lambda = 1.36\times10^{-56}\,\mathrm{cm}^{-2}$ is the cosmological constant.

existence of preassigned primordial density fluctuations. They may perhaps give place to a fractal distribution, provided the parameters of the simulation are properly adjusted. But to the author's knowledge, this has not been done so far.

In the QSS Universe, no primordial density fluctuations occur. Therefore gravitational effects may not play a key role. On the contrary, processes of *creation* (not ab nihilo ! No theology or metaphysics is implied here) of matter occur, at the time of the minimum size of the Universe, as ejecta from strong condensations (collapsed massive objects, black holes?).These ejecta act as creation centers at the period of increasing size of the universe.This is what can be called *new matter*. So, in the model, new creation centers are added to keep the model stationary. Hoyle's *toy model*, advocated by Narlikar, consists in the use of a computer to simulate that idea. Without entering into details, a clustering appears in the simulation. And, for an ad hoc but reasonable rate of creation, one obtains a fractal distribution of matter quite similar to the observed one, with a dimension 1.2; the work is in progress. A better choice of the rate of creation will allow to predict a better value of the fractal dimension, still closer to the observed one.

6. CONCLUSIONS

It seems now that the Universe being actually not continuous, being even quantized, and having a fractal structure, accompanied by an hierarchy of structures, one should incorporate these facts in any model of the Universe; Nottale's scale invariant relativity can probably be adapted to either solution of the Einstein equations - the standard and the QSS cosmologies. It is perhaps one of the most fruitful ways of research now, together with simulations of the evolution of the distribution of galaxies in time, either forwards, or backwards toward the alleged origins.

References

[1] Burbidge, G., Hoyle, F., Narlikar,J.V., 1999, *Phys. Today* **52**, 38.

[2] Charlier, C.V.L.,1886,*Arkiv Syst.Phil.* **2**, 477.

[3] ————————,1908, *Arkiv Mat.Astron.Fys* **4**, n 24.

[4] ————————, 1922, -d-, 16, n22.

[5] Einstein, A., 1917, *Sitz. Preuss. Akad. Wiss.*, 142.

[6] Gödel, K., 1949, *Rev. Mod.Phys.* **21**, 447.

[7] Narlikar, J.V., 1992, *Philosophy of Science* **59**, 361.

[8] Nottale, L., 1993, *Fractal Space-Time and Microphysics*, World Scientific, Singapore.

[9] Olbers, H.W.M.,1826, *Bode's Jahrbuch*, 111.

[10] Pecker, J.-C., Vigier, J.-P., 1976, *Astrofizika* **12**, 315.

[11] Pecker, J.-C., 1988, Cosmology, in *Grard de Vaucouleurs, a life for Astronomy*, Capaccioli, Corwin ed., World Scientific, 273.

[12] ——————,1997, (Bangalore Symposium, in press).

[13] —————-, 1998, *Understanding the Heavens*, Springer Verlag Publ., in press).

[14] ——————., 1999, in preparation (The Seeliger's effect).

[15] —————, 1894, *Astr.Nachr.* **137**,129.

[16] ————————————,1896,*Sitz.Math.Phys.Cl.,Akaz.Wiss.z.Mnchen* **26**, 373.

[17] Solomon, Iancu, ca 1970, undated and unpublished communication. Can be communicated on request by the author of this paper.

[18] Tolman,R.C., 1934, *Relativity, Thermodynamics, and Cosmology*, Clarendon Press, Oxford, U.K.

[19] de Vaucouleurs, G.,1970, *Science* **167**,1203.

[20] ————————————-1971, *Publ.Astron.Soc.Pacific* **83**, 113.

[21] Weinberg, S., 1972, *Gravitation and Cosmology*, Wiley, New York.

Chapter 25

ELECTROMAGNETIC WAVE PROPAGATION IN GENERAL SPACETIMES WITH CURVATURE AND/OR TORSION (U_4)

A. R. Prasanna and S. Mohanty
Physical Research Laboratory
Ahmedabad 380 009, India

1. INTRODUCTION

Till now the only window that we have, to view the Universe around us, is the electromagnetic window which indeed has given birth to astronomies from Radio to γ-rays. The information that one gets through receiving and analysing radiation in these frequencies has given us quite a bit of understanding about the large scale structure of space time. However, there are still many questions unanswered and this calls for as accurate an analysis as possible of the electromagnetic waves coming from distant sources to get a clearer picture of the sources as well as the material distribution through which these waves propagate before reaching us on Earth.

With the advent of general relativity as the correct theory of gravitation, the understanding of the large scale structure of the Universe - Space, Time and Matter, turned cosmology into a Science from its metaphysical status. The main aspect of general relativity *viz.* the curvature of space time produced by the matter distribution revealed several important effects on the behaviour of trajectories of both particles and photons, which are indeed well simulated into the geometry of space time. By studying these trajectories one could thus understand clearly the geometry of the Universe which is reinterpreted in terms of the Physics of the Universe. As is well known, the affine connection associated with Einstein's theory is by definition symmetric. On the other hand, if one considers a general space time manifold U_4 the connection is asymmet-

N. Dadhich and A. Kembhavi (eds.), The Universe, 277–284.

ric and the antisymmetric part, the torsion could have some influence if appropriately taken into account. Cartan (1922) generalising Einstein's theory to include torsion related it to the spin density of the matter distribution as curvature is related to the energy density. However, this approach did not get any momentum till the early seventies. On the other hand, Hehl and coworkers (1976) developed a theory of gravitation with torsion, which is more known by the name Poincaré gauge theory, while Trautman (1972) re-established the Einstein-Cartan theory in the language of differential forms, starting from an action Lagrangian which is a function of the frames and the asymmetric connection. In fact, as has been shown the connection with torsion arises naturally when gravity is treated as a gauge theory wherein the Poincaré symmetry is made a local symmetry group. The metricity condition $\nabla_\mu g_{\alpha\beta} = 0$ yields the generalised connection to be

$$\Gamma^\alpha_{\mu\nu} = \left\{ \begin{array}{c} \alpha \\ \mu\nu \end{array} \right\} + \frac{1}{2}\left(T^\alpha{}_{\mu\nu} - T_\mu{}^\alpha{}_\nu - T_\nu{}^\alpha{}_\mu\right)$$

where $\left\{ \begin{array}{c} \alpha \\ \mu\nu \end{array} \right\}$ is the Christroffel connection determined completely by the metric, whereas the torsion is regarded as a characteristic of the space-time independent of the metric. In general, T has 24 components. However, if one considers space times where $T^\alpha{}_{\mu\nu}$ is antisymmetric in all the three indices, the geodesic equation written in terms of the generalised connection also describes the shortest distance trajectories. In this context, it is worth mentioning that string theories also predict the existence of a totally antisymmetric three index tensor field which is identified with the completely antisymmetric torsion field (Green, Schwarz and Witten).

The completely antisymmetric torsion field $T_{\alpha\beta\gamma}$ has only four independent non-zero components that can also be expressed in terms of its pseudotrace part T^δ defined as $T_{\alpha\beta\gamma} = \frac{1}{3!}\epsilon_{\alpha\beta\gamma\delta}T^\delta$. In the present discussion we restrict our attention to space times with completely antisymmetric torsion wherein the generalised connection is given by $\Gamma^\alpha_{\mu\nu} = \left\{ \begin{array}{c} \alpha \\ \mu\nu \end{array} \right\} + \frac{1}{2}T^\alpha_{\mu\nu}$. In fact, as Trautman (1975) has pointed out, by splitting the torsion tensor into three parts, it can be measured only if it is purely antisymmetric or when it is due to a spinning fluid of the Wyessenhoff type.

A standard result of Einsteinian gravity is that the trajectories of all massless particles are null geodesics. It would be worthwhile to check whether there is a deviation from the null geodesics when particles have spin due to the interaction of the spin with curvature and torsion. As

Hehl et al. point out, if one tries to use the minimal coupling for the Maxwell field in a manifold with torsion, one would get the spin angular momentum tensor $\tau^{ijk} = A^{[i}F^{j]k}$ which is not $U(1)$ gauge invariant. However, Maxwell's equations can be expressed in terms of exterior derivatives which are generally covariant on any manifold.

Prasanna (1975) has written the Maxwell's equations on U_4 formally which however yield the usual equations of the Riemannian manifold as torsion terms do not appear explicitly. This is generally interpreted as the fact that the causal structure of a U_4 based on light signals is completely determined by the metric structure of the manifold.

It is interesting to note that even though the Lagrangian and the equations of motion for the electromagnetic field do not have torsion couplings, the covariant second order wave equation does have torsion couplings, due to the fact that the commutator of the covariant derivatives is proportional to both curvature and torsion. Our aim in the present work is to see the effect of curvature and torsion terms on the wave propagation in U_4. We do find two important results *viz.* (1) the presence of the curvature terms in the wave equation does not permit superluminal velocities, as claimed by some, when the background matter satisfies the strong energy condition, and (2) the effect of the torsion background is to rotate the plane of polarisation of electromagnetic waves, with the angle of rotation being independent of the wave length.

2. FORMALISM

The interaction of electromagnetic fields with gravity or equivalently the action Lagrangian describing electromagnetic fields in a general space time manifold is given by

$$S = \int d^4x \sqrt{-g} F_{\mu\nu} F^{\mu\nu} \tag{1}$$

and the corresponding equations of motion by the covariant Maxwell's equations

$$\partial_\mu \left(\sqrt{-g} F^{\mu\nu} \right) = 0 \tag{2}$$

and the Bianchi identity

$$\partial_\mu \left(\sqrt{-g} \tilde{F}^{\mu\nu} \right) = 0 \tag{3}$$

where $\tilde{F}^{\mu\nu}$ is the dual of the field tensor given by $\frac{1}{2\sqrt{-g}} \epsilon^{\mu\nu\alpha\beta} F_{\alpha\beta}$. Equations (2) and (3) may be re-expressed in a manifestly covariant form on U_4 as given by Prasanna (1975)

$$\nabla_\mu F^{\mu\nu} = \frac{1}{2} T^\nu{}_{\mu\lambda} F^{\mu\lambda} \tag{4}$$

and

$$\begin{aligned}\nabla_\mu F_{\alpha\beta} + \nabla_\alpha F_{\beta\mu} + \nabla_\beta F_{\mu\alpha} &= T^\lambda_{\alpha\beta} F_{\lambda\mu} + T^\lambda_{\beta\mu} F_{\lambda\alpha} \\ &\quad + T^\lambda_{\mu\alpha} F_{\lambda\beta}\end{aligned} \tag{5}$$

Here ∇ stands for the covariant derivative with the generalised connection.

In order to obtain the second order wave equation for electromagnetic fields on U_4, operate on equation (4) with ∇_α to obtain

$$\nabla_\mu \nabla^\alpha F^{\mu\nu} + [\nabla^\alpha, \nabla_\mu] F^{\mu\nu} = \frac{1}{2} \nabla^\alpha \left(T^\nu{}_{\lambda\mu} F^{\lambda\mu} \right) \tag{6}$$

The commutator identity is given by

$$\begin{aligned}[][\nabla_\mu, \nabla_\nu] F^{\alpha\beta} &= -T^\lambda{}_{\mu\nu} \nabla_\lambda F^{\alpha\beta} \\ &\quad + R^\alpha{}_{\lambda\mu\nu} F^{\lambda\beta} + R^\beta{}_{\lambda\mu\nu} F^{\alpha\lambda}\end{aligned} \tag{7}$$

Operating (5) by ∇^μ and rearranging the terms, one gets

$$\begin{aligned}\nabla^\mu \nabla_\mu F_{\alpha\beta} &+ [\nabla^\mu, \nabla_\beta] (F_{\mu\alpha}) + \nabla_\beta \nabla^\mu F_{\mu\alpha} \\ &+ [\nabla^\mu, \nabla_\alpha] (F_{\beta\mu}) + \nabla_\alpha \nabla^\mu F_{\beta\mu} \\ &= T^\lambda{}_{\alpha\beta} \nabla^\mu F_{\mu\lambda} + T^\lambda{}_{\mu\alpha} \nabla^\mu F_{\beta\lambda} + T^\lambda{}_{\beta\mu} \nabla^\mu F_{\alpha\lambda}\end{aligned} \tag{8}$$

wherein the derivatives of T are omitted. Using the cummutator and the Maxwell's equations (6) and (4) appropriately and simplifying, the final wave equation is obtained to be

$$\begin{aligned}\nabla^\mu \nabla_\mu F_{\alpha\beta} &= \tfrac{1}{2} \left(T^{\lambda\mu}{}_\beta \nabla_\alpha - T^{\lambda\mu}{}_\alpha \nabla_\beta \right) F_{\lambda\mu} \\ &\quad - R_{\alpha\beta r\delta} F^{r\delta} - R_{\alpha\lambda} F_\beta{}^\lambda + R_{\beta\lambda} F_\alpha{}^\lambda\end{aligned} \tag{9}$$

(One has used the cyclic identity for the curvature tensor while simplifying the above equation).

2.1 CURVATURE WITHOUT TORSION ($T^{\alpha\beta\gamma} = 0$)

The wave equation (8) immediately reduces to the form

$$\nabla^\mu \nabla_\mu F_{\nu\lambda} = R_{\rho\mu\nu\lambda} F^{\mu\rho} + R^\rho{}_\lambda F_{\rho\nu} - R^\rho{}_\nu F_{\rho\lambda} \tag{10}$$

Using the geometric optics approximation one can describe the photon trajectories by the eikonal solutions of the wave equation as given by

$$F_{\mu\nu} = e^{iS(x)} f_{\mu\nu} \tag{11}$$

where $S(x)$ the phase is a rapidly varying function of the space-time coordinates as compared to the amplitude $f_{\mu\nu}$. The gradient of the phase $\nabla_\mu S$ represents the wave number k_μ and one has

$$\nabla_\mu F_{\alpha\beta} = ik_\mu F_{\alpha\beta} \tag{12}$$

Using this one can write the wave equation in the form

$$-k^\mu k_\mu f_{\nu\lambda} + R^{\rho\mu}{}_{\nu\lambda} f_{\rho\mu} - R^\rho{}_\lambda f_{\nu\rho} + R^\rho{}_\nu F_{\lambda\rho} = 0 \tag{13}$$

which yields the dispersion relation

$$k^2 = \left(R^{\rho\mu}{}_{\nu\lambda} - R^\rho{}_\lambda f_{\nu\rho} + R^\rho{}_\nu f_{\lambda\rho}\right) \frac{f^{\nu\lambda}}{\hat{f}^2} \tag{14}$$

The solution (11) when used in the Bianchi identity (5) with $T^{\alpha\beta\gamma} = 0$, shows that of the six components of $F_{\mu\nu}$ only three are independent as expressed by the relation

$$k_0 f_{ij} + k_i f_{jo} + k_j f_{oi} = 0 \tag{15}$$

Using (15) in (13) one finds the set of three equations

$$\left(k^2 \delta_i{}^j + \epsilon_i{}^j\right) f_{oj} = 0 \tag{16}$$

where

$$\begin{aligned} \epsilon_i{}^j &= \left(R^0{}_0 + R^a{}_0 \tfrac{k_a}{k_0}\right) \delta_i{}^j \\ &\quad + \left(-2R^{oj}{}_{oi} + 4R^{aj}{}_{oi} \tfrac{k_a}{k_0} + R^j{}_i - \tfrac{1}{k_0} R^j{}_0 k_i\right) \end{aligned} \tag{17}$$

Let us consider the example of electromagnetic wave propagation in the expanding universe as depicted by the Friedmann-Robertson-Walker geometry.

The general homogeneous, isotropic universe described by the line element

$$ds^2 = dt^2 - R^2(t)\left(dr^2 + r^2 d\theta^2 + \sin^2\theta d\phi^2\right) \tag{18}$$

has for the ϵ matrix the non-zero components

$$\epsilon^1{}_1 = \epsilon^2{}_2 = \epsilon^3{}_3 = -2\left(\frac{\ddot{R}}{R} + \frac{\dot{R}^2}{R^2}\right) = -\frac{8\pi G}{3}(\rho - 3p) \tag{19}$$

with ρ and p denoting the density and pressure respectively.

Using (19) in the dispersion relation to be obtained from the matrix equation (16) for the electric field components one finds the relation

$$\omega^2 - k_i^2 = \frac{8\pi G}{3}(\rho - 3p) \tag{20}$$

and thus the photon velocity is given by

$$v_i = \frac{\partial \omega}{\partial k_i} = \left[1 + \frac{8\pi G}{3k_i^2}(\rho - 3p)\right]^{-\frac{1}{2}} \tag{21}$$

which is clearly < 1 for $\rho \geq 3p$. In the radiation dominated era when $\rho = 3p$ the photon velocity is 1. Hence we find that following special relativity we say that photons of any polarisation cannot exceed the value $c(= 1)$, then it is necessary that the material distribution through which the waves are propagating *has to satisfy the strong energy condition* $\rho \geq 3p$.

Recently Olum (1998) has discussed the issue of superluminal velocity and using the space-time diagramatic analysis has shown that 'Superluminal travel requires negative energy' in the sense that the weak energy condition has to be violated.

2.2 TORSION WITHOUT CURVATURE

¿From (8) we get after neglecting the curvature terms and restricting the torsion terms to only linear order, the wave equation

$$\nabla^\mu \nabla_\mu F_{\alpha\beta} = \frac{1}{2}\left(T^{\lambda\mu}{}_\beta \nabla_\alpha F_{\mu\lambda} - T^{\lambda\mu}{}_\alpha \nabla_\beta F_{\mu\lambda}\right) \tag{22}$$

By defining the wave vector K_μ of the propagation mode by the eikonal condition

$$\nabla_\mu F_{\alpha\beta} = iK_\mu F_{\alpha\beta} \tag{23}$$

one finds

$$K^\mu K_\mu F_{\alpha\beta} = \frac{1}{2}\left(T^{\lambda\mu}{}_\beta K_\alpha - T^{\lambda\mu}{}_\alpha K_\beta\right) F_{\mu\lambda} \tag{24}$$

Substituting for the totally antisymmetric torsion tensor, its pseudotrace $T^{\alpha\beta\mu} = \frac{1}{3!}\epsilon^{\alpha\beta\mu\nu}T_v$, the wave equation for the transverse electric field components of a wave with wave vector $K_\mu = (K_0, \theta, 0, K_3)$ propagating through space-time with torsion pseudovector $T_v = (T_0, \mathbf{T})$ is given by

$$\begin{pmatrix} K^2 & -\frac{i}{6}(T_0K_3 - T_3K_0) \\ \frac{i}{6}(T_0K_3 - T_3K_0) & K^2 \end{pmatrix} \begin{pmatrix} f_{01} \\ f_{02} \end{pmatrix} = 0 \tag{25}$$

The dispersion relation is given by the determinant of the K matrix and the wave vectors of the two propagating modes denoted by subscripts $\pm$ are

$$K_{3\pm} = K_0 \mp \frac{1}{12} (T_0 - T_3) \tag{26}$$

The solution for the electric field components of the waves are

$$E_{\pm} = E_{01} exp \{ i (K_0 t - K_{3\pm} Z) \} \tag{27}$$

The phase difference between the two modes after propagating over a distance $Z = L$, will be observed as the rotation of the plane of polarisation of the plane polarised modes $E^{(1)}$ and is given by

$$\Delta\varphi(L) = (K_+ - K_-) L = -\frac{1}{6} (T_0 - T_3) L \tag{28}$$

For light signals propagating at angle θ with respect to the of direction $\mathbf{T}$ the optical rotation angle (28) is given by

$$\Delta\varphi(L) = \frac{1}{6}(T_0 - \mid \mathbf{T} \mid \cos\theta) L \tag{29}$$

We note that this optical rotation by torsion is independent of wavelength and this effect can therefore be distinguished from Faraday rotation by galactic magnetic fields where the optical rotation is proportional to the square of the wavelength.

Nodland and Ralston [10] using observations from 90 sources with $z < 0.3$ and 71 sources with $z > 0.3$ found an anisotropic effect in the wavelength independent optical rotation which in our formalism could be accounted for by a non-zero spacelike component of the torsion $\mid \mathbf{T} \mid \simeq H_0 \simeq 10^{-42}$ GeV.

Recently Carroll and Field [11] analysed data from both nearby ($z < 0.3$) and distant ($z > 0.3$) sources to conclude that the signal of optical rotation is consistent with zero and one can at best put upper bounds from such observations. Using the Carroll and Field results [11] and using equation (29) we put upper bounds on the time and spacelike components of the torsion pseudovector given by

$$T_0 = (1.74 \pm 2.40) H_0 = (3.72 \pm 5.10) \times 10^{-42} h_o \, Gev \tag{30}$$

$$\mid \mathbf{T} \mid = (3.36 \pm 4.2) H_o = (7.20 \pm 9.0) \times 10^{-42} h_o GeV \tag{31}$$

We have thus seen that analysing the propagation of electromagnetic waves using the wave equation (8) has given some new insights regarding both curvature and torsion effects on signal propagation, intrinsic to the space time structure. It is indeed a great pleasure for us to dedicate this article to Professor Jayant V. Narlikar, on his sixtieth birthday. Jayant has been a source of inspiration to several younger generations of relativists and astrophysicists. We wish him a very happy and fruitful next sixty years keeping a steady state of interaction with the community of scientists all around.

Acknowledgments

One of us (ARP)would like to thank the ICTP,Trieste and the Max Planck institute for Astrophysics, Garching for the hospitality during the autumn of '98,when part of the work was carried out.

References

[1] Cartan, E., 1922, *Comptes Rendus*, **174**, 322.

[2] Hehl, F.W., Von der Hyde, P., Kerlick, G.D. & Nester, J.M., 1976. *Rev. Mod. Phys* **48**, 393.

[3] Trautman, A., 1973. *On the Structure of Einstein-Cartan Equations*, Symposia Mathematica, Bologna, **XII** (Monograf).

[4] Green, M.B., Schwarz, J.H. and Witten, E., 1987. *Superstring Theory*, Cambridge University Press, Cambridge.

[5] Admowicz, W. and Trautman, A., 1975. *Bull. del'Academic Polonaise des Sciences*, **XXIII (3)**, 339.

[6] Prasanna, A.R., 1975. *Phys. Letts. A.*, **54 (1)**, 17.

[7] Mohanty, S. and Prasanna, A.R., 1998. *Nuclear Phys. B. 526* 501-508,1998 .

[8] Mohanty, S. and Prasanna, A.R., 1998. *Optical Activity by Spacetime Torsion*, Phys. Rev. Lett. (submitted).

[9] Olum, K.D., 1998. *Phys. Rev. Letts.* **81**, 3567.

[10] Nodland B. and Ralston, J.P., 1997.*Phys. Rev. Lett.*, **78**, 3043.

[11] Carroll S.M. & Field, G.B., 1997. *Phys. Rev. Lett.*, **79**, 2397; Carroll, S.M., Field G.B., & Jackiw, R., (1990). Phys Rev **D 41** 1231.

Chapter 26

A FRESH LOOK AT THE SINGULARITY PROBLEM

A. K. Raychaudhuri
Relativity and Cosmology Centre,
Physics Department,
Jadavpur University,
Calcutta - 700 032, India

Abstract The slant of singularity theorem was towards proving the non-existence of singularity free cosmological solutions. The recent discovery of singularity free solutions demands a fresh look at the problem to find out the characteristics and limitations of singularity free solutions. The present article shows that a characteristic of almost all these solution is the vanishing of spatial averages of important scalars that govern the dynamics of models.

Jayant Narlikar has somewhere related how he came across my paper when he was working for his doctoral thesis in 1960. He was interested in finding the role of rotation in cosmology - in particular whether it could prevent the collapse singularity that occurs in Friedmann models. About three year later, we met at Dallas during the first Texas symposium. That was in December, 1963. Since then a somewhat intimate and strange sort of relationship has developed between us - call it friendship if you like, although there is a great disparity in age and position. During all these years, the versatility of his genius and numerous qualities of his head and heart have made a great impression on me.

It was the singularity problem that introduced me to him and so I think it fit to have a look at that problem on this occasion. In fact the discovery of some singularity free solutions demands a fresh look.

In a way, the problem is simple to understand. Consider two bodies interacting gravitationally. Apparently they must meet sometime in the future or the past. If they are approaching or receding with a kinetic energy less in magnitude than their gravitational potential energy, they

N. Dadhich and A. Kembhavi (eds.), The Universe, 285–288.

will meet in future while if they are receding with a velocity higher than a critical value, an extrapolation back in time leads to a meeting of the two bodies. If this simple argument be carried over to a distribution of matter, finite or infinite; discrete or continuous, may we not expect a similar meeting together resulting in an infinite density.

So far we have used Newtonian language and one may raise some questions. Is it not possible that in a distribution, as the gravitational attraction on a body comes from different directions, there will be a cancellation and gravitation will be effectively removed. Indeed if the distribution be homogeneous and isotropic, then there should not be any non-vanishing gravitational field as the field is a vector field and selects out a preferred direction. However Poisson's equation shows that a non-vanishing density of matter must have a non-vanishing gravitational field. These and some other difficulties merely highlight the logical fallacy of using Newtonian ideas in cosmology.

Going over to general relativity, we can reconcile homogeneity and isotropy with non-vanishing gravitational field basically because it is now a tensor field. But the symmetry assumptions severely restrict the possibilities. There remain only an unknown function of time, usually called the scale factor and a parameter determining the nature of curvature of the space sections. But conclusions regarding the presence of a singularity where physical and geometrical entities blow up persist in these simple solutions.

Recognised as a basic difficulty, it became important to investigate whether singularities are inescapable in general relativity. One might think of non-gravitational forces (recall that pressure gradient maintains the equilibrium in stars and centrifugal force maintains the planets in stationary orbits). Could they not stop the collapse? What seemed to be a final answer to these questions came in the seventies. These theorems relied on several conditions, which apparently seemed to be generally valid and then concluded that a singularity is inevitable. However there came a new definition of singularity - a finiteness of the life-history of free particles, massive or massless. It is not our purpose to question the legitimacy of this definition but we like to emphasise that this definition and the underlying conditions were often overlooked. Thus statements were frequently made that with general relativity, there must be a big bang type of singularity involving a blow up of physical and geometrical variables. The reasons for such over-simplified and incorrect statements were two fold. An exact statement would have been less sensational and so not quite salable to the public and secondly the proofs of the theorems were so complicated, that very few even amongst professional physicists made a critical study to realise the importance of the definition and the underlying conditions.

In this background singularity free solutions seemed somewhat puzzling but it was soon found out that they did not satisfy one of the underlying conditions of the singularity theorems. So the singularity problem somewhat changed its complexion - one had to recognize the existence of singularity free solution; the task was to spell out the peculiarities (or limitations) of such solutions.

We recall that non-gravitational forces may play an important role. Non-gravitational forces cause a departure from geodesicity technically termed as acceleration. The influence of non-gravitational forces is given by the divergence of the acceleration. As the acceleration is to be univalued and bounded at all points including those at spatial infinity, it easily follows that the spatial average of the divergence of acceleration must vanish. Using this result, one obtains in case rotation is absent, the following inequality:

$$-\frac{\partial}{\partial s} < \theta > \geq \frac{1}{3} < \theta^2 > + < \chi > . \tag{1}$$

where θ is the rate of volume expansion and χ is the gravitational interaction term and $< \alpha >$ signifies the spatial average of α and $\frac{\partial}{\partial s}$ is the differential coefficient with respect to the proper time.

The gravitational term χ involve the energy stress tensor and for ordinary matter one has $\chi \geq 0$. However χ may not be positive in exceptional case — thus in the inflationary scenario, the false vacuum has $\chi < 0$. We shall restrict to the case $\chi \geq 0$ (the strong energy condition) and thus obtain

$$-\frac{\partial}{\partial s} < (\theta > \geq \frac{1}{3} < \theta^2 > \geq \frac{1}{3} < \theta >^2 . \tag{2}$$

The above inequality shows that if $< \theta > \neq 0$, then $< \theta >$ will blow up either in the past or the future at a finite proper time. This will mean a singularity. Consequently for the non-singular solution, the space average of both the acceleration term and the expansion must vanish. When this is the case $< \chi >$ also vanishes.

The above results hold good for both open and closed spaces. But for closed spaces, the vanishing of the spatial average of a positive definite quantity signifies the vanishing of the quantity itself. Thus for closed spaces, there is no non-trivial singularity free solution if rotation be absent.

The situation is somewhat complicated when rotation is present. In some cases of rotating solutions, closed timelike curves (CTCs) appear

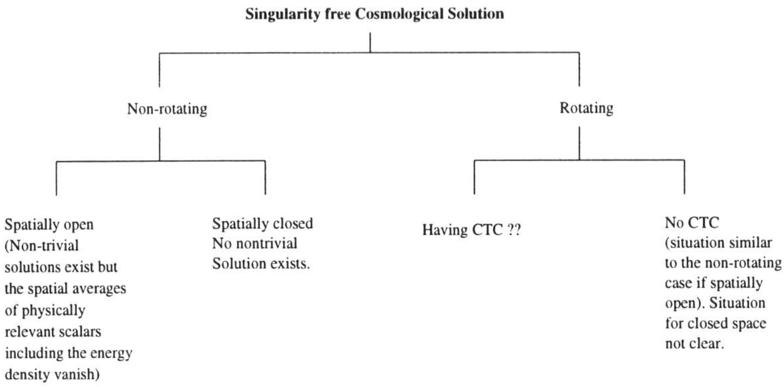

Figure 26.1

such that there is a breakdown of causality. The precise condition for this to occur is not as yet clear - one can only say that in such cases the space time does not admit a foliation into space sections and consequently that the pattern of the discussions that we have made so far becomes untenable. In any case spacetimes having closed time like curves are usually considered physically unacceptable.

However all rotating solutions do not contain closed time curves. In that case one can make a foliation into space sections and show that if the space sections are open, then the vorticity vanishes sufficiently rapidly at infinity and we recover the vanishing of spatial averages of all relevant physical scalars. However if the space sections be closed, no clear conclusion seems possible. Summing up we may present our results in Figure 26.

Chapter 27

PROBING BEYOND THE COSMIC HORIZON

Tarun Souradeep

Department of Physics, Kansas State University, Manhattan, KS 66506, U.S.A.

Abstract There are reasons to believe that the region of uniform curvature that we observe within our horizon is perhaps a tiny patch, much smaller than the length scales of inhomogeneity and global connectivity of an extremely complicated manifold. However, the recent supernova (SN) results suggest that the horizon scale could be comparable to or even much larger than curvature radius. Non trivial global structure too tends to be of the order of the local curvature scale. Probing (slightly) beyond the cosmic horizon can potentially reveal nontrivial global structure lurking around or just beyond the horizon scale. The cosmic microwave background (CMB) anisotropy is a sensitive probe of the universe on length scales up to and somewhats beyond the horizon scale and is perhaps poised to detect or put interesting limits on non trivial features the global structure. This point is exemplified by high CMB anisotropy signal of the Elliptical topology in a large fraction of the revised parameter space of the closed FRW universe (spherical geometry) currently preferred by SN observations.

... How can I
even for an instant understand the beginning, the end,
the meaning, the theory – of something outside of which
I can never go? Only this I know –
that this thing is beautiful, great, terrifying,
various, unknowable, my mind's ravisher...
– *Rabindranath Tagore*, (1901),
translated from Bengali[1]

1. INTRODUCTION

I am delighted to contribute to the festschrift honouring an illustrious scientific career such as that of Jayant Narlikar and to pay my tribute to a great teacher. It is indeed an amazing and fortunate coincidence

N. Dadhich and A. Kembhavi (eds.), The Universe, 289–300.

that this festschrift comes at a time when my research interests have wandered back into the interests of my early graduate student days under Jayant's supervision. I owe my continued interest in cosmic topology to his whole hearted encouragement and tolerance towards rather crazy and even incorrect ideas at the start of my research career.

One of the first scientific papers on cosmology that I read during that phase was Jayant's paper (with Seshadri) addressing the observability of Elliptical universes through the 'counter-images' of cosmologically distant light sources (say, quasars or very high redshift galaxies) [2]. In discussing the prospects of probing beyond the horizon, it seems appropriate for the occasion to present results from an ongoing work of mine that addresses the same question using a different probe and, more importantly, in the light of some very exciting recent observations.

The recent results from the high redshift supernova searches are pointing to the presence of a cosmological constant, Λ[3]. A non-zero Λ term [1] allows for values of the 'horizon' size relative to the curvature scale, τ_0/d_c of order unity in spherical models for acceptable values of Ω_D (see figure 27.1). *An important and exciting ramification of this result is that nontrivial features in the global spatial structure of universe may be within the grasp of observations.* The argument has two parts. First, the length scale in the simplest forms of multiple connectivity (nontrivial topology) or breakdown of homogeneity is of the order of the curvature radius, d_c. Examples demonstrating this fact are plenty: the connecting stalk in a bubble universe (see figure 27.2) can be at most at a distance πd_c from any observer on the smooth spherical region; compact hyperbolic spaces have diameters [2] of the order of d_c; the diameter of an elliptical universe is $\pi d_c/2$, etc. . The second part is the fact that the observed (Cosmic Microwave Background) CMB anisotropy probes the spatial structure of the universe up to scales comparable to and somewhats beyond the horizon. In this article, I illustrate this point with the example of Elliptical universe models.

There are theoretical motivations to believe that the spatial section of the universe is indeed a complex manifold such as shown in figure 27.2. That is not tractable in full generality. A more tractable and well defined problem would be to restrict attention to homogeneous (uniform) curvature manifolds with non trivial topology [4]. The CMB anisotropy and the constraints from data have been studied in many multiply connected Euclidean and Hyperbolic universe [5]. Spherical geometry did not get attention possibly because in a dust filled universe, $\tau_0/d_c \ll 1$ for reasonable values of its cosmological density. However, in the presence of a non zero Λ, that is no longer true and the observability of an Elliptical universe is an interesting question.

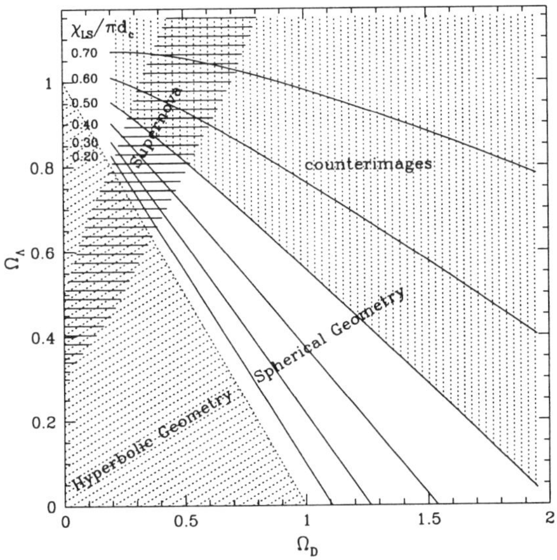

Figure 27.1 The contours of constant ratio between the horizon radius and curvature radius, τ_0/d_c are shown on the $\Omega_D - \Omega_\Lambda$ plane for models with spherical spatial section. The models with hyperbolic geometry in the hatched lower left corner are not considered here. The dividing line represents Euclidean models ($d_c \to \infty \Rightarrow \tau_0/d_c \to 0$). In the absence of a cosmological constant ($\Omega_\Lambda = 0$), counter-images lie within the horizon (i. e. , satisfies $\tau_0/d_c > \pi/2$) when $\Omega_D > 2$, whereas, in the presence of a Λ term the condition can be satisfied for all Ω_D (vertically hatched region). Recent high redshift supernovae measurements point to the presence of a cosmologically significant Λ term; the horizontally hatched region in the parameter space is a rough depiction of the range of models preferred by the Supernova results (part of the 68% likelihood region that lies within the range of the plot; the peak lies outside further into the Spherical and high τ_0/d_c region at around $(\Omega_D, \Omega_\lambda) \approx (0.75, 1.4)$). For most of the preferred spherical models, the possibility of Elliptical topology can be easily verified or refuted using the CMB anisotropy data

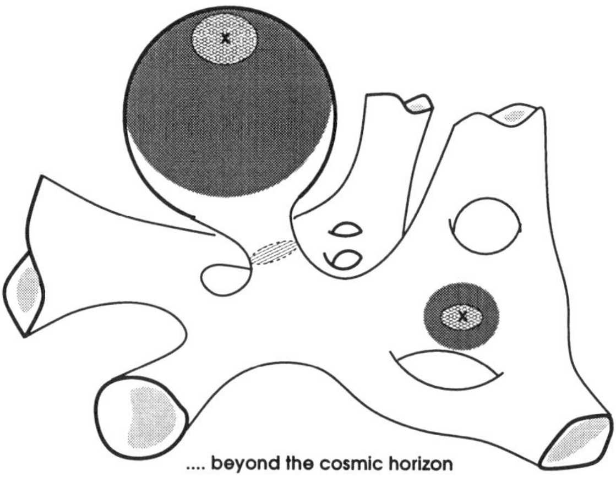

Figure 27.2 A cartoon of the possibly rich global structure of universe on ultra-large scales is depicted. The 'X' marks denote two observers in regions with locally spherical and hyperbolic geometry, respectively. In cosmological settings such that $\tau_0/d_c << 1$, i. e., observable volume shown by the cross hatched discs around the observer is small), the observer will never detect the rich global structure and be satisfied with a standard "open" or 'closed' FRW model of the universe. However, in a cosmology such τ_0/d_c is of order unity (the larger shaded discs around each 'X'), certain signatures of global structure may be detectable. Note region within the observable volume need not encompass any significant curvature variation, or be multiply connected. The non-trivial global structure could reveal itself in the CMB anisotropy through the modification of the spectrum of fluctuations; e. g. the narrow throat in the spherical case (boundary of the hatched ellipse) and the 'toroidal' regions in the hyperbolic case.

2. CMB ANISOTROPY IN A ELLIPTICAL UNIVERSE

The realization that the universe with the same local geometry has many different choices of global topology is as old as modern cosmology – De Sitter was quick to point out that Einstein's closed static universe model with spherical geometry $\mathcal{S}^3$ could equally well correspond to a multiply connected *Elliptical* universe model where the antipodal points of $\mathcal{S}^3$ are topologically identified [5].

In order to keep within the suggested size for the article, I restrict my text to points of direct relevance to the problem at hand. The reader is directed to [2] for a good description of Elliptical universe; excellent recent reviews [5] and some classic papers [6] on cosmic topology.

Elliptical universe is unique in being the only multiply connected Friedmann-Robertson-Walker (FRW) cosmology that preserves global isotropy and homogeneity. This in terms of global symmetries of the universe is the mildest possible deviation from a trivial global structure. As in a simply connected FRW universe, the CMB anisotropy here is statistically isotropic; the angular correlation function, $C(\theta)$ is simply a function of the separation implying that the angular power spectrum $\mathcal{C}_\ell$ is a complete description.

Figure 27.3 shows sample plots of $\mathcal{C}_\ell$ and $C(\theta)$ expected in Elliptical universe. The corresponding results for the spherical universe is also plotted for comparison. The signature of Elliptical universe is described in the caption. In this article I present a brief explanation of the basic effect. (A more detailed explanation will be given in a future publication [7].)

In the standard picture, the CMB that we observe is a Planckian distribution of relic photons which decoupled (last scattered) from matter at a redshift ≈ 1100. These photons have freely propagated over an affine distance $\chi_{ls} \approx \tau_0$ from this two-sphere of last scattering (SLS) to the observer at its center. The CMB anisotropy arises from the variations in the local properties on the SLS such as the gravitational potential fluctuations $\rightarrow$ Surface Sachs-Wolfe effect (SSW), velocity of baryons $\rightarrow$ acoustic peaks, etc. . and from an integral over the evolving gravitational fluctuations encountered by the photons along its path $\rightarrow$ the Integrated Sachs-Wolfe effect (ISW).

The CMB anisotropy signature of an Elliptical universe is on large angular scales where the Sachs-Wolfe effect dominates the CMB anisotropy. The angular correlation function of the CMB anisotropy then depends only on the spatial correlation functions on a equal-time spatial hypersurfaces (see [8]). The method of images implies that for Ellip-

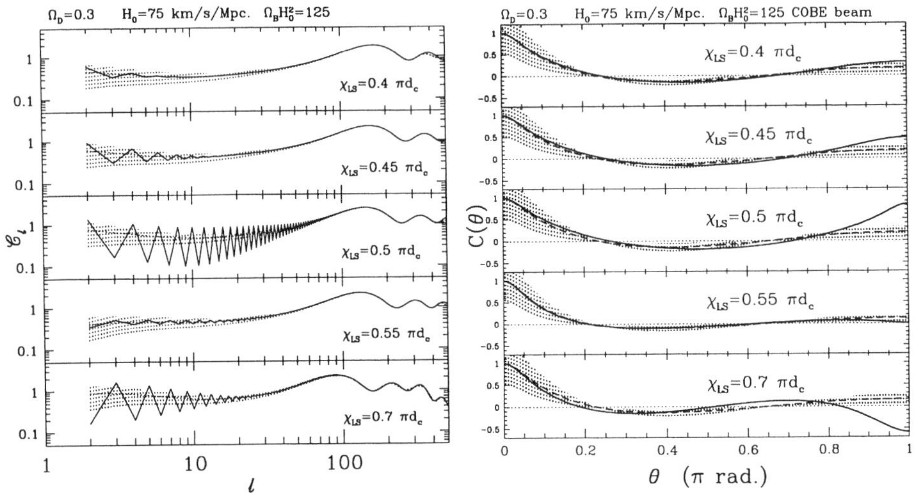

Figure 27.3 The figure presents a comparison of the CMB anisotropy between the multiply connected Elliptical universe and the corresponding simply connected ('closed') FRW universe with spherical geometry. The density of pressure-less "Dust" component is fixed at the $\Omega_D = 0.3$ and the contribution of cosmological constant, Ω_Λ, is selected to give the stepped values of $\chi_{ls} \approx \tau_0/d_c$ marked on the panels. The *left panels* show the angular power spectra, $\mathcal{C}_\ell$, for the Elliptical (jagged curve) and 'closed' model (smooth) curve. The shaded band around the closed model is an estimate of the cosmic variance. The $\mathcal{C}_\ell$ are normalised to unity at $\ell = 250$ to facilitate a good visual comparison; a COBE normalisation generally introduces an offset which is important to bear in mind. The $\mathcal{C}_\ell$'s for the Elliptical models show an marked difference in amplitude between the odd and even values of the harmonic, ℓ, up to $\ell \lesssim 100$. For $\tau_0/d_c \leq \pi/2$, the even values of $\mathcal{C}_\ell$ are enhanced relative to the odd one while the trend tends to reverse for $\tau_0/d_c > \pi/2$. This alternating pattern of $\mathcal{C}_\ell$'s is reflected strongly in the angular correlation function, $C(\theta)$. The *right panels* plot $C(\theta)/C(0)$ for the same set of models measured with a COBE beam. The solid and dashed curves are for the Elliptical and 'closed' models, respectively; the shaded band estimates the error bar. The ratio $C(0)/C(\pi)$ seems to be a good measure to distinguish locally (geometrically) identical universes with and without Elliptical topology.

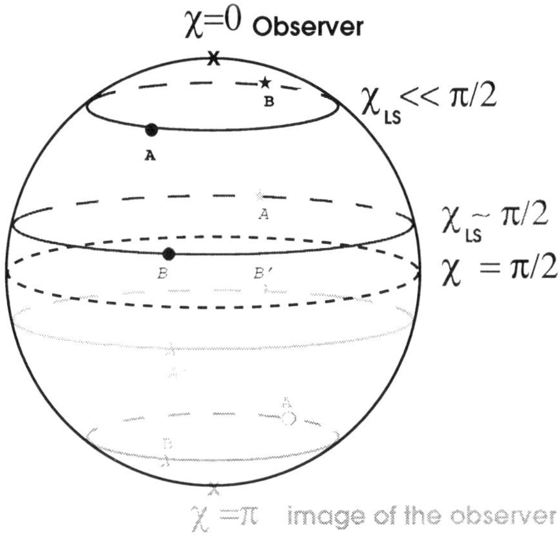

Figure 27.4 The figure supplements the explanation based on the method of images provided in the text for the CMB anisotropy signatures of an Elliptical universe. The CMB in an Elliptical universe can be interpreted as that on a simply connected spherical ('closed' FRW) universe but with a doublet of sources. The spatial section of an Elliptical universe is represented as a spherical manifold with antipodal points identified. The polar angle χ measures the radial light-travel distance from the observer, 'X'; every circle of constant χ denotes a physical two-sphere at that distance. The spheres of last scattering (SLS) is shown for a cosmological setting with $\tau_0 = \chi_{ls} \ll \pi/2$ and one for which $\tau_0 \approx \chi_{ls} \lesssim \pi/2$. The points A and B with three coordinates $(\chi_{ls}, \hat{q})$ and $(\chi_{ls}, -\hat{q})$ represent two CMB source points on the SLS along the lines of sight in two diametrically opposite directions $\hat{q}$ and $-\hat{q}$ in the sky. Each source point of the CMB anisotropy A and B has an image A' and B' at the antipode of the 3-sphere.

tical spaces, the correlation function, $\xi^e(\mathbf{x}_A, \mathbf{x}_B)$ between two points $\mathbf{x}_A \equiv (\chi_A, \hat{q}_A)$ and $\mathbf{x}_B \equiv (\chi_B, \hat{q}_B)$ can be expressed as

$$\xi^e(\mathbf{x}_A, \mathbf{x}_B) = \xi^s(\mathbf{x}_A, \mathbf{x}_B) + \xi^s(\mathbf{x}_A, \mathbf{x}_{B'}) \tag{1}$$

in terms of the correlation function, ξ^s, on the spherical space where $\mathbf{x}_{B'} \equiv (\pi - \chi_B, -\hat{q}_B)$ is the antipode of $\mathbf{x}_B$. The only feature of ξ^s required for this explanation is that it falls off with increasing separation.

Consider the correlation between the CMB temperature in two diametrically opposite directions in the sky, $\pm\hat{q}$, and assume for simplicity that the surface Sachs-Wolfe effect dominates the CMB anisotropy. The CMB correlation is then just the correlation between the potential fluctuation on two diametrically opposite points on the SLS.

Two cases of the above situation is illustrated in figure 27.4. For the case, when $\chi_{ls} \ll \pi/2$, the correlation $\xi^e \approx \xi^s$ between A and B because the second image term in eq.(1) is much smaller (A is far from B'). Hence, CMB anisotropy in the Elliptical universe is essentially indistinguishable from that in the corresponding closed FRW universe. This is to be expected since the scale of global connectivity is $\pi/2$ in units of the curvature radius, d_c.

In contrast, for the case when $\chi_{ls} \sim \pi/2$, the correlation between the point A and B given by eq.(1) is much higher than what one get in a 'closed' universe because the second term is large owing to the proximity of B' to A. This explains the enhanced angular correlation between diametrically opposite directions in the sky in the Elliptical universe in the $\chi_{ls} = 0.45\pi$ and $\chi_{ls} = 0.5\pi$ panels of fig. 27.3. Such a correlation function translates to a jagged C_ℓ curve where the odd multipoles are suppressed relative to the even ones. As $\chi_{ls} \to \pi/2$, the pattern gets stronger and persists up to progressively larger values of ℓ.

In figure 27.3, the $\chi_{ls} = 0.7\pi$ panels show that the CMB anisotropy spectrum and correlation functions for the Elliptical universe also differs dramatically from its spherical counterpart for $\chi_{ls} > \pi/2$. However, the deviation is reversed: there is now a strong anticorrelation diametrically opposite directions in the sky, and consequently, the even harmonics in C_ℓ are now suppressed relative to its odd neighbours. This cannot be understood solely in terms of the surface term of the Sachs-Wolfe effect which predicts identical CMB for models with equal $|\pi/2 - \chi_{ls}|$. The effect arises because the photon path from the SLS to the observer now crosses the image of the SLS leading to a large negative interference term between the SSW and ISW effects. This generic correlation feature, seen in multiply connected universes where ISW is important, was pointed out in [8] for compact hyperbolic universes. There is a value of χ_{ls} just beyond $\pi/2$ where the positive contribution of the pure SSW correlation to $C(\pi)$ is cancelled by the negative contribution from the interference term, e. g. see the $\chi_{ls} = 0.55\pi$ panel in fig. 27.3.

An important signature of the Elliptical universe relative to its spherical counterpart is the anomalously large magnitude of $C(\pi)/C(0)$. Fig. 27.5 presents a contour plot of $C(\pi)/C(0)$, measured with a Cosmic Background Explorer-Differential Microwave Radiometer (COBE–DMR) beam, in the $\Omega_D - \tau_0$ plane. Models in the bright (high positive values) and dark regions (high negative values) in the contour plot predict a potentially detectable signal in the COBE–DMR data. This overlaps with a large portion of the preferred (1σ) region of the recent high redshift supernova results in the $\Omega_D - \Omega_\Lambda$ plane (see fig. 27.1). Roughly, only a few levels in gray corresponding to $C(\pi)/C(0)$ between -0.1 to 0.4 will

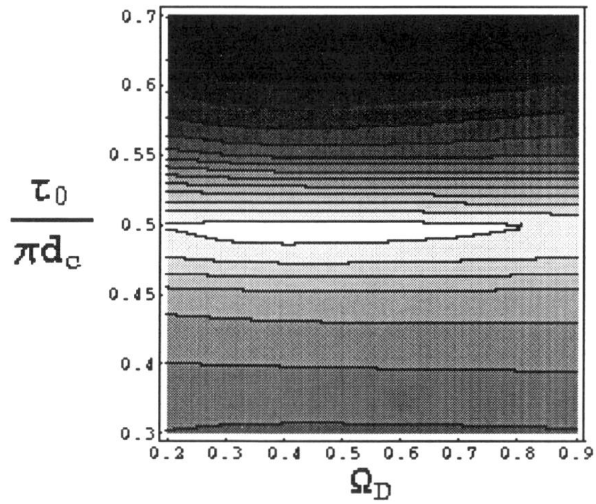

Figure 27.5 The extreme levels in contour plot of the ratio $C(\pi)/C(0)$ for Elliptical universes in the $\Omega_D - \tau_0$ parameter space are the regions most readily probed by the CMB anisotropy data. The correlation is computed assuming the COBE–DMR beamwidth. The contours are in steps of 0.1, ranging from $C(\pi)/C(0) = -0.5$ to 0.8. Since the magnitude of ratio is $\sim 0.1 - 0.2$ in simply connected models, a large region of the parameter space can be probed for Elliptical topology using the COBE–DMR data.

be inaccessible. This roughly corresponds to the models where either the horizon size is too small $\tau_0/d_c < 0.45\pi$, or which correspond to the thin band just above $\tau_0/d_c = \pi/2$ where two competing effects annul the signal. This provides only a rough guide to the results expected from the complete likelihood analysis using the full COBE–DMR data that is currently underway [7].

It should be noted that future CMB data covering large portions of the sky at higher resolution will be more sensitive to the signature of an Elliptical universe. Ideally, the contrast between $C(\pi)/C(0)$ in a Spherical and Elliptical universe can be enhanced by filtering out the first few multipoles (this ensures the $C(\pi)/C(0)$ in spherical model is very small) and filtering out power on $\ell \gtrsim 100$ (since difference between the elliptical and spherical model is negligible on small angular scales). Since COBE–DMR beam corresponds to a Gaussian cutoff around $\ell \sim 18$ there is scope for improvement with future 'all-sky', higher resolution data.

3. ADDRESSING GENERAL ULTRA-LARGE SCALE POSSIBILITIES

Nontrivial topology is just one aspect of the possible nontrivial global spatial structure of the universe. In general one could envisage bizarre possibilities such as shown in fig. 27.2. How should one address the detectability of a general breakdown of homogeneity and connectivity on scales just beyond the horizon? One such general idea is the Grischuk-Zeldovich effect which uses the small values of the detected quadrupole in the CMB anisotropy to constrain the length scale on which universe could have density fluctuations of order unity [9].

Exploiting the general connections between the geometry of manifolds and the spectrum of eigenvalues of the Laplacian is perhaps an interesting approach [10]. The constraints from CMB anisotropy data on the lowest eigenvalues of the Laplacian could have a lot to say . Let me try to motivate this approach with a toy example. Consider the bubble universe shown in fig. 27.2 connected to the rest of the universe by the throat. Consider the throat to have area A, the volume of the bubble region to be V and the curvature radius in negatively curved throat region is greater than d. Ignore the complication of the rest of the manifold and simply assume that it is compact and has much volume greater than V, and that there are no sections narrower than the throat nor regions more negatively curved. Given these conditions, the first eigenvalue of the Laplacian, k_1^2, can be bounded using known mathematical results to be in the range $h_C^2/4 \le k_1^2 \le 4h_C/d + 10h_C^2$, where $h_C = A/V$ is the

Cheeger's constant. Sharper bounds and less restrictive conditions may be possible, but even with what we have, one can, in principle, constrain the size of the throat. Very narrow throat would imply small values of h_C and can force the value of k_1^2 to be below what is expected in a simply connected spherical universe with curvature d_c. If the horizon is sufficiently large, the presence or absence of a supercurvature mode will be discernible in the CMB anisotropy.

4. CONCLUSIONS

The recent high redshift supernova results which point to the existence of a non zero cosmological constant also have encouraging news for the detectability of nontrivial features in the global spatial structure of our universe using the CMB anisotropy measurements. This general statement is borne out by this study of CMB anisotropy in the multiply connected Elliptical universe models.

5. ACKNOWLEDGEMENTS

I thank Naresh Dadhich and Ajit Kembhavi for the opportunity to contribute to this festschrift. Some of the ideas expressed here have originated from discussions with Dick Bond and Dmitry Pogosyan during our collaboration on the CMB anisotropy in CH universes. The CMB calculations were based on a 'closed' universe code created by appropriately modifying the 'open' universe code in the CMBFAST package. Part of this work was done while the author was at CITA, Toronto. At present the author is supported by NSF grant EPS-9550487 with matching support from the state of Kansas.

Notes

1. It implies, more generally, some exotic form of matter that accelerates the expansion of the universe. In this article, all my comments regarding the possible Λ term also applies to these other exotic possibilities.
2. The diameter of a manifold is the maximum of distances between all possible pairs of points.

References

[1] Dyson, K. K., 1992, *I Won't Let you Go –selected poems of Rabindranath Tagore*, (UBS Publishers).

[2] Narlikar, J. V., and Seshadri, T. R., 1985, Astrophys. J., **288**, 43.

[3] Perlmutter, S. et al. , 1999, preprint (astro-ph/9812133) (*To appear in* Astrophys. J. **516**).

[4] Wolf, J. A., 1994, *Space of Constant Curvature (5th ed.)*, (Publish or Perish, Inc.); Vinberg, E. B., 1993, *Geometry II – Spaces of constant curvature*, (Springer-Verlag).

[5] Lachieze-Rey, M. & Luminet, J.-P., 1995, Phys. Rep. **25**, 136; Starkman, G., Class. Quantum Grav. **15**, 2529, (1998); and other articles in the same issue: Proceedings of *Topology and Cosmology*, CWRU, Cleveland, Oct. 17-19, 1997.

[6] Ellis, G. F. R., 1971, Gen. Rel. Grav. **2**, 7; Sokolov, D. D., and Shvartsman, V. F., 1974, Zh. Eksp. Theor. Fiz. **66**, 412; Gott,J. R., 1980, Mon. Not. R. Astr. Soc. **193**, 153.

[7] Souradeep, T., *in preparation.*

[8] Bond, J. R., Pogosyan, D., and Souradeep, T., Class. Quant. Grav., **15**, 2671, (1998); *ibid., to appear in Phys. Rev. D.*

[9] Grishchuk, L. P. and Zeldovich, Ya. B., 1978, Astron. Zh., **55**, 209.

[10] Berard, P. H., 1980, *Spectral Geometry: Direct and Inverse Problems*, Lec. Notes in mathematics, **1207**, (Springer-Verlag, Berlin); Chavel, I., 1984, *Eigenvalues in Riemannian geometry*, (Academic Press).

Chapter 28

THE KERR-NUT METRIC REVISITED

P. C. Vaidya and L. K. Patel
Department of Mathematics,
Gujarat University,
Ahmedabad 380 009, India

Abstract Using Galilean time and retarded distance as coordinates, the combined Kerr-NUT metric is investigated in connection with Einstein field equations for a perfect fluid plus a pure radiation field. Two particular solutions of the field equations are discussed. One of them describes the field of a Kerr particle embedded in flat Robertson-Walker universe.

1. INTRODUCTION

We [1] have investigated the combined Kerr - NUT metric earlier for pure radiation fields. In addition to the Kerr [2] and NUT [3] solutions of Einstein equation, three other types of solutions were obtained. They were (i) the radiating Kerr solution (ii) the radiating NUT solution satisfying $R_{ik}- = \sigma\xi_i\xi_k, \xi_i\xi^i = 0$ and (iii) the associated Kerr solution satisfying $R_{ik} = 0$ [4]. For the derivation of these solutions, we have considered the Kerr - NUT metric in the form.

$$ds^2 = 2(du+g\sin\alpha d\beta)dt - M^2(d\alpha^2+\sin^2\alpha d\beta^2) - 2L(du+gsin\alpha d\beta)^2 \quad (1)$$

where

$$g = g(\alpha), L = L(t, u, \alpha), M = M(t, u, \alpha) \quad (2)$$

We introduce the tetrad

$$\begin{aligned} \theta^1 &= du + g\sin\alpha d\beta, \theta^2 = M d\alpha, \\ \theta^3 &= M\sin\alpha d\beta, \theta^4 = dt - L\theta^1 \end{aligned} \quad (3)$$

So that the metric (1) becomes

$$ds^2 = 2\theta^1\theta^4 - (\theta)^2)^2 - (\theta^3)^2 = g_(ab)\theta^a\theta^b \quad (4)$$

N. Dadhich and A. Kembhavi (eds.), The Universe, 301–308.

Here and in what follows the bracketed indices denote tetrad components with respect to the tetrad (3). The tetrad components $R_{(ab)}$ of the Ricci tensor for the metric (1) have been given by us in the reference [1]. They are reproduced in the appendix for ready reference. From the expressions listed in the appendix, a lengthy but straight forward calculation leads to

$$[R_{(44)}]_y - 4C\frac{M}{g}R_{(34)} + \frac{2M}{g}\frac{C_t}{C}R_{(24)} - \left[\frac{2M}{g}R_{(24)}\right]_t = 0 \tag{5}$$

and

$$[R_{(44)}]_u - 4C\frac{M}{g}R_{(24)} + \frac{2M}{g}\frac{C_t}{C}R_{(34)} - \left[\frac{2M}{g}R_{(34)}\right]_t = 0 \tag{6}$$

where

$$C = \frac{f}{M^2},\ 2f = g_\alpha + g cot\alpha \tag{7}$$

and a suffix denote partial derivatives,e.g. $g_\alpha = \frac{\partial g}{\partial \alpha}$, $L_{uu} = \frac{\partial^2 L}{\partial u^2}$ etc. The variable y is defined by a differential relation

$$g d\alpha = dy \tag{8}$$

If $R_{(24)} = 0, R_{(34)} = 0$, then (5) and (6) imply that $R_{(44)}$ is a function of cosmic time t. Again by direct calculation we have verified that

$$[R_{(14)} - LR_{(44)}]_u = \frac{C_t}{C}\frac{M}{g}\{R_{(13)} - LR_{(34)}\}$$
$$+2C\frac{M}{g}\{R_{(12)} - LR_{(24)}\} - \left[\frac{M}{g}\{R_{(13)} - LR_{(34)}\}\right]_t \tag{9}$$

and

$$[R_{(14)} - LR_{(44)}]_y = -\frac{C_t}{C}\frac{M}{g}\left\{R_{(12)} - LR_{(24)}\right\}$$
$$+2C\frac{M}{g}\left\{R_{(13)} - LR_{(34)}\right\} + \left[\frac{M}{g}\{R_{(12)} - LR_{(24)}\}\right]_t. \tag{10}$$

If $R_{(24)} = R_{(34)} = 0, R_{(12)} = R_{(13)} = 0$, then (9) and (10) indicate that $R_{(14)} - LR_{(44)}$ is also a function of t alone. The above discussion leads us to suspect that our Kerr-NUT metric (1) is more likely to describe cosmological situations. We shall now check that this is really the case.

2. THE FIELD EQUATIONS

We take the usual perfect fluid distribution traversed by unidirectional radiation flow. The energy momentum tensor for such a distribution is given by

$$T_{ik} = (p+\rho)v_i v_k - p g_{ik} + \sigma w_i w_k \tag{11}$$

with

$$v_i v^i = 1, w_i w^i = 0, v^i w_i = 1 \tag{12}$$

Here p, ρ and σ are repsectively the fluid pressure, the matter density and the radiation density. The field equations are

$$R_{ik} - \frac{1}{2} R g_{ik} = -8\pi T_{ik} \tag{13}$$

where T_{ik} is given by (11) and (12). These field equations can be expressed in tetrad basis as

$$R_{(ab)} = -8\pi[(p+\rho)v_{(a)}v_{(b)} - \frac{1}{2}(\rho - p)g_{(ab)} + \sigma w_a w_b] \tag{14}$$

where $v_{(a)}$ and $w_{(a)}$ are the tetrad components of the flow vector v_i and the null vector w_i respectively.

For the metric (1) we take $v_{(a)}$ and $w_{(a)}$ as

$$v_{(a)} = (\lambda, 0, 0, \frac{1}{2\lambda}),\ w_{(a)} = (2\lambda, 0, 0, 0) \tag{15}$$

where λ is a function of coordinates to be determined from the field equations. In view of (15) the field equations (14) give rise to the following relations :

$$R_{(23)} = 0, R_{(24)} = 0, R_{(34)} = 0 \tag{16}$$

$$R_{(12)} = 0, R_{(13)} = 0 \tag{17}$$

$$8\pi p = -R_{(14)} \tag{18}$$

$$8\pi\rho = -[2R_{(22)} + R_{(14)}] \tag{19}$$

$$2\lambda^2 = \frac{R_{(22)} + R_{(14)}}{R_{(44)}} \tag{20}$$

and

$$16\pi\sigma = R_{(22)} + R_{(14)} - \frac{R_{(11)}R_{(44)}}{R_{(22)} + R_{(14)}} \tag{21}$$

where $R_{(ab)}$ are given by the expressions listed in the appendix.

3. SOLUTIONS OF THE FIELD EQUATIONS

For the metric (1), $R_{(23)}$ is zero identically, the other two equaitons $R_{(24)} = 0$ and $R_{(34)} = 0$ of (16) give

$$M^2 = \frac{f}{Y\dot{\varphi}}(X^2 + Y^2), X = u - \varphi(t), Y = -y \tag{22}$$

where $\varphi(t)$ is an undetermined function of time t. Now onwards an overhead dot indicates differentiation with respect to t. Instead of $\varphi(t)$, it is more convenient to use the function $H(t)$ defined by $\dot{\varphi} = e^{-H(t)}$. Now the equations (17) determines the metric function L as

$$L = A(t) - \frac{2B(t)X}{X^2 + Y^2}),\ \dot{B} + Ae^{-H} = \frac{1}{2} \tag{23}$$

where A and B are functions of t. Thus the three functions H, A and B of time t are related by only one relation. Therefore, for explicit solutions, two of them can be chosen arbitrarily. It is interesting to note that the field equations (16) and (17) do not put any restriction on the metric potential $g(\alpha)$. For Kerr metric we have$g = ksin\alpha$ where k is a constant related with the angular momentum of the body. So, for simplicity we take

$$g = k\sin\alpha,\ f = k\cos\alpha,\ y = -k\cos\alpha \tag{24}$$

Consequently we have

$$M^2 = e^H(X^2 + k^2\cos^2\alpha), X = u - \int e^{-H}dt \tag{25}$$

Using the above results, the remaining $R_{(ab)}$ for our metric can be determined. They are given by

$$\begin{aligned} R_{(44)} &= \ddot{H} + \frac{1}{2}\dot{H}^2, R_{(14)} = LR_{(44)} + \ddot{A} + \dot{A}\dot{H}, \\ R_{(22)} &= -A(\ddot{H} + \dot{H}^2) - \dot{A}\dot{H} + \frac{1}{(X^2+Y^2)}[2Ae^{-H} - 2B\dot{H} - 1] \\ &+ \frac{X}{(X^2+Y^2)}[2\dot{A}e^{-H} - \dot{H} + 2B(\ddot{H} + \dot{H}^2)], \\ R_{(11)} &= L^2 R_{(44)} + \frac{2}{(X^2+Y^2)}[2\dot{B} - B\dot{H}] \\ &+ \frac{2X}{(X^2+Y^2)}[A\dot{H} - \dot{A}] \end{aligned} \tag{26}$$

where L is given by (23).

Using (25) in the equations (18) - (21), we can find the physical parameters ρ, p, σ and λ^2.

We shall discuss the explicit solutions for two particular cases: Case I : $B = me^{nH}$ and Case II : $\ddot{A} + \dot{A}\dot{H} = 0$ where m and n are constants.

Case I: $B = me^{nH}$. For this case (3.5) give

$$\begin{aligned} R_{(44)} &= \ddot{H} + \frac{1}{2}\dot{H}^2, \\ R_{(14)} &= \left[A - \frac{2BX}{X^2+Y^2}\right](\ddot{H} + \frac{1}{2}\dot{H}^2) + \ddot{A} + \dot{A}\dot{H}, \\ R_{(22)} &= -\frac{1}{2}e^H(\ddot{H} + 2\dot{H}^2) + mne^{(n+1)H}\dot{H}\left\{2\ddot{H} + (n+2)\dot{H}^2\right\} \\ &+ \frac{2mXe^{nH}}{(X^2+Y^2)}\left[(1-n)\ddot{H} - \dot{H}^2(n^2+n-1)\right] - \frac{2me^{nH}\dot{H}(n+1)}{(X^2+Y^2)}, \\ R_{(11)} &= \left[A - \frac{2BX}{X^2+Y^2}\right]^2(\ddot{H} + \frac{1}{2}\dot{H}^2) + \frac{2m(2n-1)e^{nH}}{(X^2+Y^2)}\dot{H} \\ &+ \frac{2mnXe^{(n+1)H}}{(X^2+Y^2)}[\ddot{H} + n\dot{H}^2] \end{aligned} \tag{27}$$

We now pick up particular subcases.

Case I (i) $m = 0$. If $m = 0$, then we obtain

$$\begin{aligned} A &= \frac{1}{2}e^H, B = 0, \sigma = 0, 2\lambda^2 = \frac{1}{2}e^H, \\ 8\pi p &= -e^H(\ddot{H} + \frac{5}{4}\dot{H}^2), 8\pi\rho = \frac{3}{4}e^H\dot{H}^2 \end{aligned} \tag{28}$$

In this case we get the flat Robertson - Walker universe. The explicit form of the line element is

$$ds^2 = 2(du + ksin^2\alpha d\beta)dt - e^H(X^2 + k^2cos^2\alpha)$$
$$(d\alpha^2 + sin^2\alpha d\beta^2) - e^H(du + ksin^2\alpha d\beta)^2 \tag{29}$$

where $X = u - \int e^{-H}dt$.
This is the flat Robertson-Walker metric in rotating coordinates.
Case I (ii) $n = 0$. In this subcase we have

$$B = m, A = \frac{1}{2}e^H, L = \frac{1}{2}e^H - \frac{2mX}{X^2 + Y^2} \tag{30}$$

The physical parameters for this case are given by

$$8\pi p = -e^H(\ddot{H} + \frac{5}{4}\dot{H}^2) + \frac{2mX}{(X^2 + Y^2)}(\ddot{H} + \frac{1}{2}\dot{H}^2) \tag{31}$$

$$8\pi\rho = \frac{3}{4}e^H\dot{H}^2 - \frac{mX}{(X^2 + Y^2)}(2\ddot{H} + 3\dot{H}^2) + \frac{4m\dot{H}}{(X^2 + Y^2)} \tag{32}$$

$$2\lambda^2 = \frac{1}{2}e^H + \frac{2m\left[\frac{X\dot{H}^2}{2} - \dot{H}\right]}{(\ddot{H} + \frac{1}{2}\dot{H}^2)(X^2 + Y^2)} \tag{33}$$

and

$$16\pi\sigma\left[\frac{1}{2}e^H(\ddot{H} + \frac{1}{2}\dot{H}^2) + \frac{mX\dot{H}^2 - 2m\dot{H}}{(X^2 + Y^2)}\right]$$
$$= \frac{m(\ddot{H} + \frac{1}{2}\dot{H}^2)}{(X^2 + Y^2)}[2\dot{H}(1 - e^H) + 2Xe^H(\ddot{H} + \dot{H}^2)]$$
$$+ \frac{4m^2}{(X^2 + Y^2)^2}[\dot{H}^2 - X\dot{H}^3 - X^2\ddot{H}(\ddot{H} + \dot{H}^2)] \tag{34}$$

When $m = 0$, we recover the case I (i) of flat Robertson-Walker universe. If we remove the expansion of the universe (i.e. $H = 0$) , the above solution reduces to the well-known Kerr empty space - time. Therefore the solution of this subcase describes the field of a rotating mass particle embedded in the flat Robertson-Walker universe. One such solution is discussed earlier by Vaidya [5]. In Vaidya [5] solution, the fluid pressure is anisotropic. Also the solution was obtained by a coordinate transformation. In our solution the fluid pressure is isotropic but there is a pure radiation field in addition to perfect fluid. The explicit metric of our solution is

$$ds^2 = 2(du + ksin^2\alpha d\beta)dt - e^H(X^2 + k^2os^2\alpha)(d\alpha^2 + sin^2\alpha d\beta^2)$$
$$-\left[e^H - \frac{4mX}{X^2 + k^2cos^2\alpha}\right](du + ksin^2\alpha d\beta)^2 \quad (35)$$

where $X = u - \int e^{-H}dt$.

We have also worked out the details of the subcases $n = 1/2$ and $n = -1$. For the sake of brevity these details are not reported here.

Case II $\ddot{A} + \dot{A}\dot{H} = 0$. If we take $e^{-H} = \dot{F}$, then we get

$$A = aF + b, \; B = \frac{1}{2}t - \frac{a}{2}F^2 - bF + c \quad (36)$$

where a, b, c are arbitrary constants. One can work out the details for the above solution (36). We shall work them out in a particular case

$$b = c = 0, F^2 = t, A = at^{1/2}, B = \frac{(1-a)t}{2} \quad (37)$$

The physical parameters for this case are given by

$$8\pi p = \frac{3}{8t^2}\left[at^{1/2} - \frac{(a-1)tX}{X^2 + Y^2}\right] \quad (38)$$

$$8\pi\rho = \frac{3a}{8t^{3/2}} + \frac{9(1-a)X}{8t(X^2+Y^2)} + \frac{3(1-a)}{X^2+y^2} \quad (39)$$

$$2\lambda^2 = at^{3/2} + \frac{(1-a)t(X+4t)}{(X^2+Y^2)} \quad (40)$$

and

$$\frac{6\pi\sigma}{t^2}\left[at^{1/2} + \frac{t(1-a)(X+4t)}{(X^2+Y^2)}\}\right]$$
$$= \frac{9(1-a)}{16t^3(X^2+Y^2)}\left[(X+2t)\left\{at^{1/2} + t^2\frac{(1-a)(X+2t)}{X^2+Y^2}\right\} + t\right] \quad (41)$$

where $X = u - t^{1/2}, Y = kcos\alpha$. When $0 < a \leq 1$, the physical requirements $p \geq 0, \rho \geq p$ and $\sigma \geq 0$ are satisfied for all $t > 0$. Here $t = 0$ is a singularity because as $t \to 0, p$ and ρ diverge. It is interesting to note that when $a = 1$, we recover a particlular case of the subcase case I (i) with $\rho = p$ (stiff-fluid).

In the above discussion we have assumed the form $g = ksin\alpha$ for the metric potential $g(\alpha)$. But the field equations do not give any condition on $g(\alpha)$. Therefore the other choices of $g(\alpha)$ are also possible. Thus the Kerr - NUT metric (1) can give rise to some other new exact solutions of Einstein equations corresponding to a mixture of perfect fluid and pure radiation.

References

[1] Vaidya,P. C., Patel L.K. & Bhatt P.V., 1976, *Gen. Rel.Grav.***7**, 701.

[2] Kerr, R. P., 1963, *Phys. Rev. Lett.* **11**, 237.

[3] Newman, E., Tamburino, L. & Unti, T., 1963, *J. Math. Phys.***4**, 915.

[4] Vaidya, P.C., 1976, *Curr. Sci.***45**, 490.

[5] Vaidya, P.C., 1977, *Pramana* **8**, 512.

Appendix

$$
\begin{aligned}
R_{(23)} &= 0 \\
R_{(44)} &= (2/M)[M_{tt} - f^2/M^3] \\
R_{(24)} &= (g/M)[(M_t/M)_y - (f/M^2)_u] \\
R_{(34)} &= -(g/M)[(M_t/M)_u - (f/M^2)_y] \\
R_{(14)} &= L_{tt} + (2/M)[M_{tu} + (LM_t)_t + (Lf^2/M^3)] \\
R_{(12)} &= LR_{(24)} + (g/M)[(L_t + M_u/M)_y + (2fL/M^2)_u)] \\
R_{(13)} &= LR_{(34)} + (g(M)[-(L_t + M_u/M)_u + (2fL/M^2)_y)] \\
R_{(22)} &= R_{(33)} = (1/M^2)\Big[g^2(M_u/M)_u + g^2(M_y/M)_y \\
&+ 2f(M_y/M) + 4f^2L/M^2 - 1 - (M^2)_{ut} - \Big\{L(M^2)_t\Big\}_t\Big] \\
R_{(11)} &= L^2R_{(44)} + (1/M^2)[g^2(L_{uu} + L_{yy}) + 2fL_y \\
&+ 2L_uMM_t + 4LMM_{ut} - 2L_tMM_u + 2MM_{uu}] \qquad (42)
\end{aligned}
$$

Chapter 29

BLACK HOLES IN COSMOLOGICAL BACKGROUNDS[1]

C. V. Vishveshwara
Indian Institute of Astrophysics,
Bangalore 560 034, India

For some two decades now, Jayant Narlikar and I have been participating in various activities together - organizing conferences and workshops, planning and teaching courses at schools for doctoral students and so on. Since its very inception, I have had the good fortune of associating myself, in some capacity or the other, with IUCAA which has blossomed into a fine academic institution under Jayant's leadership. Over the years, I have enjoyed reading his books, articles and stories. It is with great pleasure that I dedicate this article to Jayant, a close friend and an esteemed colleague.

—

1. MOTIVATION

Black hole physics has been one of the most active areas of research in general relativity. A great deal of information has been gathered on the structure of black holes and physical phenomena that take place in their spacetimes. These spacetimes, such as those associated with the Schwarzschild and Kerr black holes, are time-independent and asymptotically flat. Time symmetry is equivalent to the requirement that the spacetime admit a global timelike Killing vector field. In a totally realis-

[1]This article is based on ongoing work of K. Rajesh Nayak, B. S. Ramachandra and C. V. Vishveshwara at the Indian Institute of Astrophysics.

N. Dadhich and A. Kembhavi (eds.), The Universe, 309–318.

tic model, however, the black hole should be imbedded in or associated with a cosmological background. In such a scenario, neither of the above two conditions would be valid. Being part of an expanding universe, the black hole would cease to be time independent, i.e., the spacetime will no longer admit a timelike Killing vector field. Furthermore, spacetime would become cosmological and non-flat at large distances from the black hole. Very little has been done in exploring such black holes. It is not at all unlikely that the structure and properties of these black holes may differ significantly, or even drastically, from the ones that have been studied. Even in the case of the latter it is well known that the introduction of rotation, i.e., the passage from the non-rotating, spherically symmetric Schwarzschild to the rotating Kerr black hole, brings about profound changes. For instance, in the case of the Schwarzschild black hole the timelike Killing vector becomes null (static limit) on the black hole which is itself a null surface (Killing event horizon)[1]. On the other hand, in the case of the Kerr black hole the stationary limit at which the timelike Killing vector becomes null does not coincide with the event horizon which is required to be a null surface. However, Kerr spacetime admits a globally hypersurface orthogonal, irrotational timelike vector field which does become null on the event horizon[2]. The separation of the stationary limit from the event horizon and the consequent existence of the ergosphere in between lead to several interesting phenomena such as the Penrose process and superradiance. Similarly, phenomena that occur in the Schwarzschild spacetime may not take place in the Kerr spacetime, for instance the generation of gravitational synchrotron radiation. In the same manner, the introduction of the cosmic background may radically transform the physics of black holes.

2. SOME BASIC ISSUES

There are several broad issues one would like to address in considering cosmological black holes. Following are some of them.

- Definition: One of the basic issues is that of defining the black hole. Tipler[3], for instance, defined a black hole for stably causal spacetimes as an object containing all small trapped surfaces. On the other hand, Joshi and Narlikar[4] offered a definition using the notion of trapping of light by the strong gravitational field of a collapsing object in a globally hyperbolic spacetime. It is essential to study such alternative definitions as well as any new viable ones and decide which of these would hopefully represent a true cosmological black hole. Furthermore, one would also like to know whether such a definition would make the black hole a null surface as in the case of regular black holes.

- Comparison: Once a proper definition of a cosmological black hole has been arrived at, the next task is to test whether the known properties of the conventional black holes are valid or not. There are several such properties some of them involving the existence of the global timelike Killing vector and some dependent on asymptotic flatness. It is possible that some of them would remain unaltered, while others may get modified or altogether violated.

- Physical phenomena: Physical phenomena in the gravitational fields of black holes, often unusual and interesting, have been well studied. Once again it is worthwhile finding out whether these phenomena continue to exist in connection with cosmic black holes. It would, of course, be important to seek new physical phenomena in this context.

3. METHODOLOGY AND RESULTS

Our approach to the problem of cosmological black holes is to consider specific examples of black hole solutions in non-flat backgrounds. We shall proceed step by step relaxing successively the conditions of asymptotic flatness and time symmetry. The aim is to investigate different issues, some of which have been outlined in the previous section within this framework.

For this purpose we consider the exact solutions found by Vaidya [5] that are supposed to incorporate black holes in cosmological backgrounds. For details regarding the derivation of these solutions we refer the reader to the above paper. Suffices here to mention that the starting point is the particular background metric in which one wishes to incorporate a black hole, e.g. flat, Einstein, or Robertson-Walker background. Then transform to ellipsoidal polar coordinates and finally make certain adjustments that yield the required exact solutions. We list the line elements of these spacetimes below.

Kerr black hole in flat background (KFB)

$$ds^2 = 2(du + a\sin^2\alpha\, d\beta)dt - (r^2 + a^2\cos^2\alpha)(d\alpha^2 + \sin^2\alpha\, d\beta^2) - \left[1 + \frac{2mr}{(r^2 + a^2\cos^2\alpha)}\right](du + a\sin^2\alpha\, d\beta)^2 \tag{1}$$

This is the usual Kerr spacetime in the new coordinates. Here $u = t - r$ is the null coordinate; α and β are the polar angles usually denoted by θ and ϕ.

Kerr black hole in Einstein universe (KEB)

$$\begin{aligned} ds^2 &= 2(du + a\sin^2\alpha\, d\beta)dt - (1+2m\mu)(du + a\sin^2\alpha\, d\beta)^2 \\ &\quad - M^2\,[(1 - a^2\sin^2\alpha/R^2)^{-1}d\alpha^2 + \sin^2\alpha\, d\beta^2), \end{aligned} \tag{2}$$

where
$M^2 = (R^2 - a^2)\sin^2(\frac{r}{R}) + a^2\cos^2\alpha$
and $\mu = (\frac{R}{M})^2\sin(\frac{r}{R})\cos(\frac{r}{R})$. Note that for $m = 0$ we recover the Einstein universe and in the limit R tending to infinity we obtain the Kerr metric.

Kerr black hole in Robertson-Walker background ((KRWB)

$$\begin{aligned} ds^2 &= e^{2F(t)}\{2(du + a\sin^2\alpha\, d\beta)dt - (1+2m\mu e^{-2F(t)}) \\ &\quad (du + a\sin^2\alpha d\beta)^2 - M^2[(1 - a^2\sin^2\alpha/R^2)^{-1}d\alpha^2 + \sin^2\alpha d\beta^2\}, \end{aligned} \tag{3}$$

where F is an arbitrary function of time t. Once again we recover the Robertson-Walker and Kerr metrics in the limits of m tending to zero and R tending to inifinity respectively. We shall now discuss specific examples.

3.1 KERR BLACK HOLE IN EINSTEIN UNIVERSE

We transform to coordinates analogous to those of Boyer and Lindquist to obtain

$$\begin{aligned} ds^2 &= (1 - 2m\mu)dt^2 - 2ma\mu\sin^2\alpha\, dtd\beta \\ &\quad -[M^2\sin^2\alpha + a^2(1+2m\mu)\sin^4\alpha]d\beta^2 \\ &\quad -M^2[m^2(1-2m\mu) + a^2\sin^2\alpha]^{-1}dr^2 \\ &\quad -\frac{M^2}{1 - \frac{a^2}{R^2}\sin^2\alpha}d\alpha^2 \end{aligned} \tag{4}$$

In this form orthogonal transitivity is manifest with respect to the time-like Killing vector $\xi^a = \delta^a_0$ and the axial Killing vector $\eta^a = \delta^a_3$. As has been shown by Greene, Schücking and Vishveshwara[2], there exists a globally hypersurface orthogonal vector field

$$\chi^a = \xi^a - \left(\frac{\xi^b \eta_b}{\eta^c \eta_c}\right) \eta^a. \tag{5}$$

Furthermore, the surface on which χ^a becomes null ($\chi^a \chi_a = 0$) is itself a null surface. In a stationary spacetime like KEB, this is indeed the event horizon and hence defines the black hole in the Einstein background. The condition $\chi^a \chi_a = 0$ is equivalent to

$$R^2 \tan^2(\frac{r}{R}) - 2mR \tan(\frac{r}{R}) + a^2 = 0 \tag{6}$$

or

$$R \tan(\frac{r}{R}) = m \pm \sqrt{m^2 - a^2} \tag{7}$$

This is the same surface identified by Vaidya as the black hole. On the other hand, the stationary limit ($\xi^a \xi_a = 0$) is given by the surface

$$R \tan(\frac{r}{R}) = \frac{1}{\left(1 - \frac{a^2}{R^2} \sin^2 \alpha\right)} \left[m \pm \sqrt{m^2 - a^2 \cos^2 \alpha \left(1 - \frac{a^2}{R^2} \sin^2 \alpha\right)}\right] \tag{8}$$

In figure 1 we show both the event horizon and the stationary limit enclosing the ergosphere for two different values of R, the parameter which incorporates the effect of the background. As R increases the situation tends to that of the usual Kerr in flat background. For lower values of R the ergosphere is larger and more distorted than in this limit.

One of the remarkable properties of the Kerr spacetime is that it admits a second rank Killing tensor which leads to a quadratic constant of geodesic motion. Further, the Killing tensor can be expressed as the 'square' of a Killing-Yano tensor. The existence of these two tensors is related to the property of the Kerr spacetime being Petrov type-D. We have found that all these features are retained in the KEB spacetime.

3.2 SCHWARZSCHILD BLACK HOLE IN EINSTEIN UNIVESRE

By setting the angular momentum parameter $a = 0$ in equation (4), we get the required line element

$$ds^2 = \left(1 - \frac{2m}{R\tan(\frac{r}{R})}\right) dt^2 - \left(1 - \frac{2m}{R\tan(\frac{r}{R})}\right)^{-1} dr^2$$
$$-R^2 \sin^2(\frac{r}{R}) \left[d\alpha^2 + \sin^2 \alpha d\beta^2\right] \tag{9}$$

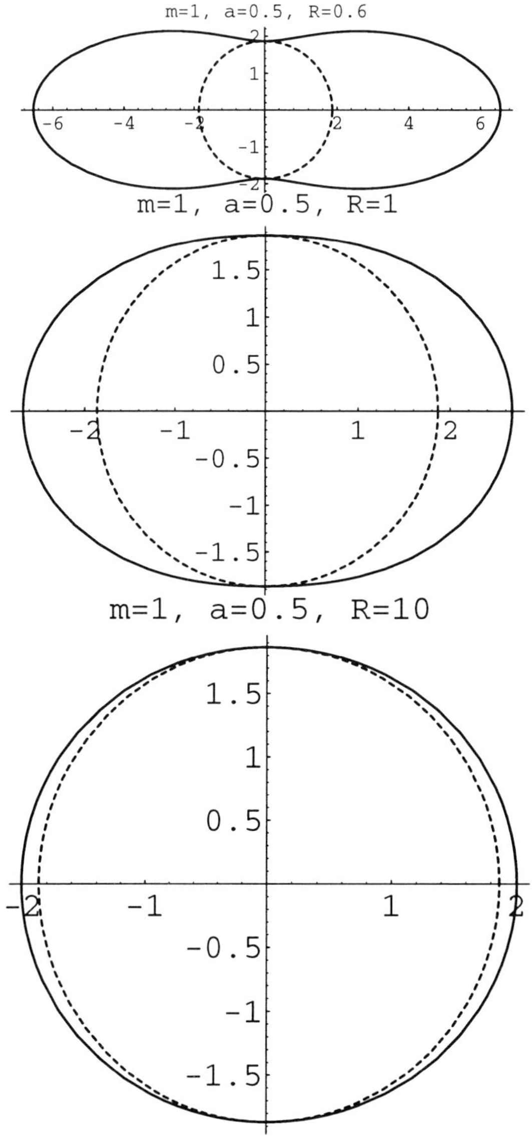

Figure 29.1 Plots of the stationary limit (solid line), event horizon (broken line) and the ergosphere in between for different values of R

As before $m = 0$ and $R \to \infty$ give respectively Einstein universe and the Schwarzschild metric. The parameter R measures the cosmological influence on the black hole given by

$$R \tan(\frac{r}{R}) = 2m \tag{10}$$

One can discuss in the present spacetime SEB all the physical phenomena known in the conventional Schwarzschild field. For instance, consider the propagation and scattering of scalar waves. By virtue of time and spherical symmetries the wave function can be written as

$$\psi = e^{i\omega t}\mathcal{R}(r)Y_l{}^m(\alpha,\beta) \tag{11}$$

Limits of the radial coordinates are given by $R\tan(\frac{r}{R}) = 2m$ to $(\frac{r}{R}) = \frac{\pi}{2}$. Further, setting the radial function $\mathcal{R}(r) \equiv \frac{u(r)}{R\sin(\frac{r}{R})}$ and defining $dr^* = \frac{dr}{1-\tan(\frac{r}{R})}$ we can derive the Schrödinger equation

$$\frac{d^2u}{dr^{*\,2}} + \left[\omega^2 - V(r)\right] u = 0 \tag{12}$$

The effective potential that controls the propagation of the scalar waves is given by

$$V(r) = \left(1 - \frac{2m}{R\tan(\frac{r}{R})}\right)\left[\frac{l(l+1)}{R^2\sin^2(\frac{r}{R})} + \frac{2m}{R^3\sin^2(\frac{r}{R})\tan(\frac{r}{R})} - \frac{1}{R^2}\left(1 - \frac{2m}{R\tan(\frac{r}{R})}\right)\right] \tag{13}$$

The effective potential vanishes at the black hole surface and goes to a constant at the other limit $\frac{r}{R} = \frac{\pi}{2}$ resembling the Eckart potential[6]. Figure 2 displays the plots of the effective potential $V(r)$ as function of $\tan(\frac{r}{R})$ for $l = 2$ and different values of R. The effective potential corresponding to the Schwarzschild metric has also been shown (dotted line) as function of r for comparison. For low values of R, i.e. high background influence, the difference is drastic, whereas for high values of R the two effective potentials approach each other.

We have also studied the three classical tests of general relativity within the framework of this spacetime. They are modified by the background as codified by the parameter R. This is true of the geodesics - the range of circular orbits, the Keplerian frequency, the effective potential and the classification scheme.

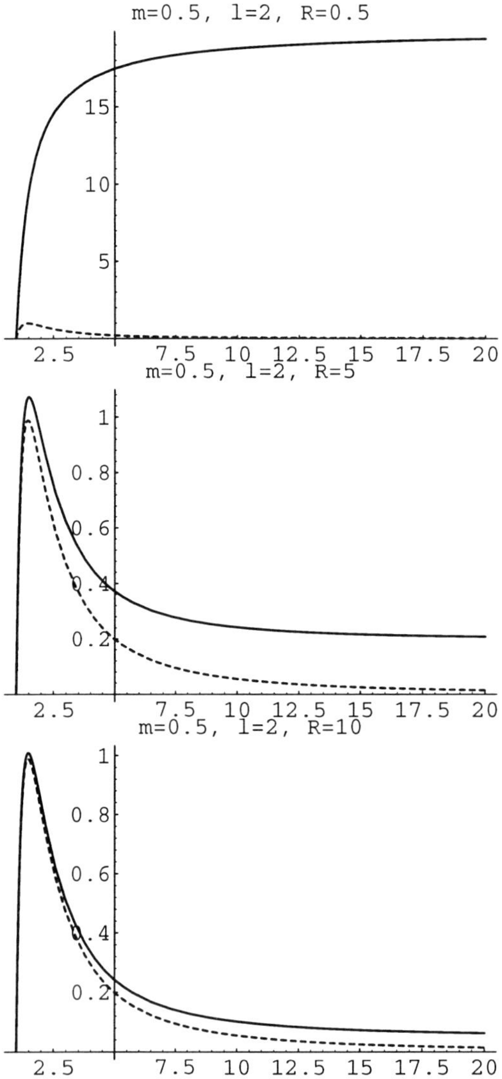

Figure 29.2 Plot of the equivalent potential for a black hole in spherically symmetric Einstein background (solid line) as a function of $R\tan(\frac{r}{R})$ compared with that for the Schwarzschild black hole (broken line) as a function of r for different values of R.

3.3 BLACK HOLE IN ROBERTSON–WALKER BACKGROUND

The spacetime is now given by the line element of equation (3). Both conditions of asymptotic flatness and stationarity have been relaxed. In the Boyer-Lindquist form, the metric may be written as

$$\begin{aligned} ds^2 \;=\; & (e^{2F} - 2m\mu)dt^2 - 2ma\mu \sin^2\alpha dt d\beta \\ & - \left[e^{2F} M^2 (a^2 \sin^2\alpha + M^2 + 2mM^2\mu e^{-2F})^{-1}\right] dr^2 \\ & - e^{2F} M^2 \left(1 - \frac{a^2 \sin^2\alpha}{R^2}\right)^{-1} d\alpha^2 \\ & e^{2F}\left[M^2 \sin^2\alpha + a^2(1 + 2m\mu e^{-2F}) \sin^4\alpha\right] d\beta^2 \end{aligned} \tag{14}$$

Now, how do we locate the black hole? Vaidya[5] wrote down the equation for this surface entirely in analogy with the Kerr metric by replacing m by me^{-2F} since e^{-2F} is the additional factor appearing in the metric now. Then the equation for this surface reads

$$R^2 \tan^2(\frac{r}{R}) - 2me^{-2F} R \tan(\frac{r}{R}) + a^2 \;=\; 0 \tag{15}$$

However, if we require of the event horizon the basic property of one way membrane then it should be a null surface with the light cone tangential to it. It is easy to verify that the surface given by equation (15) does *not* satisfy this condition. Therefore it may fail to qualify as candidate for being a black hole! This is indeed a basic problem that has to be remedied.

4. FUTURE STUDIES

We have made just a beginning of our investigation on black holes in cosmological backgrounds. In the case of Kerr spacetime in the stationary background of Einstein universe, the black hole can be well defined as a null surface on which the globally hypersurface orthogonal vector field, i.e. the projection of the global timelike Killing vector field orthogonal to the axial Killing vector, becomes null. In this case the structures of the horizon and the stationary limits can be properly studied as a function of the cosmological background influence. On the other hand, in the case of Vaidya's black hole metric in the time dependent Robertson-Walker background one fails to obtain a similar null event horizon. Either the metric should be suitably modified to accommodate a comoving black hole of this type or the definition of the black hole has to be modified in such a way that its behavior resembles that of the conventional one. This is an open problem.

Another important consideration is that of the energy-momentum tensor T_{ab}. Tensor T_{ab} is obtained from R_{ab} for a given metric. We are investigating the behaviour of T_{ab} in the spacetimes we have considered. For perfect fluid source, there seems to be a generic problem of running into negative energy density or negative pressure in certain regions of spacetime. Nevertheless, the weak energy condition is satisfied by the prefect fluid acting as the source for the spacetime thereby indicating that the situation is not pathological.

As we have seen in the forgoing discussions, several significant questions arise in considering black holes in cosmological backgrounds and more are bound to appear. We expect to be able to answer the questions that have already arisen and find out more about black holes in cosmological backgrounds.

References

[1] C. V. Vishveshwara, *J. Math Phys.*, **9**, 1319(1968).

[2] R. D. Greene, E. L. Schüking and C. V. Vishveswara, *J. Math Phys.*, **16**, 153(1975).

[3] F. J. Tipler, *Nature*, **270**, 500(1977).

[4] P. S. Joshi and J. V. Narilkar, *Pramana*, **18**, 385(1982).

[5] P. C. Vaidya, *Pramana*, **8**, 512(1977).

[6] J. M. Aguirregabiria, and C. V. Vishveshwara, *Phys. Lett.* **A 210**, 251(1996).

Chapter 30

ELEMENTARY PARTICLE INTERACTIONS AND NONCOMMUTATIVE GEOMETRY

Kameshwar C. Wali
Physics Department, Syracuse University
Syracuse, NY 13066, USA

It is indeed a great privilege to contribute to this volume celebrating the sixtieth birthday of Jayant Narlikar and to wish him many more productive years of good health and creativity. I have had the good fortune of knowing him over four decades, and have always followed his career with great pride and affection, and admiration for his tremendous accomplishments both in science, arts and public service. IUCAA is a splendid example of his dedicated effort to provide aspiring astromers and astrophysicists in India a place for education, research and training.

Abstract Brief review of the general framework of noncommutative geometry proposed by Alain Connes and its application to the Standard Model and a discretized version of Kaluza-Klein theory is presented.

1. INTRODUCTION

It has become increasingly evident that our present concepts of space and time are inadequate for a unified description of all elementary particle interactions, particularly if one wants to include gravity. In spite of great deal of effort over the decades, we find the twin pillars of twentieth

N. Dadhich and A. Kembhavi (eds.), The Universe, 319–325.

century physics, namely, General Relativity (governing the dynamics of classical space-time coupled to the dynamics of matter moving in it) and Quantum Field Theory (with rules of quantization to be applied in principle to all degrees of freedom including gravity) are found to be incompatible. Attempts at quantizing general relativity over the past several decades have yet to meet with success. String theory, as *the* candidate for a consistent quantum theory of gravity along with a unified description of all elementary interactions, has so far remained only a promise. In spite of its recent new discovery through duality of an underlying unity among a diversity of string theories, it is as yet far from being a convincing physical theory with predictable and experimentally verifiable consequences. We need new ideas.

Both general relativity and quantum field theories are constructed on a continuum picture of space-time. A pseudo-Riemannian manifold endowed with a metric structure based on a continuum picture underlies general relativity. Likewise, quantum fields and their interactions are local operators that are continuous functions of commuting space-time coordinates.

There are several reasons to believe why such a continuum picture of space-time is inadequate at all distance scales. The problem of singularities in the curvature tensor in general relativity and the ultra-violet divergences in quantum field theories are too well known to merit any discussion. To these we might add two other problems that have received considerable attention in recent years. One is the problem of black hole entropy and the enumeration of the black hole degrees of freedom and the other the problem of localization consistent with quantum mechanics. To elaborate briefly on the latter, we note that when we perform accurate measurements of the space-time localization of an event, up to uncertainties ΔX^0,ΔX^1, ΔX^2, ΔX^3, we must transfer energy of the order $E \approx 1/\Delta X$. This energy creates a gravitational field and assuming spherical symmetry for simplicity, the associated Schwarzschild radius $R \approx E \approx 1/\Delta X$. Consequently signals originating inside R cannot reach an outside observer. From such arguments, one may infer the existence of a set of fundamental space-time uncertainty relations(STURS)[1],

$$\Delta X^0 \Delta X \approx l_p^2; \qquad .\Delta X^1 \Delta X^2 \approx l_p^2$$

Such fundamental STURS are incompatible with classical commuting coordinates of a Lorentzian or a Riemannian manifold. They seem to imply[2] that the quantum theories of space, time and matter are interwoven; they are not, a priori, a property of space-time, but a property of space-time in which quantum mechanical matter triggers events. Said in different words[3], "Curvature oscillations may tend to become uncontrollable at short distances. Is there a way to 'smoothen out' short scale curvatures? In some sense nature must become regular there. It is suggestive to speculate that space-time might cease to be continuous but become 'quantized' into some sort of space-time lattice."

¿From such considerations, it appears that the ultimate goal of a fundamental theory should be a generalized quantum theory, a theory that does not, at the outset, begin with a continuum space-time as an input, but gives rise to the classical continuum of space-time in an appropriate limiting regime just as classical behavior of quantum systems emerges in an appropriate limit. However, at present , we have no real candidate for such a generalized quantum theory of matter, space and time, although there are promising guideposts and indications of progress from several different points of view. Superstring theory claims to have a quantized theory of all interactions including gravity. A version of pure quantum gravity suggests 'loops,' whereas causal sets of discrete points are the basis of an alternate approach.

In recent years, Alain Connes has proposed an approach based on noncommutative geometry that is regarded by some, if not many, as providing a new and suitable framework for a geometrical description of all elementary particle interactions [4]. It is, according to Fröhlich [2], a mathematical tool that looks promising for constructing a notion of differential geometry compatible with quantum theory. Connes' ideas hinge upon the well known theorem due to Gelfand and Naimark, which states that the classical manifold based on a continuum can be equivalently described by the abelian or commutative and associative algebra of smooth functions defined on that manifold. Connes' starting point is an extension and a generalization of the idea expressed in the above mentioned theorem to noncommutative spaces, by adopting associative noncommuting algebras as describing such spaces. Connes has reconstucted the standard objects of differential geometry in a purely algebraic way, providing the framework of a noncommutative geometry that permits one to deal with more general spaces with both continuous and discrete degrees of freedom.

In the next section, I discuss briefly the general framework of Connes, followed by a its application to the Standard Model. In Section 3, I present a qualitative discussion of a discretized version of Kaluza-Klein theory. The final section is devoted to some concluding remarks.

2. GENERAL FRAMEWORK; STANDARD MODEL

Connes' formalism consists of three basic elements, called the SPECTRAL TRIPLE,

$$(\mathcal{A}, \mathcal{H}, \mathcal{D}),$$

where $\mathcal{A}$ is an involutive, associative, commutative or noncommutative algebra, $\mathcal{H}$ is a Hilbert space which acts as the carrier space for a representation of the algebra $\mathcal{A}$, and $\mathcal{D}$ is a self-adjoint operator acting on $\mathcal{A}$ with the property that

$$[\mathcal{D}, \mathcal{A}] \quad \text{is bounded} \quad \forall a \in \mathcal{A}.$$

The algebra $\mathcal{A}$ generalizes the commutative algebra of smooth functions. The operator $\mathcal{D}$ allows one to build a differential structure associated with any associative algebra. To construct a Lagrangian and an action based on such a spectral triplet, one first constructs the so called universal differential algebra $\Omega^*(\mathcal{A})$ based on $a \in \mathcal{A}$ and a symbol δ that obeys

$$\delta(1) = 0, \delta(ab) = (\delta a)b + a(\delta b) \forall a, b \in \mathcal{A}.$$

and a representation $\Pi\Omega^*(\mathcal{A})$ of this universal algebra on the Hilbert space $\mathcal{H}$ consisting of bounded operators,

$$\Pi : \Omega^*(\mathcal{A}) \longmapsto \mathcal{L}(\mathcal{H}) \Pi(a_0 \delta a_1 \ldots . \delta a_n) = \rho(a_o) \Pi[\mathcal{D}, \rho(a_i)],$$

where $\mathcal{L}(\mathcal{H})$ is the space of bounded operators on $\mathcal{H}$ and ρ is a representation of of $\mathcal{A}$ on $\mathcal{H}$.

Using such a general framework, one defines a hermitian connection one-form ∇ and the curvature ∇^2. A suitably chosen scalar product then helps to construct a Lagrangian and action.

In Connes-Lott [5] formulation of the Standard Model, the spectral triple $(\mathcal{A}, \mathcal{H}, \mathcal{D})$ consists of

$$\begin{aligned} \mathcal{A} &= \mathbf{C}^\infty(\mathbf{M}) \otimes \mathbf{A_F}; \qquad \mathbf{A_F} = \mathbf{C} \oplus \mathbf{H} \oplus \mathbf{M_3}(\mathbf{C}) \\ \mathcal{H} &= L^2(M, S) \otimes H_F \\ \mathcal{D} &= \gamma^\mu \otimes I + \gamma^5 \otimes \mathcal{D}_\mathcal{F}, \end{aligned} \tag{1}$$

(2)

where $C^\infty(M)$ is the algebra of smooth functions on a real manifold and A_F is a finite matrix consisting of a direct sum of complex C, quaternion

H and 3*3 complex matrices $M_3(C)$. H_F is the Hilbert space that is spanned by particles and antiparticles of the standard model. Finally D_F contains block matrices of the form

$$D_5 = \begin{bmatrix} 0 & M \\ -\bar{M} & 0 \end{bmatrix}$$

,

where M and $\bar{M}$ are matrices signifying coupling constants that eventually result in mass parameters of fermions.

Using this as input, Connes and Lott[5], and subsequently many others [6] have derived the Standard Model with some amazing successes that go beyond the conventional model. In spite of the spectacular agreement of the Standard Model predictions with experiments, the model has many shortcomings, notably the spontaneous symmetry breaking mechanism through the introduction of the Higgs scalar meson. As is well known, unlike the gauge sector, the form, the content and the couplings of the scalar field to fermions are not prescribed by gauge principles alone. Additional seemingly ad hoc assumptions are needed in model building. In contrast, in the approach based on noncommutative geometry, one finds a geometric origin for the scalar just like the gauge mesons. They appear on equal footing and *a spontaneous symmetry breaking mechanism appears naturally*. One can predict, *in principle*, some of the parameters that are arbitrary in the Standard Model, such as the Weinberg angle, and the quark and lepton masses However, such predictions in reality are based on certain approximations and are subject to uncertainties.

3. INCLUSION OF GRAVITATIONAL INTERACTIONS

The noncommutative geometrical approach also lends itself naturally to include gravity along with other interactions. In the minimal Standard Model of electro/weak interactions, parity violation in weak interactions is incorporate by attributing different symmetries to left-handed and right-handed fermions. One may imagine that the right-handed and left-handed components live on two different copies of space-time. this leads to the concept of an extended space-time that includes two discrete points. Such a space-time may be looked upon as a discretized version of Kaluza-Klein theory in which the continuous fifth dimension is replaced by two points. Riemannian geometry on such an extended space-time will inevitably lead to a different picture of gravity. In the background of such a geometry, elementary particle interactions are modified. Nguyen Ai Viet and myself have studied such a geometry [7] and find it intriguing in that it gives rise to several rich and complex models. Standard Model in the background of such a geometry is currently under investigation [8].

Without going into details, let me summarize some basic results;

- The extended space-time permits one to introduce a generalized vielbein consisting of a pair of tensor, a pair of vector, and a pair of scalar fields.

- In the general case, one component of each pair has zero mass while its partner is massive.

-Metric compatible, torsion free connection one-forms lead to constraints on the vielbeins in the form of dynamical dilaton fields that imply new and intersting consequences on gravity.

- In the conventional Riemannian geometry non-vanishing torsion does not lead to unique determination of the connection one-form coefficients in terms of the metric components and their derivatives. In contrast, it is possible in the extended geometry to determine both nonvanishing torsion components and one-form coefficients in terms of the assumed vielbeins.

- With the unique determination of the connection coefficients, the Lagrangian and the action one obtains appears as a sum of two terms, each consisting of all the six independent fields and each representing a generally covariant action.

- These resulting actions present rich and complex structures that suggest various physical models for further study and investigation. In such models, in contrast with the conventional Kaluza-Klein theories, mass spectrum is finite, needing no truncation of the mass spectrum.

4. CONCLUDING REMARKS

Connes' recent development of noncommutative geometry has provided new ideas and new tools for a unified description of elementary particle interactions. In the context of the Standard Model, it is fair to say that the model in the framework of Connes' noncommutative geometry has reached a heightened status of a fundamental theory.. With spontaneous symmetry breaking arising naturally, with fundamental Higgs and gauge fields appearing on the same footing and with the prospect of predicting the values of some parameters arbitrary in the original model, one has the hope that the approach based on noncommutative geometry will provide a deeper understanding of the successes of the Standard Model.

A continuum space extended by the inclusion of discrete points provides an example of a noncommutative space. Geometry based on

Connes' ideas leads to new version of gravity that includes interactions of other subsidiary vector and scalar fields with complex, but well specified linear and nonlinear interactions. Will such modified action lead to a renormalizable theory of gravitation? This is an intriguing question that certainly warrants study.

References

[1] S.Doplicher, K. Fredenhagen, and J.E. Roberts, *Physics Letters* B 331 (1994) 39-44.

[2] J. Fröhlich, in *PASCOS 94-Proc.IV Int. Symp. on Particles, Strings and Cosmology(Syracuse, 1994)*, ed. K.C.Wali (World Scientific, Singapore, 1995),p.443.

[3] 't Hooft, G. High energy Physics, *Physics Reports*, 104, Nos 2-4, 129-142.

[4] A.Connes, NonCommutative geometry, Academic Press

[5] A.Connes and J. Lott, *Nucl.Phys.* **B18**(Proc.Suppl.), 29 (1990).

[6] B.Iochum, D. Kastler, and T. Schücker, CPT-95/P.3260 hep-th/9511011.

[7] Nguyen Ai Viet and Kameshwar C. Wali, *Int. J. Mod. Phys. A, Vol. 11, 533 (1996); ibid Vol. 11, 2403(1996).* Please see references to other related work in these papers.

[8] James A. Javor and Kameshwar C. Wali, *Construction of Action Functionals in Non-Commutative Riemannian Geometry, preprint SU 4240-692*

Chapter 31

FROM INTERSTELLAR GRAINS TO PANSPERMIA

N.C. Wickramasinghe
School of Mathematics
Cardiff University
PO Box 926, Cardiff CF2 4YH, UK

Abstract Data relating to interstellar dust could be interpreted in terms of a widespread distribution of bacterial material in the galaxy. The present contribution reviews the main arguments supporting a modern version of panspermia.

1. INTRODUCTION

Lyman Spitzer Jr. prophetically referred to interstellar dust particles as "grains" many years ago. It is scarcely conceivable that Sptizer would consciously have foreseen modern debate concerning the possible biological nature of grains, nor the recent trends that will be discussed in this contribution. Fred Hoyle and the present writer first approached the subject of panspermia not from a biological point of view but from an attempt to understand the nature of interstellar dust (Hoyle and Wickramasinghe [12]). The first relevant point to consider is that these particles of interstellar dust or grains appear to be much the same in all directions, as we look outwards from the Earth. They are of a size that would be typical for bacteria, a micrometre or less.

Another fact relevant to panspermia is that the total mass of interstellar dust in the galaxy is as large as it possibly can be if all the available carbon, nitrogen and oxygen in interstellar space is condensed as grains. The amount is about three times too large for the grains to be mainly made up of the next commonest elements, magnesium and silicon, although magnesium and silicon could of course be a component of the particles, as would hydrogen, and also many less common elements in comparatively trace quantities.

N. Dadhich and A. Kembhavi (eds.), The Universe, 327–343.

If one now asks the question: what precisely are the interstellar dust grains made of, a number of inorganic molecules composed of H,C,N,O present themselves as possible candidates. These would include water-ice, carbon dioxide, methane, ammonia, all such materials being easily condensible into solids at temperatures typically of about 20-50 degrees Kelvin, which is the usual temperature of the grains.

During the decade starting from the early 1960's the properties of a wide range of inorganic grain models were studied, and their electromagnetic properties checked against the number of new observations that were beginning to emerge. Such models stubbornly refused to fit the available data, particularly relating to interstellar extinction, to anything like the precision that was required. The correspondences between predictions for assemblies of inorganic particles and the observations could be lifted to a certain moderate level of precision but never beyond that, no matter how hard one tried.

It was certainly a milestone in progress towards panspermia when the present writer realised that there is another very different class of materials that can be made from the same four commonest elements - C,N,O,H, namely organic materials, possibly of a polymeric type (Wickramasinghe [26]). Of course there are a vast number of possible organic compositions, making for a great number of further investigations that could be made. By the mid-1970's, the astronomical observations were spanning a large range in wavelength, from $30\,\mu\,\mathrm{m}$ in the infrared, through the near infrared, into the visible spectrum, and further into the ultraviolet. So a satisfactory theory of the nature of grains had by now to satisfy a very large number of observational constraints. Figure 31.1 shows the so-called extinction curve of starlight, the way that starlight is dimmed as it traverses clouds of interstellar dust. A puzzle here relates to how the visual part of this curve (over the $1 - 3\,\mu\,\mathrm{m}^{-1}$ range) could be reproduced almost exactly in all directions of the sky. For inorganic condensation models one requires a rather precise definition of particle sizes, and that is difficult to justify. The puzzle remained unresolved until we first began to consider organic particles, particularly organic particles that were hollow. Particles that have about 70 percent hollow space gave very good results. This is what bacteria become when they are fully dried out. The solid curve in Figure 31.1 (heavy line) combines the effects of hollow bacterial particles with clusters of aromatic molecules that result naturally from the inevitable degradation of bacteria, along with a small admixture of silica-iron particles of submicron sizes that could explain the rise in the extinction into the far ultraviolet.

The excellent fit shown in Figure 31.1 follows from the assumption of the grains being mostly comprised of bacterial material. The invariance

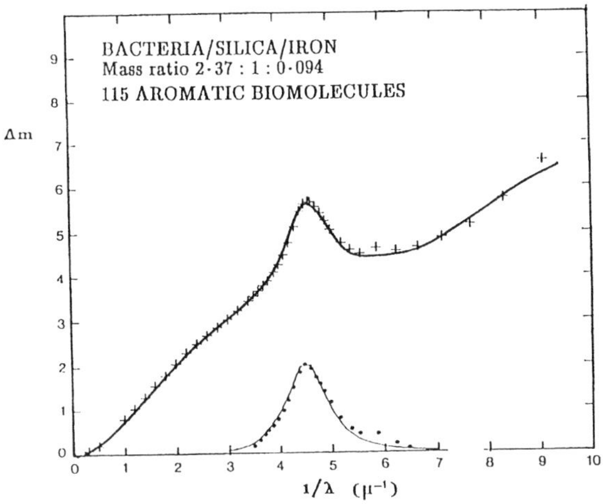

Figure 31.1 The filled circles (points) are excess interstellar absorption values over and above a scattering curve for hollow bacteria. Crosses are the mean interstellar extinction data. The heavy curve is calculated for hollow bacteria with an admixture of bioaromatic molecules and trace quantities of silica and iron in the form of submicron sized grains. The thin line is the absorption profile for an ensemble of bioaromatic molecules (full references and credits in Hoyle and Wickramasinghe [12] and Wickramasinghe [27]).

of the visual extinction curve follows from this assumption, with no additional *ad hoc* hypotheses being required.

2. THE 10 MICRON FEATURE IN INTERSTELLAR GRAINS AND SILICATES

Woolf and Ney [28] made the first detections of infrared emission from interstellar grains in the 8 to 13 μm waveband. The grains responsible for the observed infrared emissions were quickly characterised as "silicates", without it being considered necessary to specify what type of silicate was involved.

Improvements in infrared techniques over subsequent years soon permitted an extended range of astronomical objects to be studied in detail - late-type stars, planetary nebulae, compact HII regions, the galactic centre, comets and the Trapezium nebula. Hot stars in this latter nebula were heating interstellar grains in the vicinity to temperatures of about 175K, thereby causing detectable infrared emission over the 8 to 12 μm waveband.

What was significant in the case of the Trapezium dust was that the emission suffered little self-absorption in the nebula itself. Under

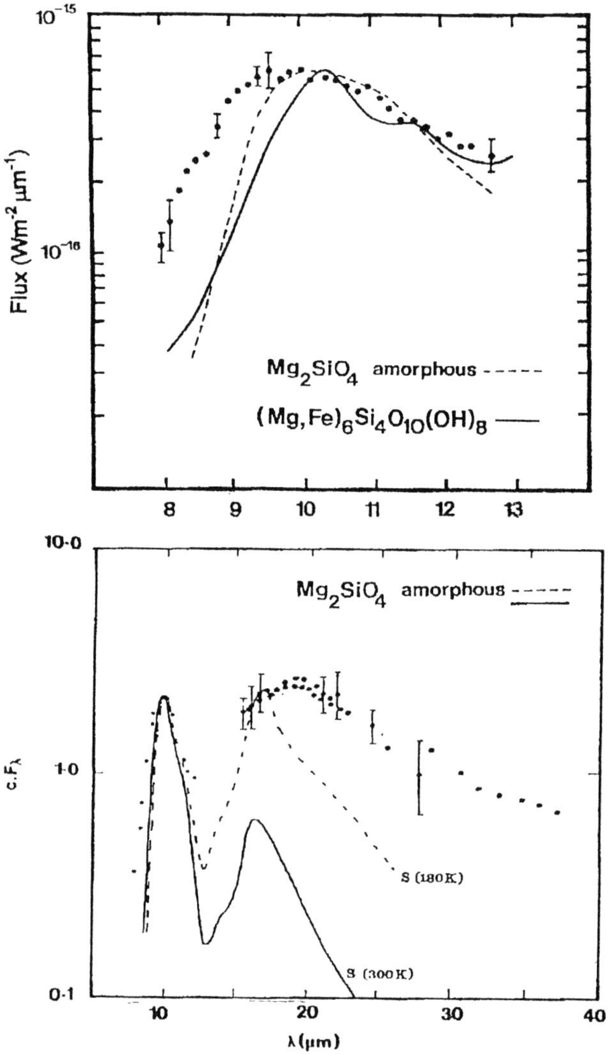

Figure 31.2 *Top panel*: The points are the flux measurements from the Trapezium nebula over the 8 to 13 μm waveband (full references in Hoyle and Wickramasinghe [12]). *Bottom panel*: Trapezium nebula fluxes over the waveband 8 to 35*mum*, compared with predictions for amorphous silicates heated to two temperatures.

such conditions of optically thin emission the flux of radiation at any particular wavelength λ is given by

$$F_\lambda = \text{constant} \times \tau(\lambda) B_\lambda(T), \tag{1}$$

where $B_\lambda(T)$ is the Planck function at the temperature T of the particles and $\tau(\lambda)$ is the opacity of grain material. This function $\tau(\lambda)$ can be measured in the laboratory for any particular material. And in the astronomical case it can be obtained observationally to within a constant factor from Equation 1, once the temperature T is specified. So with the left hand side of Equation 1 determined by astronomical observations at various values of λ, the observed opacity function $\tau(\lambda)$ is obtained by an easy calculation. Thus if we think the particles in the Trapezium nebula consist of a certain type of silicate we can readily verify our belief, or otherwise, by comparing the resulting observed $\tau(\lambda)$ with the $\tau(\lambda)$ obtained for the material in question in the laboratory. The two ways of finding $\tau(\lambda)$ must agree to within a constant factor, which necessarily must be expected because the amount of the sample in the laboratory is unlikely to be the same as in the Trapezium.

When such a comparison was made for all silicates that anybody cared to try out in the laboratory the results were appallingly bad as can be seen in Figure 31.2. The curves are calculated by using Equation 1 with a temperature of $T = 175\,\text{K}$ and the opacity values of silicates $\tau(\lambda)$ measured in the laboratory. The points are the actual flux observations for the Trapezium nebula. The situation only got worse when the observations of the Trapezium were extended further into the infrared (See references in [12]. The bottom panel of Figure 31.2 shows what happened for the best amorphous silicates that actually exist.

A remarkable resolution of the difficulty eventually came to be offered. Using the observed points of Figure 31.1 on the left hand side of Equation 1, the astronomically required function $\tau(\lambda)$ was worked out, which can be called $\tau_{\text{obs}}(\lambda)$. Then instead of looking for an actual substance with $\tau_{\text{lab}}(\lambda) = \tau_{\text{obs}}(\lambda)$, such a substance was invented by hypothesis. And the proposed so called "astronomical silicate" solution to the problem was to consider the hypothetical substance actually to exist.

The scientific validity of this procedure leaves much to be desired. Of course it could not be asserted that real silicates, amorphous or hydrated, did not exist anywhere in the Universe. They certainly exist on Earth and elsewhere in the solar system as well. All we could say from the Trapezium nebula data is that anything remotely resembling a silicate cannot contribute any significant fraction to the mass of the dust.

3. THE 10 AND 20 MICRON FEATURES IN BIOLOGICAL GRAINS

Absorption and emission features at 10 and 20 μm are by no means a prerogative of mineral grains. Many complex organic materials, including biopolymers, exhibit broad features arising from C-O, C=C, C-N, C-O-C bonds centred on wavelengths close to 10 and 20 μm. As our thoughts began to turn in the direction of cosmic biology it occurred to us that there is a possible contribution from biogenically generated silica, as for instance are found in a class of algae known as diatoms, a class that appears to have made a sudden appearance on the Earth some 65 million years ago.

Shirwan Al-Mufti, who was making laboratory measurements for all manner of possible candidate substances, managed after some searching around, to obtain a mixed culture of diatoms taken from waters of the River Taff [14, 2]. Here both 10 and 20 μm absorptions arise from a combination of biologically generated carbonaceous and siliceous material.

Going back to Equation 1 and using this measured $\tau_{\text{lab}}(\lambda)$ for $\tau(\lambda)$ on the right hand side of Equation 1, together with the same temperature of 175 K as before, permits the expected emission of diatoms to be worked out at each wavelength λ. Thus the curve in the upper panel of Figure 31.3 shows the expected curve for diatoms. When this curve is compared with the observed points the agreement is seen to be most impressive indeed. And when the comparison was subsequently extended further into the infrared up to 40 μm, the agreement still remained good, as can be seen in the lower panel of Figure 31.3.

Recent data As with the introduction of every new observing technique the use of ISO (Infrared Space Observatory) launched by ESA on 17 November 1995 provided new opportunities for testing astronomical theory. Particulates in localised regions, for example, planet-forming regions around young stars, would be expected to contain a fair proportion of silicates,and this expectation was indeed borne out in some recent investigations. Spectral features near 19, 24, 28, and 34 μm that have been attributed to hydrated silicates have been observed in several such sources including HD100546 and also Comet Hale-Bopp ([4, 23]). The uniqueness of these assignments is still in some doubt, and even on the basis of a silicate identification in the case of Hale-Bopp such material appears to make up only some few percent of the mass of the dust, the rest being Trapezium-type grains [13]. In all cases where grains in the general interstellar medium or in extended

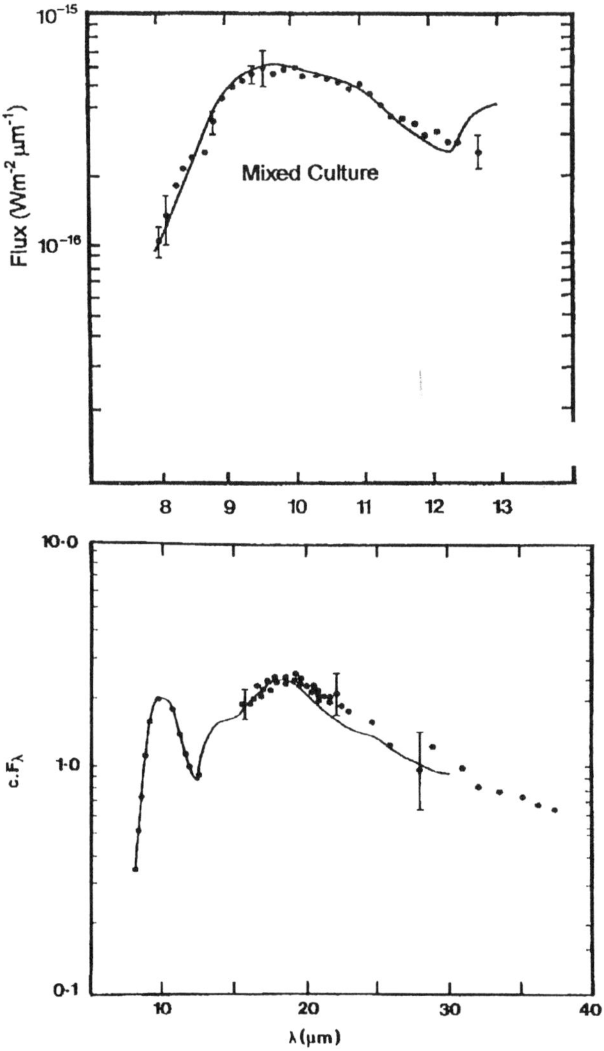

Figure 31.3 Points are same as for Figure 31.2. The curves show calculated emission behaviour of diatoms at 175 K.

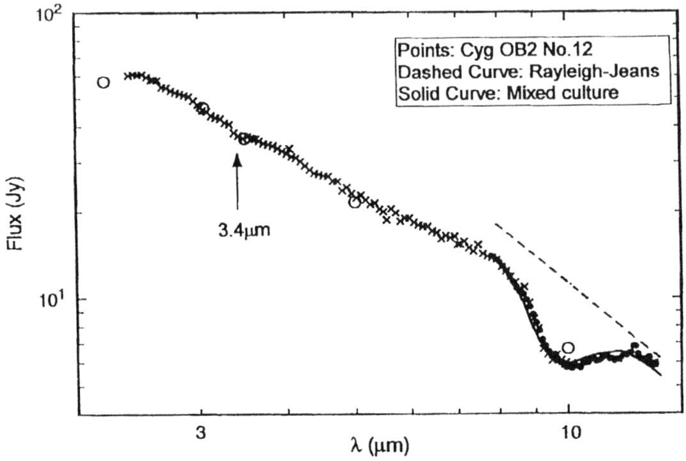

Figure 31.4 Spectrum of VI Cyg OB2 No.12, combining ground based observations and satellite data points (adapted from Whittet and Tielens [25]). The dashed line is the Rayleigh-Jeans tail of stellar emission. The segment of solid curve is calculated assuming extinction by diatom-type material. regions of space have been studied the situation is exactly as we have discussed earlier - no real silicate can explain the observations over the 8-14 μm waveband.

An object that has recently been re-examined and one that is interesting in the present context is the highly reddened B star VI Cyg No.12. This star has a normal interstellar extinction curve with a total visual extinction of some 10 mag. So it could be inferred that its reddening is due to dust over an extended path length in the diffuse interstellar medium. W.A. Stein and F.C. Gillett [21] first examined this star to search for a 3.1 μm water ice band that was expected for the then popular ice grain theory. The results for the ice grain theory were disappointingly negative as it eventually turned out. Now the same star has been studied at high spectral resolution using both ground-based telescopes and satellite observations ([5, 3, 25]). This data is reproduced in Figure 31.4. The filled and open circles are ground-based data and the crosses represent SWS ISO observations. We note first that a hint of a feature occurs at 3.4 μm amounting when measured accurately to an extinction of 0.12 mag. This data is also s een to be consistent with the earlier data which implies that there is little or no evidence for water-ice absorption at 3.07 μm in the general interstellar medium.

The most striking feature of the spectrum of VI Cyg No.12 is the broad smooth absorption feature over the 8-12 μm waveband, which must be due to grains in the general interstellar medium. The dip below a continuum level near 9.5 μm corresponds to an extinction of about 0.8mag.

The dashed curve displayed in Figure 31.4 corresponds to a Rayleigh-Jeans spectrum for the longwave emission from the star. The expected reduction of flux at the Earth due to absorption by interstellar dust is now given by the simple formula

$$|\Delta F_\lambda| = \text{constant}\, \tau(\lambda), \tag{2}$$

where $\tau(\lambda)$, as before, refers to the opacity of a candidate grain material as measured in the laboratory. With an appropriate choice of the constant scaling factor the resulting diminished flux, using $\tau(\lambda)$ for our mixed diatom culture model, is plotted as the solid curve in Figure 31.4.

4. THE 3.4 MICRON ABSORPTION BAND

The earliest evidence of organic matter in a condensed form occurring in interstellar space had been greeted with strong scepticism from the mid-1970's through much of the 1980's. The first relevant data pointing in this direction turned up in spectra of protostellar sources such as the BN object as well as in dense clouds like the Taurus dust clouds [24, 17]. The evidence was in the form of a long-wave wing in the 3.1 μ m absorption band due to water-ice. The circumstance that the 2.9-3.3 μ m ice band with a mass absorption coefficient at its band centre of some 30,000 cm2 g-1 could mask a very much weaker CH-stretching absorption band invariably left only a residual hint of a 3.4 μ m feature to be seen. This was true wherever water-ice was able to condense on grains even in relatively small quantities. Hoyle and Wickramasinghe were the first to recognise this hint of 3.4 μ m absorption in many sources such as the BN. It was pointed out that even in these instances such as this the mass of organics exceeded the mass of ice by more than a factor of ten (Hoyle and Wickramasinghe [8, 9, 10, 11]).

The first direct evidence of complex organic molecules associated with interstellar dust came with observations of the galactic centre source GC-IRS7 [1]. Their observations, using instruments on the Anglo Australian Telescope, with possibly optimal observing conditions, showed unequivocal evidence of a broad absorption band centred at about 3.4 μ m that could be attributed mostly to CH stretching within a mixture of aliphatic and aromatic functional groups. The absorption was to be clearly detected against the background of thermal emission in a source radiating at a temperature of 1100K. Quantitatively the absorption amounted to 0.3 mag at the centre of the 3.4 μ m band. Figure 31.5 shows the spectra several similar sources distributed over an extended 3 cubic parsec volume around IRS7 which were subsequently observed by Okuda et al. [18, 19]. The circumstance that all these sources display approximately the same central optical depth (0.3 mag) at the 3.4 μ m band centre, relative to the underlying black-body continuum, makes it certain that most of the absorption arises from the diffuse distributed interstellar medium rather than from local circumstellar regions. It is therefore safe to infer that this C-H stretching absorption is characteristic of interstellar grains over an extended path length to the galactic centre of some 10kpc or so. It is also clear from Figure 31.5 and from the original observations of Allen and Wickramasinghe [1] that there is no ice band at $3.1\,\mu$m to any significant extent, at any rate none that exceeds the optical depth of the $3.4\,\mu$m band. This result is consistent with the ISO observations of VI Cyg No. 12 to which we have already referred. The points in the upper panel of Figure 31.6 shows the detailed absorption profile in GC-IRS7, combining the data of Allen and Wickramasinghe [1] with that of Okuda et al. [18]. The absorption occurs over wavelength ranges characteristic of OH stretching, CH aromatic and aliphatic stretching and NH stretching. It is immediately clear that a complex mixture of organic materials is involved, but the precise combination of functional groups within plausible models is difficult, perhaps impossible to specify. However, for any given organic substance, or mixture of organic substances, one could determine whether a fit to the astronomical data is possible or not. The general argument is exactly the same as that for the Trapezium nebula that we have discussed in an earlier section.

A laboratory sample of candidate material could give an experimentally measurable transmittance $T(\lambda) = 100 \exp^{[-\tau(\lambda)]}$, whilst the spectrum of GC-IRS7 (e.g. Allen and Wickramasinghe [1]) gives a flux

$$F_(\lambda) = A\, B_{\lambda)}(T) \exp^{-\alpha\tau(\lambda)} \tag{3}$$

A, α being constants and $B_{\lambda)}(T)$ being the Planck function. Thus we can regard the astronomical observations as determining the quantity $\tau(\lambda)$ via Equation 3, at any rate to within a constant factor.

Historically, the first organic model that was considered, and found to match the data to a remarkable degree of precision, was the material represented by the common bacterium E-coli. A spectroscopic KBr disc was prepared with a carefully measured mass of 1.5 mg of dry E-coli. The KBr disc was then heated in an inert gas upto a temperature of 350 C and the quantity $\tau(\lambda)$ for this system was measured using a standard Perkin-Elmer spectrometer. The raw spectrum showing $\tau(\lambda)$ for this case is displayed as the lower panel of Figure 31.5 [2]. The mass absorption coefficient at the peak of the $3.4\,\mu$m absorption was found from this experiment to be close to $500\,\text{cm}^2\,\text{gm}^{-1}$. The curve in the upper panel of Figure 31.6 shows the closeness of the fit that ensued with a choice of $\alpha = 1.3$ used in Equation 3. To obtain this fit, which implies an extinction value of 0.3 mag at the centre of the $3.4\,\mu$m band, we

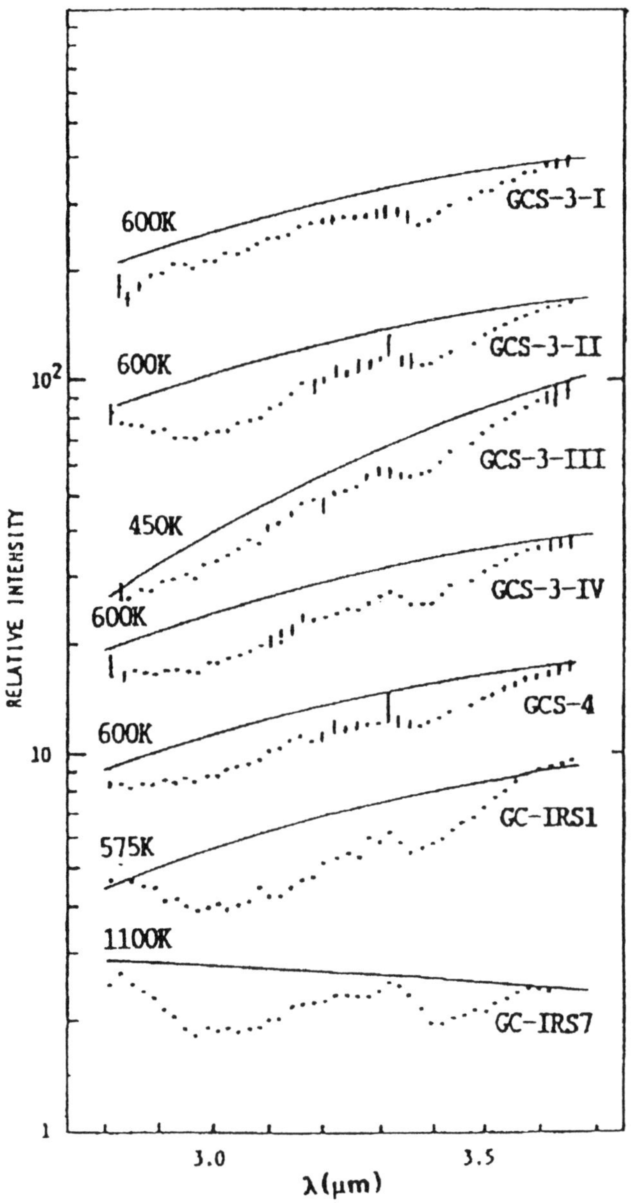

Figure 31.5 Relative flux data for several infrared sources in the galactic centre region [18].

require a distributed mass density of "bacteria-like" organic dust grains amounting to about 10^{-26} gm cm^{-3} - a large fraction of all the mass of interstellar dust.

There have also been new attempts to measure the spectrum of GC-IRS7 using better instruments than before, although not necessarily at superior observing sites with regard to ambient atmospheric water. It should be noted in this context that even minute amounts of atmospheric H2 O would introduce a 3.1 μ m feature in spectra that would be inconsistent with the original AAT observations of GC-IRS7 a nd also the ISO observations of VI Cyg No.12. The generally favoured modern spectrum of GC-IRS 7 appears to be one attributed to Pendleton et al. [20] which is reproduced as the points in Figure 31.7. We see immediately that this spectrum differs from the original spectrum of Allen and Wickramasinghe [1] (dashed line) to the extent of an excess absorption over the 2.8-3.3 μ m waveband that is generally consistent with the presence of water-ice. Our original conclusion concerning the E-coli - GC-IRS7 opacity correspondence would remain valid provided we adopt one of the following two procedures:

(1) Subtract the excess absorption in this waveband, attributing it to spurious atmospheric water (2) Add a component of water-ice to our proposed bacterial grains, an amount as little as 2sufficient for this purpose [15].

Despite the astonishingly modest nature of requirement (2), the present writer would prefer the former of these alternatives, option (1), and propose to adopt the relative flux curve of Figure 31.6 as having the correct overall shape, subject only to refinements of detail over the 3.4 μ m band profile arising from improvements in astronomical spectroscopy.

5. Uniqueness of bacterial solution

One might now ask: what other chemical system besides biology can be invented to match the data for GC-IRS7? We can use Equation 3 to invert the relationship between τ and $F(\lambda)$ and obtain the $\tau_{\rm obs}(\lambda)$ curve just as was done for the Trapezium. In view of the closeness of the fit seen here $\tau_{\rm obs}(\lambda)$ should be considered to all intents and purposes as being necessarily identical to the E-coli opacity. This is of course true only we accept the observations represented by the points in Figure 31.6 as being substantially correct.

Since 1982 many attempts have been made to match the GC-IRS7 spectrum in the 2-4 μ m waveband using abiotically generated mixtures of organic materials. Irradiation of suitably constructed mixtures of inorganic ices have been shown to result in organic residues possessing spectra that fitted the astronomical spectra to varying degrees [22]. But all these arguments and comparisons have begged the important question

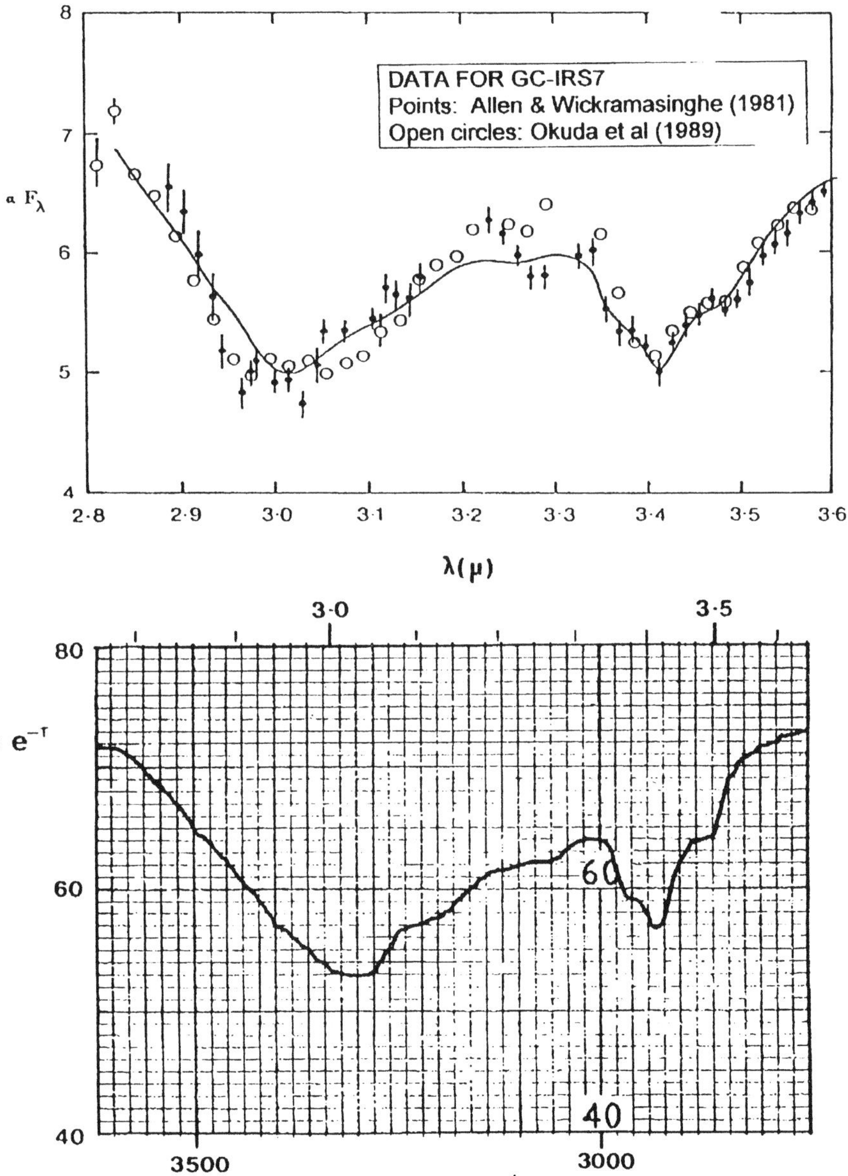

Figure 31.6 *Top panel*: Points show data in [1, 18] for GC-IRS7. The curve is the calculated behaviour of the E.coli model. *Bottom panel*: Transmittance data for dry E.coli [2].

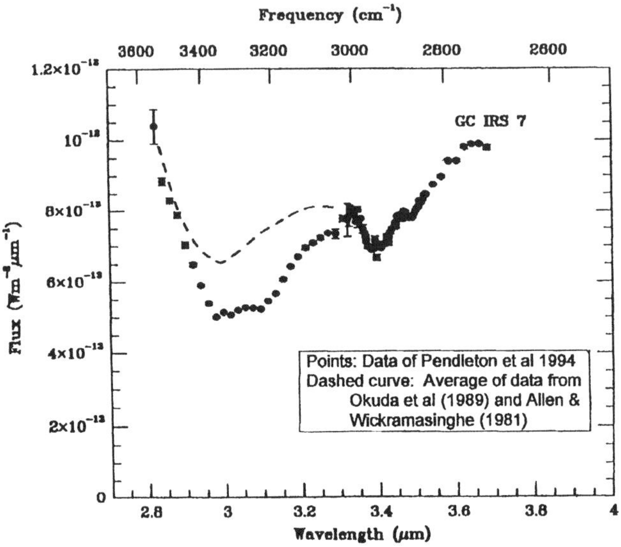

Figure 31.7 High resolution data for GC-IRS7 (Pendleton et al. [20]) (points). Dashed curve is the average relative flux values from the data in [1, 18].

as to how the precise conditions under which the laboratory experiments were conducted could be reproduced with such unerringly precision on a galaxy-wide scale.

6. Other indications of biological grains

Another remarkable development in recent years has been the discovery of independent evidence for vast quantities of aromatic molecules occuring on a cosmic scale [12]. These molecular structures appear to be distributed quite extensively on a galactic as well as an extragalactic scale, and once again a large fraction of the available interstellar carbon seems to be tied up in this form. Needless to say, such molecules are part and parcel of biology, and their occurrence in interstellar space is readily understood as arising from the break-up of bacterial cells.

Even much earlier, in 1962, the presence of aromatic molecules in space might have been inferred from the so-called diffuse interstellar absorption bands. It has been known for over half a century that some 20 or more diffuse absorption bands appear in the spectra of stars, the strongest being centred on the wavelength 4430A. Despite a sustained effort by scientists over many years no satisfactory inorganic explanation for these bands has emerged. F.M. Johnson had first shown that a molecule related to chlorophyll - magnesium tetrabenzo porphyrin - has many of the required spectral properties [16].

There is yet another property of biological pigments such as chlorophyll that persistently shows up in astronomy. Many biological pigments are known to fluoresce, in the fashion of pigments in glow worms. They absorb blue and ultraviolet radiation and fluoresce over a characteristic band in the red part of the spectrum. For some years astronomers have been detecting a broad emission feature of interstellar dust over the waveband 6000-7500 Angstroms. Chloroplasts containing chlorophyll, when they are cooled to temperatures appropriate to interstellar space fluoresce precisely over the same waveband [13].

7. Concluding remarks

In this contribution I have discussed only a small subset of the astronomical data that since the 1980's have pointed consistently in the direction of panspermia. At the most conservative the astronomical data show decisively the overwhelming dominance of highly complex organic molecules in a condensed state. This condensed particulate matter must have spectroscopic properties over ultraviolet, optical and infrared wavebands that make them indistinguishable from freeze-dried bacteria. On this there is no longer any disagreement. Also isotropy of visual extinction curve of starlight shows that these organic grains must be substantially the same in one direction from the Earth as in another. By far the simplest way to produce a vast quantity (1040 gm of small organic particles everywhere of the sizes of bacteria is from a bacterial template. The power of bacterial replication is immense. Given appropriate conditions for replication, a typical doubling time for bacteria would be two to three hours. With a continuing supply of nutrients, a single initial bacterium would generate some 240 offspring in four days, yielding a culture with the size of a cube of sugar. Continuing for a further four days and the culture, now containing 280 bacteria would have the size of a village pond. Another four days and the resulting 2120 would have the scale of the Pacific Ocean. Yet another four days and the 2160 bacteria would be comparable in mass to a molecular cloud like the Orion Nebula. And four days more still for a total since the beginning of 20 days, and the bacterial mass would be that of a million galaxies. No abiotic process remotely matches this replication power of a biological template. Once the immense quantity of organic material in the interstellar material is appreciated, a biological origin for it becomes an almost inevitable conclusion.

References

[1] Allen, D.A. & Wickramasinghe, D.T., 1981, *Nature* **294**, 239.

[2] Al-Mufti, S., 1994, PhD Thesis, University College, Cardiff.

[3] Bowley, J.E., Adamson, A.J. & Whittet, D.C.B., 1999, *Mon. Not. Roy. astr. Soc.* , in press.

[4] Crovisier, J. et al: 1997. *Science* **275**, 1904.

[5] Gezari, D.Y., Schmitz, M., Pitts, P.S. & Mead, J.M., 1993. *Catalogue of Infrared Observations*, NASA Reference Publ. 1294.

[6] Hoyle, F. & Wickramasinghe, N.C., 1977a, *Nature* **268**, 610.

[7] Hoyle, F. & Wickramasinghe, N.C., 1977b, *Nature* **270**, 323.

[8] Hoyle, F. & Wickramasinghe, N.C., 1980a, *Astrophys. Sp. Sci.* **68**, 499.

[9] Hoyle, F. & Wickramasinghe, N.C., 1980b, *Astrophys. Sp. Sci.* **69**, 511.

[10] Hoyle, F. & Wickramasinghe, N.C., 1980c, *Astrophys. Sp. Sci.* **72**, 183.

[11] Hoyle, F. & Wickramasinghe, N.C., 1983,*Nature* **305**, 161.

[12] Hoyle, F. & Wickramasinghe, N.C., 1991, *The Theory of Cosmic Grains*, Kluwer Academic Publishers.

[13] Hoyle, F. & Wickramasinghe, N.C., 1996 *Astrophys. Sp. Sci.* **235**, 343.

[14] Hoyle, F., Wickramasinghe, N.C. & Al-Mufti, S. : 1982, *Astrophys. Sp. Sci.* **86**, 63.

[15] Hoyle, F., Wickramasinghe, N.C. & Jabir, N. : 1983. *Astrophys. Sp. Sci.* **92**, 439.

[16] Johnson, F.M., 1972. Ann. NY Acad Sci, 187, 186.

[17] Merrill, K.M., Russell, R.W. & Soifer, B.T., 1976. *Astrophys. J.* **207**, 763.

[18] Okuda, H. et al. , 1989, *IAU Symp.***136**, 281.

[19] Okuda, H. et al. , 1990, *Astrophys. J.* **351**, 89.

[20] Pendleton, Y.J. et al. , 1994, *Astrophys. J.* **437**, 683.

[21] Stein, W.A. & Gillett, F.C., 1971, *Nature Phys.Sci.***233**, 72.

[22] Tielens, A.G.G.M. et al. , 1996. *Astrophys. J.* **461**, 210.

[23] Waelkens, C.. & Waters, L.B.F.M., 1997, in *From Stardust to Planetissimals* Eds. Y.J.Pendleton & A.G.G.M. Tielens *PASP Conference Series*, 67.

[24] Whittet, D.C.B. et al. , 1983, *Nature* **303**, 218.

[25] Whittet, D.C.B. & Tielens, A.G.G.M: 1997, in *From Stardust to Planetissimals* Eds. Y.J.Pendleton & A.G.G.M. Tielens *PASP Conference Series*, p.161.

[26] Wickramasinghe, N.C. 1974, *Nature* **252**, 462.

[27] Wickramasinghe, N.C. 1993, in *Infrared Astronomy* Eds. By A. Mampaso et al. , Cambridge University Press, Cambridge., p303.

[28] Woolf, N.J. & Ney, E.P., 1969, *Astrophys. J.* **L155**, L181.

Books Authored by Jayant Narlikar

Professor Jayant Narlikar has been a very prolific writer of scientific papers, which number over 250, and books which include research monographs, introductory and expository text books, popular science and science fiction books. Almost all of the popular and science fiction writings have been translated into many Indian and foreign languages. Jayant also writes in his mother tongue Marathi and in Hindi, a language that he has known since childhood. Undoubetdly, his books have been of great influence world wide. Given below is a list of non-fiction books.

- ASTROPHYSICS (Co-authors: R.J. Tayler, W. Davidson and M.A. Ruderman), W.A. Benjamin, London, 1969.
- ACTION AT A DISTANCE IN PHYSICS AND COSMOLOGY (Co-author: Fred Hoyle), W.H. Freeman and Company, San Francisco, 1974.
- THE STRUCTURE OF THE UNIVERSE, Oxford University Press, Oxford, 1977.
- GENERAL RELATIVITY AND COSMOLOGY, The Macmillan Company of India Ltd., New Delhi, 1978.
- THE PHYSICS ASTRONOMY FRONTIER (Co-author: Fred Hoyle), W.H. Freeman and Company, San Francisco, 1980.
- VIOLENT PHENOMENA IN THE UNIVERSE, Oxford University Press, Oxford, 1982.
- THE LIGHTER SIDE OF GRAVITY, W.H. Freeman and Company, San Francisco, 1982; Second Edition, Cambridge University Press, 1996.

- FROM BLACK CLOUDS TO BLACK HOLES, World Scientific Publishing Company, Singapore, 1985; Second Edition, 1995.
- A JOURNEY THROUGH THE UNIVERSE, National Book Trust's Nehru Bal-Pustakalaya, New Delhi, 1986.
- GRAVITY, GAUGE THEORIES AND QUANTUM COSMOLOGY (Co-author : T. Padmanabhan), D. Reidel, Dordrecht, 1986.
- THE PRIMEVAL UNIVERSE, Oxford University Press, Oxford, 1988.
- THE FRONTIER BETWEEN PHYSICS AND ASTRONOMY (IIT Madras Series in Science and Engineering), Macmillan India Limited, Delhi, 1989.
- COSMOLOGY AND ACTION AT A DISTANCE ELECTRODYNAMICS (Co-author: Fred Hoyle), World Scientific Publishing Co.,Singapore, 1996.
- ELEMENTS OF COSMOLOGY, Educational Monograph of the Jawaharlal Nehru Centre for Advanced Scientific Research, Bangalore, Universities Press, Hyderabad, 1996.
- MOTION AND GRAVITY, Book in the Exploratory Series, Shekhar Phatak & Associates, Pune, 1999.
- QUASARS AND ACTIVE GALACTIC NUCLEI : AN INTRODUCTION (Co-author: Ajit K. Kembhavi), Cambridge University Press, Cambridge, 1999.
- SEVEN WONDERS OF THE COSMOS, Cambridge University Press, Cambridge, 1999.